T0360710

PHYSICS OF
ELECTRONS
IN SOLIDS

Advanced Textbooks in Physics

ISSN: 2059-7711

The *Advanced Textbooks in Physics* series explores key topics in physics for MSc or PhD students.

Written by senior academics and lecturers recognised for their teaching skills, they offer concise, theoretical overviews of modern concepts in the physical sciences. Each textbook comprises of 200–300 pages, meaning content is specialised, focussed and relevant.

Their lively style, focused scope and pedagogical material make them ideal learning tools at a very affordable price.

Published

Physics of Electrons in Solids
  by Jean-Claude Tolédano

Astronomical Spectroscopy: An Introduction to the Atomic and Molecular Physics of Astronomical Spectroscopy (Third Edition)
  by Jonathan Tennyson

A Guide to Mathematical Methods for Physicists: Advanced Topics and Applications
  by Michela Petrini, Gianfranco Pradisi & Alberto Zaffaroni

Quantum States and Scattering in Semiconductor Nanostructures
  by Camille Ndebeka-Bandou, Francesca Carosella & Gérald Bastard

An Introduction to Particle Dark Matter
  by Stefano Profumo

Studying Distant Galaxies: A Handbook of Methods and Analyses
  by François Hammer, Mathieu Puech, Hector Flores & Myriam Rodrigues

Trapped Charged Particles: A Graduate Textbook with Problems and Solutions
  edited by Martina Knoop, Niels Madsen & Richard C Thompson

Advanced Textbooks in Physics

PHYSICS OF ELECTRONS IN SOLIDS

Jean-Claude Tolédano

École Polytechnique, France

World Scientific

NEW JERSEY · LONDON · SINGAPORE · BEIJING · SHANGHAI · HONG KONG · TAIPEI · CHENNAI · TOKYO

Published by

World Scientific Publishing Europe Ltd.
57 Shelton Street, Covent Garden, London WC2H 9HE
Head office: 5 Toh Tuck Link, Singapore 596224
USA office: 27 Warren Street, Suite 401-402, Hackensack, NJ 07601

Library of Congress Cataloging-in-Publication Data
Names: Tolédano, Jean-Claude, author.
Title: Physics of electrons in solids / Jean-Claude Tolédano, École Polytechnique, France.
Description: New Jersey : World Scientific, [2021] | Series: Advanced textbooks in physics,
    2059-7711 | Includes bibliographical references and index.
Identifiers: LCCN 2020051745 | ISBN 9781786349729 (hardcover) |
    ISBN 9781786349736 (ebook) | ISBN 9781786349743 (ebook other)
Subjects: LCSH: Solid-state physics.
Classification: LCC QC176 .T65 2021 | DDC 530.4/16--dc23
LC record available at https://lccn.loc.gov/2020051745

British Library Cataloguing-in-Publication Data
A catalogue record for this book is available from the British Library.

For any available supplementary material, please visit
https://www.worldscientific.com/worldscibooks/10.1142/Q0286#t=suppl

Desk Editors: Vishnu Mohan/Michael Beale/Shi Ying Koe

Typeset by Stallion Press
Email: enquiries@stallionpress.com

Printed in Singapore

This book is dedicated to my beloved wife Karin, to my fantastic daughter Muriel, to her dear companion Simon, and to their adorable and very cute children Olympe and Oscar.

Contents

Preface

The subject of this course, entitled "Physics of electrons in solids" is only part of the wider subject of *solid-state physics*, on which many important treatises have been written. With a more restricted scope, this book mainly concerns the properties of solids which are essentially based on *non-interacting electrons*. It is an extended version of a physics course which has been taught, in French, to the students of École Polytechnique. It was aimed at giving to students of undergraduate level an introduction to solid-state physics. The two central chapters (Chapters 5 and 6) develop in detail the situation of the free- and quasi-free electron in a periodic potential, the tight-binding approximation method, and the equilibrium physical properties which derive from the resulting electron states. Subsequent chapters focus on the dynamical and transport properties of electrons as well as on semiconductors properties. The first chapter contains the crystallographic and quantum prerequisites. Finally, the two last chapters reexamine the simplifying assumptions made and give some insight on interactions between electrons in the phenomena of ferromagnetism and superconductivity. For many of the subjects, the qualitative presentation of the physical ideas behind the considered phenomena, precedes the detailed calculations supporting these ideas. More elaborate considerations, contained in every chapter of the book, are adapted to a teaching to graduate students. We have included the texts of three final examinations on which the students were tested (texts in Appendix A and solutions in Appendix B). Certain parts of these long problems, inspired by research articles, can serve as independent short exercises.

At the end of Chapter 1, a figure explains the logical ordering of the chapters, their relationship, and their prerequisites. As most considerations are based on the knowledge of quantum mechanics and statistical physics, we have included a short reminder of a few ideas and of the formalism of quantum mechanics at the end of Chapter 3. Likewise notions of statistical physics are recalled at the beginning of Chapter 6.

In completing this book I must acknowledge the help of several colleagues from École Polytechnique. Teaching the subject was suggested to me by Yves Quéré who had taught an earlier course on the same matter. Jean-Louis Basdevant examined the projected table of contents and encouraged me to pursue the writing. During the teaching, I benefited from the expertise of my colleagues Claude Weisbuch, and of the late Heinz Schulz and Michel Voos. The book of my colleagues Claudine Hermann and Bernard Sapoval was very inspiring to write the chapter on semiconductors. I also acknowledge permanent exchanges with my brother Pierre. Finally, I thank the representatives of the publisher, Stephen Soehnlen, Michael Beale, and V. Vishnu Mohan for their kind encouragement and efficient help.

About the Author

 **Jean-Claude Tolédano** is graduated from École Polytechnique-Paris, and École Nationale Supérieure des Télécommunications-Paris, and PhD in Physics from the University of Paris. He was Research Scientist at the National Center for Research in Telecommunications, and Professor in Physics at the École Polytechnique. He was Guest Professor at the University of Illinois, the University of Nijmegen, and the University of Porto. He has researched in various fields of solid-state physics, mainly on the phase transformations in crystalline solids, but also on the mechanisms of crystal growth and on high-T_c superconductors. He is also Laureate of the French Academy of Sciences and of the French Telecommunication Association. He has authored more than 100 research papers and several books: *The Landau Theory of Phase Transitions* (with Pierre Tolédano; World Scientific, 1987), *Symmetry and Microscopic Physics* (with Jean-Paul Blaizot, in French, Editor Ellipse, 1997) *Physical Basis of Plasticity in Solids* (World Scientific, 2012).

Chapter 1

Solids as Quantum Systems

1.1 Introduction

The content of this textbook can be considered as an elaborate answer to the following question: why does copper, the metal commonly used in electrical connections, possess a value of electrical resistivity as low as $\rho \approx 1.6\,\mu\Omega\,\mathrm{cm}$ at room temperature.

For a physics textbook, such an objective will appear as excessively narrow. It is, instead, sufficiently ambitious to justify that the two hundred and fifty pages of the eleven chapters will only allow the reader to understand that only an incomplete answer can be given here. Indeed, solids are complex systems composed of a large number of interacting constituents. Although it is believed that the general principles governing their properties are understood, the methods for determining the accurate value of a specific property as the electrical resistivity of copper are still in the state of improvement by current research. An introductory text, such as the present one, can therefore only aim at presenting general principles. Even so, a number of considerations are required to reach the stated objective.

Let us outline briefly in this introductory chapter the topics and steps discussed in the subsequent chapters.

Since the electrical resistivity is a property pertaining to the flow of electric current in a metal, and that this flow is believed to be due to the motion of electrons, one has to understand, in the first place, the collective properties of the electrons which are, together with the nuclei of atoms, the constituents of a solid. In this view, one will have to distinguish between "core-electrons" and "active electrons". One has then to analyse the quantum spectrum of the latter electrons, which will appear to depend on the chemical nature and on the geometrical configuration of the constituting atoms. One has also to take into account the quantum and thermodynamical constraints resulting from Pauli's principle (Chapters 5–11).

Secondly, one has to recognize that the interaction of electrons with the collective oscillations of atoms about their equilibrium positions are the main cause of a non-zero resistivity in metals. One has therefore to study the spectrum and statistics of these oscillations. Finally, one must consider the methods relevant to the study of the non-equilibrium properties of solids, since such methods alone can account for a *dissipative* phenomenon such as the resistance of a metal to the flow of electrical current (Chapter 8).

Going through the preceding corpus of knowledge has the advantage of giving access to a much broader field than the single phenomenon of resistivity. Actually, it is the key to the general understanding of *the physical properties of solids*. Although the latter field has not played in the history of physical concepts a role as prominent as quantum mechanics, or as relativity, it is the subject of a major and diversified scientific activity directed both towards the investigation of physical phenomena and the development of objects of industrial interest as summarized by Table 1.1.

Some of the specific properties of solids are part of everyone's personal experience, e.g. the permanence of their external shape, their mechanical hardness, or (for metals) their "metallic" reflection of light. Solids are also the almost exclusive hosts of certain remarkable phenomena such as high electrical and thermal conductivities, semiconducting properties, ferromagnetism, superconductivity, all of which are bases of industrial applications.

Interest in the microscopic origin of these properties has begun at the end of the 19th century with the elaboration of theories of the electrical conductivity of metals and of their magnetic properties. A crucial step was made towards 1930 when it has become clear that understanding the principles governing the properties of solids required the use of quantum mechanics. However, the study of solids only concerned a marginal fraction of the scientific and industrial community until a major impetus was given, towards 1950, by the elaboration of the first semiconductor-devices. Following that achievement, solid-state physics has progressively become one of the

Table 1.1.   Types of solids and their technological uses.

Type of solid	Examples	Technological applications
Semiconductors	Silicon Si	Electronics, processors, solar cells
	Gallium Arsenide GaAs	Telecommunications, lasers
	Cadmium Telluride HgTe	Infrared sensors
Magnetic materials	Iron, Nickel, Ferrites, Garnets	Recording, hard disks
Superconductors	Nb–Ti, Nb_3Sn, $YBa_2Cu_3O_7$	Magnetic resonance imaging Particles accelerators
Metals, alloys	Copper, Aluminum	Electric industry
	Alloys, Titanium, Steels	Planes and other manufacturing
Laser crystals	Ruby Al_2O_3– Cr	Microsurgery, ophtalmology
	Rare earth doped crystals	Target localization in weapons
Piezoelectrics	Quartz SiO_2	Watches
	Lead Zirconate $ZrPbO_3$	Sonars, tunnel microscopes
Amorphous solids	Fused silica SiO_2	Optical fibers
	Amorphous silicon	Solar cells
Liquid crystals	Twisted nematics	Flat screens (TV, computers)

largest field of activity of physicists and its technological counterpart the largest sector of industrial innovation.

Two types of mechanisms underly the physical effects which will be discussed. Those which are based on a direct and dominant role of the electrons, and those in which this role is hidden behind that of more complex assemblies of particles such as atoms or groups of atoms. *The present course is an introduction to the former type of physical phenomena.* Besides, it concerns exclusively the property of *crystalline* solids, i.e. of solids whose atomic spatial configuration display a specific regularity. However, the methods and results which will be discussed are often starting points for the study of more complex systems, more recently investigated by physicists, such as glasses, polymers, liquid crystals, quasi-crystals, nanolayers.

Another restriction, which might seem surprising if it was not to be justified in Chapter 10, is that, in most chapters, *it will be assumed that electrons are the only constituents of a solid.* The degrees of freedom of the remaining constituents, the atomic cores (i.e. the nuclei and the most strongly bounded electrons) will be ignored in the first place. They will only manifest their presence through the occurrence of an electrostatic potential acting on the electrons. In the central chapters (3–9), it will also be assumed that the electrons *do not interact with each other.*

Hence, certain important properties of solids pertaining either to the degrees of freedom of the atomic cores (essential part of the specific heat at room temperature, mechanical properties, phase transformations, infrared emission and absorption, etc.) or to the interaction between these degrees of freedom and electrons, or between electrons (e.g. magnetism and superconductivity) have a marginal place in this course. They will be studied more briefly, mainly in Chapters 7 and 11.

1.2   Basic Principles of the Physics of Electrons in Solids

A number of ideas, all important, have marked the progressive elaboration of the theory of electrons in solids. In this introductory chapter, it is worth reviewing briefly the formulation and the experimental substrate of the three main ideas.

The first one is the assumption that, *in metals,* only a fraction, generally small, of the electrons present in the solid, the so-called *conduction electrons,* is active in the mechanism underlying the flow of current through a metal. The second idea, already mentioned in the preceding paragraph, is the fact that the state of the electrons can only be described through the use of *quantum mechanics.*[1] The third

---

[1] A brief introduction to the ideas of quantum mechanics and to the Schrödinger equation will be found at the end of Chapter 3.

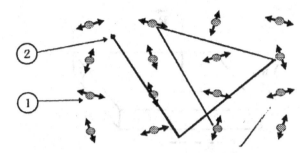

Fig. 1.1. Schematic model of a metal. (1) Ions constituted by "core" electrons and nuclei. The ions oscillate with very small amplitudes around their equilibrium positions (the amplitudes have been exaggerated for the sake of visibility). This oscillation is not taken into account in the Drude model (cf. Chapter 8). (2) Less bounded electrons circulate almost freely within the volume of the metal. Their trajectories are only modified by collisions with the ions.

idea refers to the essential role of the *spatial periodicity* of the atomic configuration in a solid.

1.2.1 Conduction electrons

The assumption that the flow of current through a metal is realized by a displacement of the electrons was formulated (Drude 1900) soon after the discovery of this particle (1897). A metal was described as composed, on the one hand, of ionized atoms M^{z+}, positively charged, and whose positions form a regular array in space, and, on the other hand, of *conduction electrons*, in number z per atom. Each conduction electron can move almost-freely within the volume of the metal, only perturbed in its motion, by collisions with the array of ionized atoms situated at fixed positions (Fig. 1.1). The collective motion of electrons ensures the displacement of electric charges, hence the flow of current. Conduction electrons are the electrons of the constituting atoms with the smallest binding energy to the nuclei of the atoms.

A confirmation of this model could be found in the results of *Hall-effect*-experiments performed in "good metals" such as copper or silver.[2] In such an experiment, a flow of current I_x is passed through

---

[2]In many substances however, complications arise which will be discussed in the light of the quantum theory in Chapters 6 and 7.

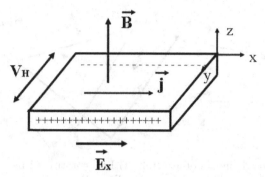

Fig. 1.2. Hall effect configuration. Directions of the magnetic field $\vec{B}$, the current density $\vec{j}$, and of the Hall voltage V_H.

the considered metal along the direction x, and a magnetic field B_z is applied parallel to the direction z (Fig. 1.2). A voltage V_H, *the Hall-voltage*, can then be detected between the surfaces of the sample perpendicular to the direction y (the dimensions of the sample are denoted L_x, L_y, and L_z).

A simple qualitative explanation of this effect can be formulated within the framework of classical mechanics and electromagnetism. The mobile particles, carrying a charge q, with velocity $\mathbf{v}_x$, which determine the flow of current, are submitted to the Laplace force $F_y = q\,\vec{v} \times \vec{B} = q\mathbf{v}_x B_z$. This force F_y, parallel to the direction y, tends to displace the mobile carriers and produce an accumulation of charges on one of the surfaces perpendicular to y (a positive charge if $q > 0$, and a negative charge if $q < 0$). Owing to the initial macroscopic electric neutrality of the material, the opposite surface, partly depleted of its mobile carriers, develops an electric charge of sign opposite to that of the mobile carriers. In consequence, this double electric layer, separated by the distance L_y, generates an electric field E_y directed, within the sample, from a y face to the opposite y face, and relatedly, of a voltage V_H, the sign of which depends on the sign of the charge of the mobile carriers.

As shown by an elementary quantitative analysis of the effect, one can deduce from the value of V_H the number of mobile carriers per unit volume of the sample as well as their *sign*.

Indeed, one can assume that the electric field E_x inducing the flow of current determines a constant average velocity for the mobile charges, proportional to the field:[3]

$$\overrightarrow{\mathbf{v}} = \alpha.q.\overrightarrow{E}_x \qquad (1.1)$$

with $\alpha > 0$. The electric current density $\overrightarrow{j}$ is equal to the number of mobile carriers crossing the unit section of the sample per unit time:

$$\overrightarrow{j} = qn\overrightarrow{\mathbf{v}} = \alpha q^2 n \overrightarrow{E}_x \qquad (1.2)$$

where n is the number of mobile charges per unit volume. As a result of the formation of a positive and negative layer of electric charges on the surfaces, the magnitude of the electric field E_y which develops in presence of a magnetic field, is such as to compensate the effect of the Laplace force. Hence:

$$q\overrightarrow{E}_y = -q\overrightarrow{\mathbf{v}} \times \overrightarrow{B} = -\alpha q^2 \overrightarrow{E}_x \times \overrightarrow{B} \qquad (1.3)$$

Referred to the direction of $\overrightarrow{E}_x \times \overrightarrow{B} \propto \overrightarrow{j} \times \overrightarrow{B}$, the sign of the Hall-voltage $V_H = L_y.E_y$ is that of $(-q)$. Its magnitude is

$$\frac{V_H}{j} = \frac{BL_y}{qn} \qquad (1.4)$$

Experiments show (cf. Table 1.2) that in a number of metals, and in particular in copper or silver, the results of the measurements[4] are in agreement with the fact that q *is negative* consistent with the assumption that *mobile carriers are electrons* $(q = -e)$. Besides, in copper and silver, n corresponds to one electron per atom, i.e. much less than the total number of electrons constituting the atom (respectively 29 and 47 in copper and silver), therefore supporting the idea that only the least bounded electrons (the so-called valence electrons) participate in the mechanism of conduction.

---

[3]This proportionality is consistent with Ohm's law, which expresses the proportionality of the current with the field. It does not derive from Newton's law of mechanics which would rather determine an accelerated motion. Its physical background will be analyzed in Chapters 7 and 8.

[4]As apparent from the formula, for given total current injected in a sample $I = jL_zL_y$ the Hall voltage is proportional to $(1/L_z)$. The measurement has therefore to be performed preferentially on a very thin sample (e.g. $L_z \approx$ a few microns) in order to maximize the measured voltage.

Table 1.2.   Calculated and measured values of the Hall voltage.

Metal	Valence electrons density	Calculated ratio V_H/jBl_y (V.A^{-1}.Tesla^{-1}.m^{-1})	Measured ratio V_H/jBl_y (V.A^{-1}.Tesla^{-1}.m^{-1})
Copper Cu	8.5×10^{28} m^{-3}	7.4	5.1
Silver Ag	5.9×10^{28} m^{-3}	10.7	8.8

1.2.2   Quantum character of the electronic properties

Specific heat of metals

One of the important experiments which has contributed to elaborate the relevant theoretical framework is the measurement of the specific heat C of metals for which experimental laws were established already in the beginning of the nineteenth century. Figure 1.3 of such measurements for two common metals as a function of the temperature. The feature which is worth considering first is the asymptotic behaviour in the upper range of temperatures, above the so-called *Debye temperature* Θ_D. The specific heat approaches a "universal" value (i.e. independent of the metal considered) equal to 6 calories.degree^{-1} ($\sim$ 25 J.deg^{-1}) for one atom-gram of metal ($N = 6.02.10^{23}$ atoms). This amount corresponds to $3k_B$ per atom, where k_B is the Boltzmann constant ($k_B = 1.38.10^{-23}$ J/ deg).

The existence of such a universal value constitutes the *Dulong and Petit law*. Its interpretation in the framework of "classical" (non-quantum) physics reveals a major inconsistency.

The model described in the preceding paragraph considers that a metal contains two types of particles: free electrons which carry the current, and N positive ions. The ions oscillate about their fixed equilibrium positions. Each type of particles is expected to contribute to the specific heat.

Consider first the contribution of the N positive ions oscillating about their average position in the solid. It is equal to the contribution of N three-dimensional oscillators. The energy of each oscillator can be shown to have the form:

$$E = \sum_{i=1}^{3} \left\{ \frac{P_i^2}{2M} + \frac{M\omega_i^2}{2}X_i^2 \right\} \tag{1.5}$$

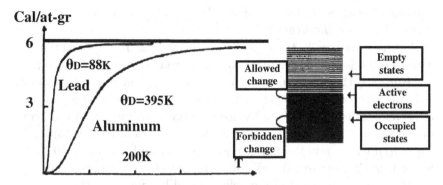

Fig. 1.3. Specific heat of lead and aluminum as a function of temperature. Quantum energy level scheme of a solid. Θ_D is the "Debye" temperature of the solid corresponding to the approximate transition between a rapidly varying temperature dependence and the approach of an asymptotic behaviour (cf. Chapter 10).

In which M the mass of the ion, the P_i are the components of its momentum, the X_i the components of its *deviation* from its average position and ω_i the angular frequency corresponding to the restoring force along direction i. This expression being the sum of six square terms as a function of the degrees of freedom (X_i, P_i). It is adapted to the application of the so-called *equipartition theorem* valid for the statistical equilibrium of *classical systems*.[5] In this framework, each square term can be shown to contribute by the amount $k_B/2$ to the specific heat. The overall contribution of the N positive ions is thus $(6Nk_B/2)$ or $3k_B$ per atom, i.e. precisely equal to the result of the Dulong and Petit experimental law.

However, one also expects a contribution of the conduction electrons. If each atom has z conduction electrons, for N atoms there will be Nz "free" electrons contributing to the specific heat as an ideal gas of Nz free particles. In the framework of classical statistical physics, the contribution to the specific heat of each particle can be calculated as $(3k_B/2)$. The contribution of the Nz particles should then be $(3Nzk_B/2)$ equal to $3z$ calories.degree^{-1} for an atom gram. *In the Dulong and Petit experiment, there is no evidence of*

---

[5] A brief reminder of statistical physics can be consulted in Chapter 6.

this additional contribution of the free electrons in the asymptotic value of the specific heat.

Let us state briefly the reason of this paradox. It is indeed reasonable to consider that the conduction electrons of a metal form a *free-electron-gas*. However, contrary to the assumption made hereabove, this gas is of *quantum nature*. In its quantum description, a particle is associated to a "cloud" whose spatial extension is of the order of the so-called De Broglie wavelength (cf. Appendix of Chapter 3) $\lambda_B = h/p$ where p is the particle momentum and $h = 6.6 \times 10^{-34}$ J.s is the Planck constant. As will be recalled in Chapter 6, to behave as a classical system, a gas of particles should be such that the "clouds" relative to the different particles do not overlap. As we now argue, this condition is not realized for electrons in metals owing to their high density. Indeed, in a solid, the constituting atoms are in contact with each other. If this solid is a metal formed by a single atomic species, the average volume occupied by an atom will therefore be approximately equal to the volume of the free atom. Assume that each atom contributes by, at least, one electron, to form the "gas" of free electrons. The density of the electron gas will be at least of one electron per atomic volume (e.g. ≈ 1 Å in radius) thus, approximately, an average distance of 2 Å $= 2 \times 10^{-10}$ m between electrons. In order for this electron gas to behave classically, the De Broglie wavelength $\lambda_B = h/p$ associated to an electron should be much smaller, and this would imply an average velocity:

$$\langle \mathbf{v} \rangle = \frac{p}{m} = \frac{h}{m\lambda_B} \gg \frac{6.6.10^{-34}}{9.10^{-31}.2.10^{-10}} \approx 10^7 \, \text{m/s} \qquad (1.6)$$

i.e. two orders of magnitude larger than the value expected for a classical gas ($\sim 10^5$ m/s). Hence, the classical description of the electron is not consistent.

In contrast, the absence of significant contribution of the electrons to the specific heat, at high temperatures, can be understood if the electron gas is a quantum gas of the same electron density.

Indeed, as quantum statistical thermodynamics indicates (cf. Chapter 6), in a *quantum gas of electrons*, not interacting with each other, the *Pauli principle* imposes a constraint in the temperature range corresponding to the Dulong and Petit law. Namely, *only a small proportion of the conduction electrons* can exchange energy with outer sources (electric or magnetic fields, heat reservoirs, etc.).

This is because the lowest electronic energy states of the conduction electrons are *neutralized*. The electrons in these states cannot modify their energy as this would imply "jumping" into a neighbouring energy state already occupied by an electron, and this is forbidden by the Pauli principle (Fig. 1.3, right). As a consequence, only the "active" electrons (approximately 1% of the free electrons at room temperature) can change their energy when the temperature of the solid is modified. Hence, their contribution to the specific heat will be ∼ 1% of the value (3p calories) calculated hereabove. In other words, electrons contribute only marginally to the specific heat of the solid at high temperatures. In the temperature range in which the Dulong and Petit law is valid, their presence is entirely hidden behind the contribution of the atomic thermal vibrations.

With respect to its *electronic properties*, a metallic solid thus appears as a *quantum system* of *macroscopic size* (*the volume of the solid*).[6]

Another important evidence of the inadequacy of the classical theory resides, as shown on Fig. 1.3, in the pronounced decrease of the specific heat, due to the oscillation of the ions, as measure as the temperature is lowered. This behaviour can be shown to be explained by the quantum treatment of the oscillations of ions about their central position.

Magnetism, a quantum effect

An other indication of the quantum nature of the physical behaviour of solids concerns magnetic phenomena. Thus, *ferromagnetism* is specific of the solid state. It is the phenomenon which underlies the existence of permanent magnets. Other effects as *diamagnetism* and *paramagnetism* exist in other media but are especially intense in solids. They have their origin in the electrons of the system.

All these effects are characterized by the onset of a non-zero value of the magnetization vector $\overrightarrow{M}$ which is the sum of the magnetic dipoles $\overrightarrow{\mu}_i$ present at microscopic level in the material. $\overrightarrow{M}$ will onset

---

[6]If the electrons of a metal always form a quantum system, it will be shown (see Chapter 9) that it is often possible to consider the electronic states of a semiconductor as those of a "classical" system. This is due to the fact that the density of free electrons in a semiconductor is much smaller than in a metal.

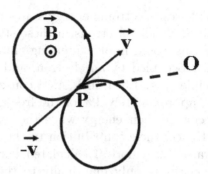

Fig. 1.4. Projection, on the plane perpendicular to the magnetic field B, of the circular paths of the free electrons in a solid. The sum of the dipoles induced by the field is equal to zero.

either by applying an external magnetic field to the medium (in the case of diamagnetism and paramagnetism) or "spontaneously" (in the case of ferromagnetism), below a certain temperature, without need of an applied field (it can then be interpreted as the result of the occurrence of an effective internal field).

All of these effects are of quantum nature, i.e. they are not compatible with a description based on classical (non quantum) mechanics and of electromagnetism as is briefly argued now.

In a classical framework, an elementary magnetic dipole $\vec{\mu}_i$ is generated by the flow of electric current associated to the displacement of an electron along a closed loop (Fig. 1.4). First, consider the system in the absence of external magnetic field. In a classical electron gas in thermodynamic equilibrium, the sum of the dipoles thus generated is necessarily equal to zero. Indeed, the distribution of electron velocities is governed by the so-called Maxwell–Boltzmann statistics. Thus, at each point of the macroscopic system, an electron has the same probability to have a given value of velocity, or the opposite value (or alternately a given value of the local electric current or the opposite value).

Now, consider the effect of a magnetic field, which acts on the electrons through the Laplace force $-e\vec{v} \times \vec{B}$. This force will induce for each electron a rotation of the electrons around the field, the sense of the rotation being the same for all the electrons (Fig. 1.4). However, for electrons located at a point P, and having equal probability of velocities $\vec{v}$ and $-\vec{v}$, of the system, the sum of the momenta

$\overrightarrow{M} = \overrightarrow{OP} \times \overrightarrow{v}$ will be equal to zero. Hence, the situation is the same as in the absence of a magnetic field, i.e. *there is no local magnetization in the system*. A more formal proof will be given in Chapter 11.

In conclusion, the onset of a non-zero value for $\langle \overrightarrow{M} \rangle$ which characterizes magnetism, cannot be described by a mechanism based on classical thermodynamics. This general result is not contradictory with the fact that the application of classical statistics to an assembly of microscopic dipoles (the so-called Langevin model)[7] can be shown to give rise to a paramagnetic behaviour. Indeed, such a model assumes as a starting point the existence of magnetic dipoles at microscopic level. It is the very existence of such dipoles which requires to be justified by quantum mechanics. These dipoles arise either from the electron spin (a pure quantum effect) or from the quantized motion of the electrons on their trajectory in the atom. In the case of ferromagnetism a second quantum aspect is involved. As will be argued in Chapter 11, ferromagnetism results from an effective interaction between dipoles deriving from the Pauli principle.

Another specific quantum effect, which involves, again the Pauli principle, is that of *superconductivity*. Among its characteristic manifestations is the lack of dissipation related to the flow of electric current in a solid material (zero electrical resistivity). Superconductivity will be discussed in Chapter 11.

1.2.3 Relevance of the spatial configuration and of the chemical nature of the atoms

A more systematic and more accurate study of solids (not only restricted to a few common metals) shows that considering the valence electrons as a free electron gas is an insufficient approximation, if one wishes to account in detail for the properties of solids. In particular this approximation does not allow to understand the large differences of electronic properties which are observed in different solids.

Hence, different metals exhibit electrical resistivities which can differ by an order of magnitude (e.g. at room temperature, in copper

---

[7]Cf. Chapter 7.

$\rho = 1.6\,\mu\Omega.\text{cm}$ while in lead $\rho = 19\,\mu\Omega.\text{cm}$). Besides, many solids are electric *insulators*, and almost prevent any flow of electric charges (e.g. in quartz at room temperature $\rho \gg 10^{10}\,\Omega.\text{cm}$).

Even if we restrict to metals, the diversity of electric properties does not seem compatible with the simple picture drawn in the preceding paragraph, i.e. that, in all metals, there is a quantum gas of free electrons formed by the valence electrons of the constituting atoms. Indeed, the electron gas of different solids should only differ by their density, which is itself determined by the valence of atoms and by their atomic volume. This would determine differences in resistivity by a factor of a few units only.

Two examples illustrate this inadequacy of the free-electrons quantum gas model.

Hence, aluminum A_L and boron B are two chemically similar atoms having the same valency $(+3)$. The two monoatomic solids constituted by these atoms should therefore have similar electric properties. In contrast with this expectation, aluminum is a "good" conductor while boron is an insulator.

On the other hand, graphite, diamond, and other forms of solids (e.g. fullerenes cf. Chapter 2) are constituted by carbon atoms. Graphite is a fairly good conductor ($\rho \sim 50\,\mu\Omega.\text{cm}$) while diamond and fullerenes are insulators. These three solids composed of carbon atoms differ by the spatial configuration of the atoms, i.e. by the geometrical configuration of their atomic positions in space (cf. Fig. 1.5).

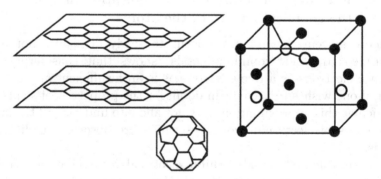

Fig. 1.5.   Upper left: Layered structure of graphite: each carbon atom is located at the summit of an hexagon. In diamond (right) the carbon atoms form regular terahedra, with vertices located at the corners and the face centers of a cube. The third structure, made up of carbon atoms, that of "fullerene", is more complex.

These two examples suggest that both the chemical nature of the constituting atoms (e.g. aluminum vs. boron) and their geometrical configuration in the solid have a major influence on the conduction properties of a solid. The reason underlying this influence resides in the fact that the action of the constituents of a solid is not only to confine the electrons within the volume of a metal by creating a uniform attractive potential. It submits them to a *spatially varying potential at the microscopic level*, whose spatial variations and magnitude depend closely on the location of the atoms and of their chemical nature.

Another peculiar effect of this sophistication of the quantum properties of electrons is revealed by the measure of the Hall effect (see section hereabove) in certain metals. In these metals (e.g. aluminum, zinc, cadmium, beryllium) the *sign of the Hall voltage*, which, as shown above, is determined by the sign of the electric charges carrying the current, is *opposite to that expected for electrons*. This *abnormal Hall effect* will also find its explanation in the influence of the chemical nature and of the spatial configuration of the constituents (cf. Chapter 7).

1.3 Microscopic Origin of the Properties of Solids

In this paragraph, we briefly review the mechanisms of other properties of solids related to the characteristics of the electrons. Some of these properties are directly related to the characteristics of the *electron gas* of the solid. Other properties imply the electrons in an indirect manner, as constituents of the atoms composing the solid. As already mentioned in Section 1.2.2 above, this is the case of the main contribution to the specific heat of solids, related to the oscillation of atoms about their equilibrium position. Certain of the following properties will be studied in detail in the subsequent chapters.

1.3.1 Thermal conduction

Thermal conduction governs the transfer of heat between two points of a system having different temperatures in such a way as to equalize their temperatures. A good thermal conductivity corresponds to a rapid transfer of the heat.

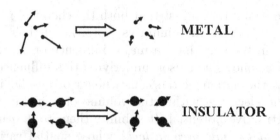

Fig. 1.6.   Mechanisms of transfer of heat in metals and insulators. In metals the average kinetic energy of the free electrons is transfered from the "hot" point, in which the average kinetic energy is high to the "cold" point where it is lower, by the motion of free electrons. In insulators the large amplitude oscillation energy of atoms at "hot" points is transfered progressively to neighbouring atoms.

In a metal, the mechanism of this transfer is related to a local increase of the *average kinetic energy* of the free electrons by an increase of the local temperature (cf. Fig. 1.6). These electrons being mobile in the system, and having a high velocity, their excess of energy will be brought rapidly at other points of the solid. An energy (i.e. heat) transfer will thus occur rapidly and determine a good thermal conductivity. One can also understand that, in metals, a relationship exists between the thermal conductivity and the electrical conductivity since both properties are determined by the same free electrons.

In electric insulators there are no mobile electrons. *In general* they have a smaller thermal conductivity than metals, as the mechanism of heat transfer is different. In these solids, a local increase of temperature induces an increase of the average *energy of oscillation* of the atoms, and it is the interaction between neighbour atoms which produces a transfer of oscillation energy from one point to another.

1.3.2   Mechanical properties

As is commonly known, metals (e.g. steel or alloys) can be deformed by exerting bending or hammering forces in order to manufacture objects of various shapes, while other solids (e.g. glass) will break before deforming, under the action of such forces. This is due to the fact that different types of binding forces insure their cohesion. The mechanism of the deformation of solids is a complex mechanism which involves the motion of specific defects called dislocations. It can be shown that this motion is made more easy in solids in which there is a certain tolerance (i.e. less increase in energy) to the change of relative orientation of the bonds between neighbouring atoms. This is precisely the case in metals, in which the binding relies mainly on the role of the electron gas of negative charges which creates an attractive potential binding together the set of positive ions. This type of binding ensures an *approximate isotropy* of the effective attractive force between ions and hence some tolerance to a relative change in orientation of the bonds between atoms.

In non-metallic solids, the cohesion is often realized by *covalent bonds* between atoms. These bonds are considerably less tolerant to a local change of orientation. As a consequence, dislocations will not be easily moved through the solid and an increase of force aimed at obtaining such a motion will rather provoke the fracture of the material.

1.3.3 Optical properties

One of the striking difference between metals and other solids is their property with respect to the *visible* part of the spectrum of *light*. Metals reflect light, thus giving their specific "metallic reflection". In contrast, insulating solids will generally be transparent, at least for part of the visible spectrum. Again, this difference is related to a difference in the quantized energies of the electrons. In metals (see Chapters 3 and 5) conduction electrons possess quantized energies belonging to a continuum of values close to each other, forming an "energy band", the width of which is $\Delta\varepsilon \geq h\upsilon$ where the frequency υ belongs to the visible part of the spectrum. Hence, if a photon of visible light impacts the surface of a metal, it will be absorbed by an electron as the electron will be able to change its energy from ε_1 to another energy ε_2 in the continuum-band, such as $h\upsilon = (\varepsilon_2 - \varepsilon_1) \leqslant \Delta\varepsilon$ (Fig. 1.7). By regaining its initial energy ε_1 the electron will emit a photon having the same energy contributing to the light reflected by the metal. In insulators (see Chapter 6) the energy spectrum of electrons involves energy "gaps", several electron-volts wide, in which there are no energy states for an electron. Consequently, part of the visible light can neither be absorbed nor reflected, and will be transmitted through the solid.

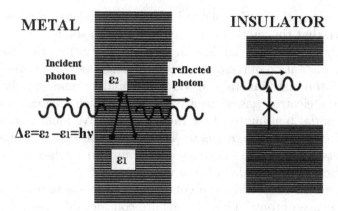

Fig. 1.7.   In a metal, which has a continuum of energy levels, an incident photon is always absorbed by electrons which will acquire its energy to occupy a higher unoccupied energy level. It will then restitute this energy, in the form of a reflected photon, by returning to a lower level. In an insulator such a mechanism is not possible due to the absence of energy levels in a wide energy band. The incident photon will not be absorbed: the solid is then transparent.

1.3.4   Optical infrared properties of insulators

The energy of photons in the infrared part of the spectrum is smaller than a tenth of an electron-volt. The interaction of these photons with insulating solids does not involve directly the electrons. The absorption of a photon determines the increase of the energy of *collective oscillation of the atoms*. Conversely, the emission of infrared light by a heated solid is achieved by a decrease of the oscillation energy of the solid. Hence, analysis of the characteristics of the infrared light emitted or absorbed by a solid is a means of studying the characteristics of the frequencies of the collective oscillation of the atoms of a solid. These frequencies provide indications on the cohesive forces binding the considered solid.

1.3.5   Dielectric permittivity of insulators

An external applied electric field $\vec{E}$ applied to an insulating solid will induce in it an electric polarization $\vec{P}$ (i.e. a macroscopic electric dipole per unit volume). The *permittivity* ϵ measures the magnitude of the polarization for a given applied field $\vec{P} = \epsilon \vec{E}$.

The induced polarization is the sum of microscopic dipoles of different origins. On the one hand, there is an *electronic contribution*. The electric field modifies the charge distribution of the electrons around each atom or ions. The center of gravity of this distribution is displaced opposite to the field while the positive nucleus of the atom is displaced in the direction of the field. Such a contribution is not specific of solids. However, the sum of the induced dipoles is larger that in a gas of atoms due to the higher density of atoms in the solid.

Another contribution comes from the relative displacement of the positive and negative ions by the electric field. This contribution is specific of the solid state. Clearly it will be larger for solids in which the binding forces between the constituting atoms are weaker, These forces are also the restoration forces bringing back the atoms to their equilibrium positions.

It is worth emphasizing that the local field acting on a given microscopic dipole (of electronic or ionic origin) differs from the external field applied to the solid sample. Indeed, the dipoles created by the external field in the entire volume surrounding the considered dipole will create an additional field superimposed on the external applied field. This additional field is of opposite sense to the external field. It is called depolarizing field.

1.4   Organization of the Book

The present book is organized as shown on Fig. 1.8.

The central question is that of studying the so-called *quantum energy spectrum of an electron* in a solid. The three following chapters (2–4) concern the introduction of the geometrical and quantum prerequisites required to deal with the quantum spectrum of electrons. As mentioned above, Chapter 3 will include, in an Appendix, a brief introduction to the quantum mechanical notions used in the book. Chapter 5 concerns the general characteristics of this spectrum, i.e. the nature and values of the quantum numbers labelling the quantum states. This problem can only be studied by approximation methods, in spite of the simplifying assumptions made. Chapter 6 is devoted to the consequences of the electronic spectrum on the physical *equilibrium* properties of solids. It contains a few indications on

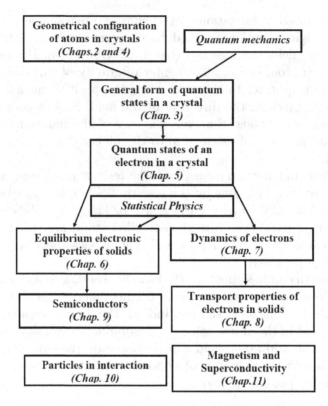

Fig. 1.8.   Relationship between the different chapters.

the results of the statistical theory of equilibrium. Chapters 7 and 8 concern the so-called *transport* properties such as the dissipative flow of current. Chapter 9 focuses on the specific electronic properties of semiconductors and describes briefly their use in making electronic devices. Chapter 10 provides the justifications and the simplifying assumptions of the central chapters, and examines the consequences of lifting these limitations. Finally, Chapter 11 is a brief overview of ferromagnetism and superconductivity, two properties which involve interactions between electrons.

Bibliography

The Physics of Metals, *F. Seitz*, McGraw-Hill (1943).   .
Introduction to Solid State Physics, *Charles Kittel*, John Wiley & Sons (First Edition, 1953).

Quantum Theory of Solids, *R.E. Peierls*, Oxford University Press (1955).

Solid State Physics, *N.W. Ashcroft, and N.D. Mermin*, Harcourt College Publishers (First Edition, 1976).

Physics of Materials, *Yves Quéré*, Gordon and Breach Science Publishers (1998).

Principles of Condensed Matter Physics, *P.M. Chalkin, and T.C. Lubensky*, Cambridge University Press, India (2004).

Introduction to Solid State Physics, *A. Aharony and Ora Entin-Wohlman*, World Scientific Publishing (2018).

Quantum Mechanics, *J.L. Basdevant, and J. Dalibard*, Springer (2005).

From Microphysics to Macrophysics, *R. Balian*, Springer (2007).

Chapter 2

The Crystalline Order

Main ideas: Algorithm of construction of a crystal (lattice + basis) (Section 2.1.3). Geometrical implications of the three-dimensional periodicity (lattice planes and rows, restrictions on the values of rotation angles and on the unequivalent types of Bravais lattices, etc.) (Section 2.2.2). Different types of unit cells (Section 2.3). Characteristics of the simple packings of atoms (Section 2.4.1). Additional matter: Classification of space groups (Section 2.5), incommensurate crystals, quasicrystals (Section 2.6).

2.1 Crystal Structure and Periodicity

2.1.1 Introduction

The solids, the electronic properties of which are studied in this book, are the *crystalline* solids generally more simply termed *crystals*. Natural mineral crystals such as rock-salt, quartz, or precious stones, have been objects of common experience for a long time. The technological developments of the last decades (cf. Table 1.1) have induced the fabrication of a variety of synthetic crystals. For instance, crystals of silicon are the "raw matter" of the microelectronic industry while rare earth doped crystals and gallium arsenide are the basis of the fabrication of solid-state lasers.

The observation of crystals, either by direct visual inspection, or by means of instruments giving access to the atomic-scale, shows the occurrence of definite *geometrical regularities*. At the macroscopic level, the surfaces of crystals display polygonal facets within planes

23

Fig. 2.1. Synthetic crystals of the superconducting material $Bi_2Sr_2CaCu_2O_8$ (courtesy of H. Noël) showing natural growth facets of dimensions $-1\,$mm.

making specific angles (Fig. 2.1). Investigations at the angström-scale (Figs. 2.2 and 2.3) reveal remarkably regular geometrical patterns of the microscopic constituents. Understanding these patterns is a necessary step in the study of the physical properties of solids. Indeed, as will be shown in Chapter 3, the nature of the *quantum states* of electrons in a crystal is strongly constrained by its microscopic geometrical characteristics. The present chapter is devoted to the specification of these characteristics which are also termed *symmetries*.

Crystal symmetries have been described during the 19th century, hence long before a direct experimental access to the atomic-scale became possible by means of X-rays or electron microscopy. "Crystallographers", having studied the regularity of the *external* shape of crystals, had *inferred* that these objects possess *an internal* regularity pertaining to the configuration of their microscopic constituents. The didactic presentation of the crystals microscopic symmetries has often followed the preceding inductive path. Today, a more straightforward introduction can rely on the multiplicity of instrumental means permitting a fairly direct display of the constituents of crystals. In the next paragraph, two examples of atomic-configurations, probed at the angströ m-scale, are discussed. On this experimental ground, the general concepts presiding over the microscopic geometry of crystals are inferred.

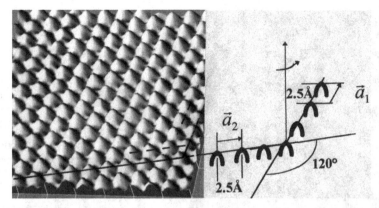

Fig. 2.2. Scanning tunneling microscope image of the surface of a graphite crystal (courtesy of J.M. Moison). The primitive translations and the rotational symmetry axis are indicated on the right.

The end of the chapter, Section 2.6, reviews briefly the geometrical description of solids which depart in a more or less pronounced manner from the case of crystalline solids.

2.1.2   Observations at the atomic scale

Figure 2.2 is the image of the surface of a graphite crystal, obtained by means of a "scanning tunneling microscope".[1] One can distinguish in it the regular repetition, in the two dimensions of the surface, of a rounded hillock. In each of the two directions marking specific alignments of hillocks, the repeat distance is approximately 2.5 Å. The angle between these directions is 120 degrees. Hence, the observed surface can be entirely constructed by replicating a given object (a hillock), by means of the repeated application of two elementary displacements $\vec{a_1}$ and $\vec{a_2}$ represented on Fig. 2.2. The two vectors have a modulus of $\sim$2.5 Å and are respectively parallel to the two marked directions. Though the nature of the observed elementary objects cannot be identified in this experiment, it is known from

---

[1]In this experimental technique, invented by Georg Binnig, a very thin probe, the tip of which is $\sim$10 Å thick, scans the surface of the sample at a distance smaller than 10 Å. The vertical displacements of the probe, imposed by the constituents of the surface are detected by the instrument, and recorded during the scanning of the probe.

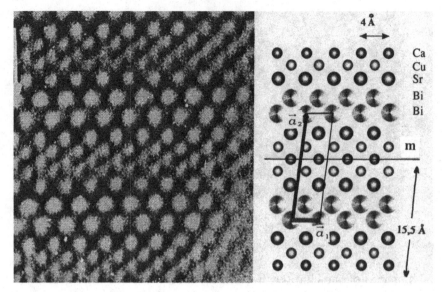

Fig. 2.3. High resolution microscope image of the superconductor $Bi_2Sr_2CaCu_2O_8$ (courtesy of Maryvonne Hervieu). On the right the primitive translations, the unit-cell containing 7 atomic columns and the horizontal symmetry plane **m** of the structure are shown.

prior X-ray diffraction experiments (see Chapter 4) that each hillock is a group of two carbon atoms.[2]

Figure 2.3 reproduces the observation of a crystal containing five different chemical species (bismuth, calcium, copper, oxygen and strontium) by means of an *electron microscope*. Each slightly-blurred disk on this figure discloses a column of identical atoms of a given species directed perpendicular to the figure. However, the oxygen columns are missing owing to the lack of sensitivity of the microscope to the lighter chemical species.

As seen on the image, the crystal can be constructed by replicating an elementary "brick" of dimension ~ 4 Å $\times$ 15 Å, through two displacements, the amplitudes and directions of which are defined

---

[2]Existing tunneling microscopes having still higher resolution (i.e. allowing observations at a smaller scale than in the picture shown) would permit to observe directly the two atoms composing one of the above microscopic units. These atoms are at a mutual distance of 1.4 Å, and aligned along a bisector of the reference axes.

by the vectors $\vec{a_1}$ and $\vec{a_2}$. This brick is composed of seven atomic columns associated to four species (Bi, Ca, Sr, and Cu). Hence, in spite of the greater complexity of the considered structure, a generation of this structure through the replication of an elementary unit can be as simply achieved. The complexity resides entirely in the elementary atomic unit which actually contains 15 atoms, if one includes the oxygen-atoms.

From the observation of the former two-dimensional views, it appears that, although crystals can display very different atomic configurations, they have the common property of being the result of the *periodic replication* in space of an elementary microscopic object, with spatial-period values in the angström range.

2.1.3 Generalization: Crystal structure space-symmetry

We can rely on the two former examples to formulate the basic concepts defining crystalline solids.

The *structure* of a solid is defined as the *configuration in space* of its constituting atoms. In principle, this configuration can be specified by the set of coordinates of the nuclei of all the atoms.[3]

An *ideal crystal* is an infinitely extended solid, the structure of which possesses a *three-dimensional spatial periodicity*. By this statement, one implies that its structure can be constructed by replicating a given set of atoms, through displacements associated to *translations* $\vec{T}$ of the form:

$$\vec{T} = n_1\vec{a_1} + n_2\vec{a_2} + n_3\vec{a_3} \tag{2.1}$$

where the coefficients n_1, n_2, n_3 are integers (positive, negative, or equal to zero), and where the three vectors $\vec{a_1}, \vec{a_2}, \vec{a_3}$ are *three*

---

[3]The correctness of the preceding statement requires mentioning that the concerned atomic positions are average positions of the atomic nuclei. Indeed, as mentioned in Chapter 1, and described in more details in Chapters 6–10, the atoms in a crystal are in constant motion, vibrating about a point which is their average equilibrium position. The microscopic observations in Figs. 2.2 and 2.3 are performed on timescales which are very large with respect to the periods of the vibrations. They therefore reveal, consistently with the above definition, the average atomic positions.

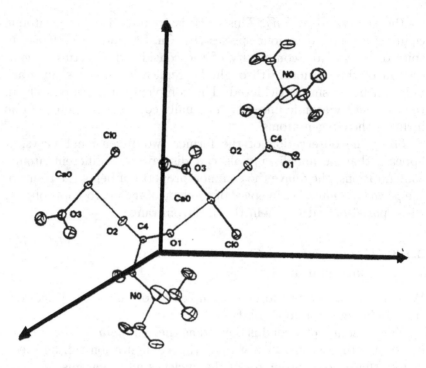

Fig. 2.4.   Basis and primitive translations of hydrated betaine calcium chloride.
The magnitude of the three primitive translations is ∼10 angströms.

non-coplanar vectors. These vectors define the basic *periodicities* of
the crystal. They are called the *primitive (or elementary) translations*
of the crystal. This construction generalizes for three dimensions the
situations represented on Figs. 2.2 and 2.3.

The set of atoms, the replication of which generates the crystal, is the *basis* of the structure. For many crystals of common use
(metals, semiconductors, simple minerals, etc.), the basis contains a
single atom or a few atoms. In contrast, in crystals made of biological molecules (e.g. protein crystals), the basis can contain several
thousands atoms or even several hundred thousands atoms. Due to
the fact that the constituents of crystals are in contact with each
other, *the lengths of the primitive translations and the linear size of
the basis are in the same range of values (Fig. 2.4).*

COMMENT. The manner of generating a given crystal structure from
a basis and a set of translations is not unique. Thus, one can either
replicate a given basis, or alternately, replicate a larger group of

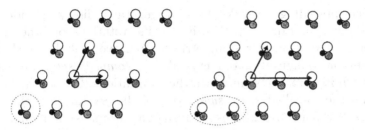

Fig. 2.5. Two possible sets (basis + translations) allowing to generate the same structure. Dotted circle (left) and ellipse (right): two bases. Arrows: corresponding primitive translations.

atoms constituted by a set of several such bases. In the latter case, certain of the three elementary translations will be multiples of the primitive translations considered initially. One can lift the ambiguity of the given definition by specifying that the basis is the smallest set of atoms allowing, by replication, the construction of the entire crystal structure[4] (Fig. 2.5).

It is worth pointing out that the translations $\overrightarrow{T}$ in Eq. (2.1) have been defined as a *tool of construction* of the structure through replication of the basis. A different point of view can be adopted. Hence, the structure being infinitely extended, its *overall displacement* by any of the translations (2.1) will consist in bringing a given set of atoms to a position where it substitutes an identical set. The structure is then *transformed* into an undistinguishable one. Each translation $\overrightarrow{T}$ in Eq. (2.1), viewed as a geometrical transformation acting in the three-dimensional space, leaves the atomic structure of the crystal globally *invariant* (unchanged). From this standpoint, $\overrightarrow{T}$ defines the *translational symmetry* of the structure.

Such a point of view can be extended to other geometrical transformations than translations. For instance, examination of the graphite surface in Fig. 2.2 shows that a rotation by an angle of 120 degrees about an axis passing through the origin, and perpendicular to the surface of the crystal, exchanges identical constituents and preserves globally the structure. Similarly, in Fig. 2.3, a reflection (mirror) symmetry through the horizontal plane **m** exchanges chemically identical atoms and is a symmetry of the structure.

---

[4]In Section 2.2, we will see that, even for the given "minimal" basis, the set of corresponding primitive translations $\overrightarrow{a_1}, \overrightarrow{a_2}, \overrightarrow{a_3}$ is not defined uniquely.

The preceding remarks have a general validity. Besides the translations in Eq. (2.1), certain specific rotations or other transformations (e.g. mirror symmetries) can leave globally unchanged the atomic configuration of a crystal. *The complete set of these symmetry transformations* (in which the translations (2.1) are included), defines the so-called *space-symmetry* of the crystal. It is this complete set which substantiates the idea of an atomic-scale regularity in crystals.

The possible space-symmetries of crystals have been classified into a finite number (230) of classes. The principles underlying this classification are outlined, in Section 2.5.

2.2 Bravais Lattices

2.2.1 Definition

The Bravais lattice associated to a given crystal structure is the three-dimensional, infinite, set of points formed by the endpoints $M(n_1, n_2, n_3)$ of all the vectors:

$$\overrightarrow{OM}(n_1, n_2, n_3) = \overrightarrow{T}(n_1, n_2, n_3) = n_1\overrightarrow{a_1} + n_2\overrightarrow{a_2} + n_3\overrightarrow{a_3} \qquad (2.2)$$

where O is an origin (Fig. 2.6). Each point M of this lattice is called a *node*. Two different meanings can be attached to the Bravais lattice of a crystal. First, this lattice can be considered as *part of the structure*, by choosing its origin O within the structural basis, e.g. on the nucleus of a specific atom of the basis. All the nodes will then be located on the nuclei of identical atoms. From this standpoint, the Bravais lattice constitutes the underlying skeleton of the structure. Putting "flesh" on this skeleton in the view of obtaining the actual structure consists in placing at each node M a set of atoms related to the basis by the displacement equation (2.2).

On the other hand, the Bravais lattice can be considered as an abstract entity, a mere set of points, separate from the structure of the crystal, but *associated* to it, and providing a *geometrical picture of the set of translations* $\overrightarrow{T}$ defining the periodicity of the crystal. As will be shown in Section 2.2.2, this picture is a convenient tool for deriving the rules governing the symmetry properties of crystals.

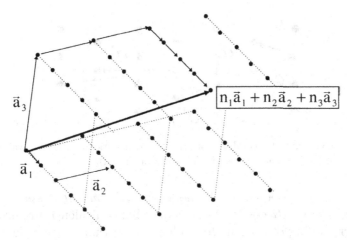

Fig. 2.6. Portion of the infinite three-dimensional lattice of points generated by Eq. (2.2).

2.2.2 Properties

Non-unicity of the generating set of primitive translations

Figure 2.7 provides an example of a two-dimensional lattice of points obtained by means of a formula of the type (2.2), involving two vectors $\vec{a_1}$ and $\vec{a_2}$. All the points of this two-dimensional Bravais lattice can also be generated by using the two basic vectors $\vec{a_1'} = (2\vec{a_1} + \vec{a_2})$ and $\vec{a_2'} = (\vec{a_1} + \vec{a_2})$ or, alternately, the two basic vectors $\vec{a_1''} = \vec{a_1}$ and $\vec{a_2''} = (-\vec{a_1} + \vec{a_2})$. Likewise, it is easy to check that the Bravais lattice of any three-dimensional crystal can be generated either by the set of vectors $(\vec{a_1}, \vec{a_2}, \vec{a_3})$ or, for instance, by the set $(\vec{a_1} - \vec{a_2}, \vec{a_2}, \vec{a_3})$. Hence, several choices of primitive translations are possible to generate the uniquely defined lattice of points underlying the structure of a crystal.

Equivalence of nodes: Lattice planes and rows

Owing to its infinite extension, a Bravais lattice is not modified by displacing its origin at any of the nodes of the lattice: indeed, in expression (2.2), one obtains the same infinite set of translations $\vec{T}$ by adding (or subtracting) fixed integers to the values of the n_i

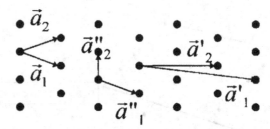

Fig. 2.7.   Existence of several sets of primitive translations for the Bravais lattice of a crystal. Each set generates all the nodes of the lattice.

indices. As a consequence, the nodes are all *equivalent since any of them can be substituted to the origin.* This equivalence implies that the surrounding of each node is identical to that of the origin, and that any property attached to a node must be realized in an identical manner for the other nodes.

Two remarkable properties stem from the equivalence between nodes:

(a) *All the nodes of a Bravais lattice can be grouped on a set of equally spaced parallel planes constituting a family of **lattice planes.** This distribution of all the nodes into a stack of parallel planes can be achieved in an infinity of manners by stacks having different orientations.*

(b) *Similarly, all the nodes of a Bravais lattice can be grouped, in an infinity of manners, on a set of equally spaced parallel lines constituting a family of **lattice rows.***

Let us establish, for the sake of illustration, property (a). Consider (Fig. 2.8), O, M, N any three non-aligned nodes of the Bravais lattice, with $\overrightarrow{OM} = \vec{v}$ and $\overrightarrow{ON} = \vec{w}$. They define a plane which we denote by **P**. Owing to the equivalence between lattice nodes, if we take as origin of $\vec{v}$ and $\vec{w}$ the node M or N (instead of O), the endpoints of these vectors define additional lattice nodes belonging to **P**. Repeating systematically this operation generates in **P** an infinity of lattice nodes. Other nodes (in infinite number) may exist in **P**, which are not generated by this procedure. The equivalence between *all the nodes* in **P** implies that their set is a two-dimensional Bravais lattice, for which one can choose two primitive translations $\overrightarrow{V}$ and $\overrightarrow{W}$.

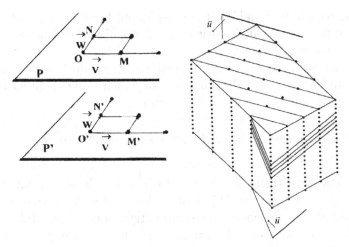

Fig. 2.8. Left: Construction of a family of lattice planes starting from three nodes of the lattice. Right: Decomposition of a Bravais lattice into two families of lattice planes.

If we now consider any node O' external to **P**, its equivalence to O implies that the plane **P′** passing through O' and parallel to **P** contains an identical two-dimensional Bravais lattice. By further repeating this procedure with points O'', etc., external to the planes already defined, an infinity of planes is generated.

Hence, as stated above, all the nodes of the lattice can be grouped on an infinite set of planes parallel to **P**. Moreover, the invoked equivalence between nodes implies that these planes are equally spaced. Let $\overrightarrow{X}$ be a vector joining two nodes of *consecutive* planes parallel to **P**. Clearly, $(\overrightarrow{V}, \overrightarrow{W}, \overrightarrow{X})$ constitutes a set of primitive translations of the crystal lattice, *adapted to the considered family of lattice planes*.

Since such a family of lattice planes can be defined by the initial choice of any three non-aligned nodes of the lattice, *a numerable infinity of families exist*. Each family, defined by the orientation of the perpendicular vector $\overrightarrow{u}$ (Fig. 2.8) corresponds to a specific spacing $d(\overrightarrow{u})$ between planes.

Lattice planes with different orientations do not *generally* contain the same planar density of nodes (unless they are symmetry related): the most dense planes are obtained by selecting three nearest-neighbour nodes in the lattice. The denser a lattice plane, the larger the associated spacing between planes, since the decomposition of the three-dimensional lattice into different families of lattice planes having different planar densities must determine the same number of nodes in a given volume.

A discussion similar to the preceding one would hold for the definition of lattice rows, the direction of which is determined by the selection of any two nodes of the lattice.

Point-symmetry of the lattice: Constraints on the rotation angles

The geometrical transformations which define the space-symmetries of crystals are (cf. Section 2.5) translations, rotations, mirror symmetries (also called reflexions), the symmetry about a point (also called inversion), as well as combinations of the former transformations.

It can be shown that any of the preceding geometrical transformation S can be denoted

$$S = \{R_O | \overrightarrow{t}\} \tag{2.3}$$

R_O is a *point-symmetry* transformation i.e. a pure rotation, a reflexion, an inversion, or a combination of these transformations. R_O leaves unmoved the origin O of the lattice (e.g. O is a point common to the rotation axes). $\overrightarrow{t}$ is a translation applied *after* R_O. In the framework of this notation, the lattice-translations $\overrightarrow{T}$ are expressed as $\{0|\overrightarrow{T}\}$ (rotation of angle $\theta = 0$ followed by the translation $\overrightarrow{T}$).

If O is a node of the Bravais lattice, this lattice is transformed into itself by the set of all the *point-symmetry* transformations R_O.[5]

For instance, if the structure of a crystal is unchanged by a transformation consisting in the succession of a $\pi/2$ rotation around the z-axis, *and* of a definite translation $\overrightarrow{t}$, the Bravais lattice is necessarily invariant by the *sole rotation*, applied about any of its nodes, and is therefore "square-shaped". The implications of this invariance property of Bravais lattices is described in Section 2.5.4. Let us only examine here the important constraint imposed to the angle θ of a rotation R_O.

---

[5]In other terms, the set of translations defining the periodicity of the crystal, must be compatible with the other symmetry transformations of the structure.

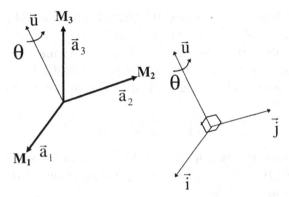

Fig. 2.9. Reference frames respectively adapted to the Bravais lattice and to the rotation axis.

Consider first in the three-dimensional space a frame of reference adapted to the Bravais lattice, and consisting of the set of primitive translations $\vec{a_1}, \vec{a_2}, \vec{a_3}$ defining the lattice (Fig. 2.9). Let $R_O(\vec{u}, \theta)$ be a rotation of angle θ around an axis $\vec{u}$ passing through the origin. Since we assume that R_O leaves the Bravais lattice globally invariant, the endpoints M_1, M_2, M_3 of the $\vec{a_i}$, which are nodes of the lattice, are transformed by the rotation into nodes M'_1, M'_2, M'_3 of this lattice. The latter statement can be expressed differently: the $\vec{a_i}$ are transformed into linear combinations of the $\vec{a_i}$ with *integer* coefficients, of the form (2.2). Hence, in this reference frame, the matrix $M(R_O)$ which represents the transformation of the basic-vectors of the frame is composed of integers. Accordingly, the *trace* of this matrix (the sum of its diagonal elements) is an integer N. The example hereunder is relative to a rotation of $2\pi/6$ around the z-axis in a hexagonal lattice:

$$(\vec{a_1}, \vec{a_2}, \vec{a_3}) \rightarrow \begin{bmatrix} 0 & 1 & 0 \\ -1 & 1 & 0 \\ 0 & 0 & 1 \end{bmatrix}$$

$$(\vec{i}, \vec{j}, \vec{u})$$

$$\rightarrow \begin{bmatrix} \cos\dfrac{2\pi}{6} & \sin\dfrac{2\pi}{6} & 0 \\ -\sin\dfrac{2\pi}{6} & \cos\dfrac{2\pi}{6} & 0 \\ 0 & 0 & 1 \end{bmatrix} \qquad (2.4)$$

Consider now an *orthogonal* reference frame (Fig. 2.9) defined by the unit vector $\vec{u}$ of the rotation axis, and two vectors $\vec{i}$ and $\vec{j}$ lying in the plane perpendicular to $\vec{u}$. In this frame, the trace of the matrix associated to a rotation R_O is $(1 + 2\cos\theta)$. The value of the trace of a matrix being independent of the reference frame chosen, one can write:

$$(1 + 2\cos\theta) = N \tag{2.5}$$

Solving this equation leads to a limited set of possible values for $\cos\theta$, namely $(0, \pm\frac{1}{2}, \pm 1)$. Thus, the possible values of the angle of a rotation R_O are as follows:

$$\theta = 0, \pm\frac{\pi}{3}, \pm\frac{\pi}{2}, \pm\frac{2\pi}{3}, \pm\pi \tag{2.6}$$

Any other value of the rotation angle such as, for instance, $\pi/6$, $\pi/5$, or $\pi/4$ is excluded.

2.3 Unit Cells

2.3.1 Primitive unit cells

Consider (Fig. 2.10) the parallelepipedic body defined by the three primitive translations $\vec{a_1}, \vec{a_2}, \vec{a_3}$. This body constitutes a *primitive* (*or elementary*) *unit cell* of the crystal. Displacing the unit cell by all the translations $\vec{T}$ of the form (2.1), generates a set of identical adjacent cells, and realizes a *paving* of the three-dimensional space, i.e. a filling, without overlap, of the entire space. Shifting the origin

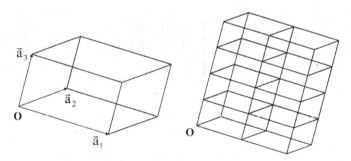

Fig. 2.10. Primitive unit cell of a crystal, and paving of space by adjacent unit cells.

of the paving by $(\vec{a_1}/2, \vec{a_2}/2, \vec{a_3}/2)$ with respect to a node, results in positioning a lattice-node at the center of each cell. Hence, there is a one-to-one correspondance between the nodes and the unit cells, the density of the nodes in space being equal to one node per volume of the primitive unit cell.

We have pointed out in Section 2.2.2 that the set of primitive translations of a crystal is not uniquely defined. Consequently, there is also an arbitrariness in defining a primitive cell. However, all the possible primitive cells have the same volume, owing to their one-to-one correspondance with the nodes of the Bravais lattice.

2.3.2　Conventional unit cells

One can generalize the definition of a unit cell by keeping two of the three properties of primitive unit cells: (a) parallelepiped based on the lattice translations; (b) paving of the space by an infinite set of identical bodies; (c) one-to-one correspondance between the cells and the nodes of the Bravais lattice. The two first properties are preserved in the definition of the *conventional unit cell*, while the two last ones are preserved in the definition of the *Wigner–Seitz unit cell*. Both the conventional cell and the Wigner–Seitz cell, unlike the primitive-cell, *are uniquely defined* for a given crystal.

In order to understand the advantage of recurring to alternate definitions, let us consider the example in Fig. 2.11, of a two-dimensional "rectangular centered" Bravais lattice. This lattice is invariant by mirror symmetries through the vertical and horizontal planes passing through any node. The conjunction of these two symmetries is characteristic of a *rectangle*. However, the primitive unit cell is not a rectangle but a diamond, the edges of which are two half diagonals of the rectangle. A rectangular cell can be constructed (Fig. 2.11) by using as edges the two lattice translations:

$$\vec{c_1} = (\vec{a_2} - \vec{a_1}) \quad \vec{c_2} = (\vec{a_1} + \vec{a_2}) \tag{2.7}$$

If one applies to this rectangular cell the translations defined by all the linear combinations of $\vec{c_1}$ and $\vec{c_2}$ with integer coefficients, one generates a set of identical adjacent rectangles which realize a paving of the plane. The rectangular cell is not a primitive-cell: the nodes generated by Eq. (2.7) coincide with half the nodes of the Bravais lattice (the centers of the cells are not accounted for). Each cell has

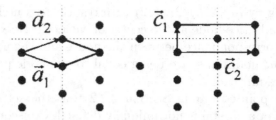

Fig. 2.11. Primitive and conventional unit cell of the two-dimensional rectangular centered Bravais lattice. The dotted lines are mirrors leaving the lattice invariant.

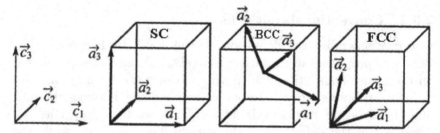

Fig. 2.12. Relationship between the edges of the conventional cell and the primitive translations of the simple cubic (SC), body centered cubic (BCC) and face centered cubic (FCC) Bravais lattices.

a surface twice that of a primitive unit cell, and is in correspondance with two nodes of the Bravais lattice.

One could define other rectangular cells by taking multiples of $\vec{c_1}$ and $\vec{c_2}$ as edges, with however the drawback of generating an even smaller fraction of the Bravais lattice nodes. The rectangular cell defined by (2.7), which is the *conventional unit cell* of the considered two-dimensional lattice, is the smallest of the rectangular cells.

More generally, the *conventional cell* of a crystal is the smallest parallelepiped, based on translations of the form (2.1), the shape of which displays in an obvious manner the symmetries (mirror reflexions, rotations) of the Bravais lattice. Figure 2.12 shows an example of conventional cell for three-dimensional Bravais lattices. In general, the conventional cell is a *multiple-cell*, associated to a density of several nodes per cell. There are cases, however, in which the conventional cell coincides with a primitive unit cell. Such a situation defines *simple Bravais lattices*.

2.3.3   Classification of the Bravais lattices: Cubic lattices

The distinction between primitive cells and conventional ones provides a means for classifying Bravais lattices. The resulting classification is a two-level one.

The first level consists in grouping together all the Bravais lattices possessing conventional cells having the *same shape*. For instance, a Bravais lattice will be classified as *cubic* if its conventional cell is a *cube*. For a cube of edge $\mathbf{c}$, the translations $(\vec{c_1}, \vec{c_2}, \vec{c_3})$ defining the conventional cell satisfy the conditions:

$$|\vec{c_1}| = |\vec{c_2}| = |\vec{c_3}| \quad \vec{c_i}.\vec{c_j} = \mathbf{c}^2.\delta_{ij} \qquad (2.8)$$

There are seven different shapes of conventional cells determining seven classes of lattices (Fig. 2.13): cube (*cubic lattices*), prism with square basis (*tetragonal lattices*), prism with rectangular basis (*orthorhombic lattices*), prism with parallelogram basis (*monoclinic lattices*), prism with a 120 degrees diamond basis (*hexagonal lattices*), rhombohedron (*rhombohedral lattices*), parallelepiped with arbitrary shape (*triclinic lattices*).

Within each of the former classes, a second level of classification distinguishes *Bravais lattices types* on the basis of the form of the relationship which exists between the vectors defining the conventional and the primitive cells.[6]

Let us consider for instance the case of cubic lattices. Three types of such lattices can be distinguished[7] (Fig. 2.12).

(a) *Simple cubic lattice (abbreviated as SC)*

For this lattice, the primitive and conventional cells coincide. The Bravais lattice is generated by the translations (2.8).

---

[6]Another base of the classification, related to symmetry-groups, is provided in Section 2.5.5.

[7]The distinction of only three distinct types of cubic Bravais lattices implies that we consider as equivalent the lattices having the same geometrical shape of conventional cell, but a different size (i.e. different value of $\mathbf{c}$).

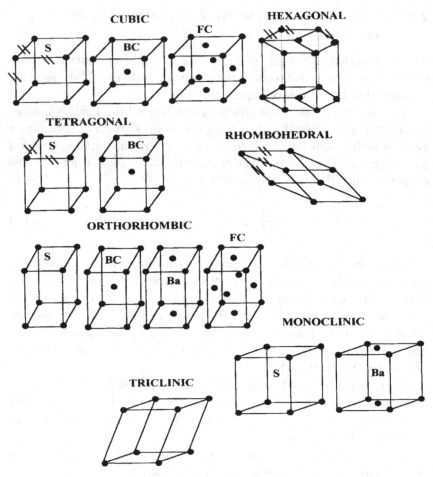

Fig. 2.13.   Fourteen types of Bravais lattices. S: simple; BC: body centered; FC: face centered; Ba: base centered.

(b) *Body centered cubic lattice (abbreviated as BCC)*

The primitive translations are related to those (2.8) defining the conventional cell by:

$$\vec{a_1} = \frac{1}{2}(\vec{c_1} + \vec{c_2} - \vec{c_3}) \quad \vec{a_2} = \frac{1}{2}(-\vec{c_1} + \vec{c_2} + \vec{c_3})$$

$$\vec{a_3} = \frac{1}{2}(\vec{c_1} - \vec{c_2} + \vec{c_3})$$

$$(2.9)$$

The mixed product $(\vec{a_1}, \vec{a_2}, \vec{a_3})$, determining the volume of the primitive unit cell, is half the volume $(\vec{c_1}, \vec{c_2}, \vec{c_3}) = \mathbf{c}^3$ of the conventional unit cell. This result is in agreement with the fact that the nodes of the Bravais lattice occupy the vertices *and* the center of the conventional cubic cell (thus justifying the terminology used). The shortest distance between nodes, equal to the length of a primitive translation, is $d_{\text{node}} = |\vec{a_i}| = \frac{1}{2}\mathbf{c}\sqrt{3}$.

(c) Face centered cubic lattice (abbreviated as FCC)

The relevant relationship is

$$\vec{a_1} = \frac{1}{2}(\vec{c_1} + \vec{c_2}) \quad \vec{a_2} = \frac{1}{2}(\vec{c_2} + \vec{c_3}) \quad \vec{a_3} = \frac{1}{2}(\vec{c_3} + \vec{c_1}) \qquad (2.10)$$

The volume of the primitive unit cell, $\frac{1}{4}\mathbf{c}^3$, is one-fourth of that of the conventional cell, in agreement with the fact that the nodes of the Bravais lattice occupy the vertices *and* the face-centers of the conventional cubic cell (consistently with the terminology). The shortest distance between nodes is $d_{\text{node}} = |\vec{a_i}| = \frac{1}{\sqrt{2}}\mathbf{c}$.

If one performs for the other shapes of conventional cells the same enumeration of distinct lattice types, one finds that there are **14** *Bravais lattice types* unequally distributed among the seven shapes of conventional cells. Their traditional labelling is indicated on Fig. 2.13. Seven of these Bravais lattice types are *simple* (i.e. the primitive cell is identical to the conventional one).

2.3.4 Wigner–Seitz unit cell

We have seen that for most of the Bravais lattices, the primitive unit cell does not display in an obvious manner the symmetry of the lattice. Besides, its one-to-one correspondance with nodes is hidden by the fact that the nodes are situated at its vertices and are shared between several cells. It is always possible to construct a "symmetric" cell associated to a single node situated at its center, and therefore free from the preceding drawbacks. However, unlike the primitive and conventional cells, it is not a parallelepiped with edges constituted by translations of the type (2.1).

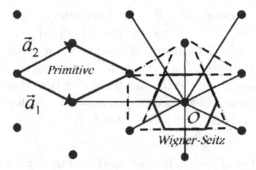

Fig. 2.14.   Primitive unit cell and Wigner–Seitz unit cell in a two-dimensional rectangular centered lattice.

Let us consider (Fig. 2.14) the segments joining a node O to all the other nodes of the lattice. The intersection of the mediator planes of these segments defines an inner "surface" (a volume in the three-dimensional case) around O. This volume is called the *Wigner–Seitz cell* of the lattice. Owing to its construction, all the symmetry-elements of the lattice about the origin O are necessarily symmetries of the Wigner–Seitz cell (mediator planes transform in the same manner as nodes). Besides, it has the obvious property of containing the points which are closer from the central node than from any other node of the lattice. This property clearly implies that the set of Wigner–Seitz cells associated to all the nodes are adjacent and fill the entire space without overlap. Besides, the one-to-one correspondance between nodes and Wigner–Seitz cells also warrants that this cell has the same volume as the primitive unit cells.

2.4   Examples of Crystal Structures

It will be shown, in the following chapters, that one can explain a number of general physical properties of crystalline solids (e.g. the form of their quantum states) by relying on a very simplified description of their atomic structure restricted to the mere specification of their Bravais lattice. However, a more detailed account of the properties of a specific solid (e.g. the recognition of its insulating or conducting character) requires the knowledge of its complete crystal structure, i.e. of the chemical nature and position of all the atoms in its primitive unit cell.

A great variety of structures is found in nature. This diversity is the source of two questions pertaining to the field of *crystal chemistry*. The first one concerns the classification of structures into "families" having similar atomic configurations and, relatedly, similar physical properties. The second one concerns the prediction of the crystal structure which will arise from the solidification of a given chemical substance.

As shown in Chapter 10 (Section 10.3), the cohesion of a crystal structure is ensured, in the same manner as the cohesion of isolated molecules, by the "valence electrons" of its constituting atoms. Conversely, in a solid, the state of the valence electrons is determined by the crystal's structure. The complexity of this mutual dependence for as large a system as a solid (which involves a macroscopic number of atoms and electrons) explains that the two underlined questions have no general answers at present.

Let us restrict here to a simplified approach, the validity of which can be argumented in the case of certain *monoatomic* crystal structures, i.e. the structures built from a single chemical element of the Mendeleiev table (Fig. 2.20). Other types of structures will only be qualitatively mentioned in Sections 2.4.2 and 2.4.3.

2.4.1 Simple monoatomic structures: Packings

One can expect that solids made from a single type of atom will possess a simpler crystal structure than other solids. This is only partly true. It is generally observed that even a single chemical element (e.g. iron or carbon) will give rise, as a function of temperature or pressure, to several different crystalline structures some of which are fairly complex. Nevertheless, in a large class of monoatomic solids, mainly those made from elements situated in the left part of the Mendeleiev table (Fig. 2.20), the predominantly observed structure resembles the one obtained by *packing together hard spherical balls*.

The existence of such simple structures relies on the fact that their cohesion is determined by *non-directional bonds between atoms:* atoms are in mutual attraction without favouring any specific direction. Accordingly, structures can be built by piling together atoms in contact, as if these atoms were hard spheres. Such an idea holds mainly for "good metals". Indeed, in this case, the cohesion of a crystal is ensured by *free-electrons* having an approximately *isotropic distribution in space*. Such a situation differs from that of isolated molecules, for which the angles between covalent bonds are often constrained to specific values (e.g. the tetrahedral configuration of bonds found in carbon compounds).

There are several manners of packing hard spheres to build a crystalline structure. Two of the resulting structures are termed *closed-packed*. This qualification means that, for a given radius of the atomic sphere, the compacity of the packing is maximum: the voids existing between the atoms occupy a smaller fraction of the total volume than in any other packing.

Cubic close-packing

This packing is obtained by placing an atom of the considered chemical species at each node of an FCC-lattice (cf. Section 2.3.3). The edge-length c of the conventional-cubic cell is such as to ensure the contact between the atomic spheres (Fig. 2.15). If r is the atomic radius, and $|a_i|$ the length of a primitive translation of the lattice, the latter condition implies the relationship $|a_i| = c/\sqrt{2} = 2r$. Each atom is in contact with 12 closest neighbours situated along the face diagonals at a distance $2r = |a_i|$.

The construction of the structure (one atom at each node) determines a density of one atom per primitive unit cell. The conventional cell contains four atoms (cf. Section 2.3.3). The compacity of the structure can be characterized by the ratio between the volume of

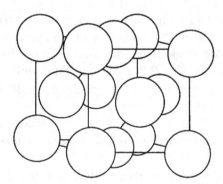

Fig. 2.15. Cubic close packing. For the sake of clarity, the atomic radius has been reduced.

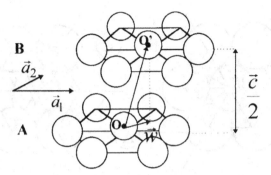

Fig. 2.16. Position of the atoms in the hexagonal close-packing. The radius of the atom has been reduced for the sake of clarity.

an atom and that of the primitive unit cell:

$$\frac{V_{\text{atom}}}{V_{\text{cell}}} = \frac{4\pi r^3/3}{c^3/4} = \frac{\pi}{3\sqrt{2}} = 0.74 \tag{2.11}$$

Hexagonal close-packing

The construction of this packing is more complex (Fig. 2.16). Consider first the hexagonal Bravais lattice (Fig. 2.13) generated by the set of primitive translations $(\vec{a_1}, \vec{a_2}, \vec{c})$ where $\vec{a_1}$ and $\vec{a_2}$ are vectors of equal length making an angle of 60 degrees, and where $\vec{c}$ is perpendicular to the $(\vec{a_1}, \vec{a_2})$ plane. In each primitive cell, place one atom at the origin O and the other at O' such as $\overrightarrow{OO'} = \frac{1}{3}(\vec{a_1} + \vec{a_2}) + \vec{c}/2$. The atoms of the structure will be in direct contact with their neighbours if the values of the atomic radius **r** and of the parameters **c** and **a** are related by:

$$\frac{c}{a} = \frac{c}{2r} = 2\sqrt{\frac{2}{3}} \tag{2.12}$$

In this structure, the basis contains *two* identical atoms (located respectively at O and at O'). Thus, the compacity is given by

$$\frac{2V_{\text{atom}}}{V_{\text{cell}}} = \frac{8\pi r^3/3}{a^2 c\sqrt{\frac{3}{2}}} = 0.74 \tag{2.13}$$

It has the same value as in the cubic close packing.

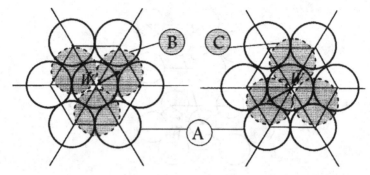

Fig. 2.17. Position of the B or C planes (shaded spheres) with respect to the A close-packed planes.

If one applies to the atom located in O the lattice translations generated by $(\vec{a_1}, \vec{a_2}, \vec{c})$, one obtains a succession of planes (which we label **A**), parallel to $(\vec{a_1}, \vec{a_2})$, spaced by $|\vec{c}|$, and containing one half of the atoms of the structure. In each of these planes, atoms form a regular lattice of equilateral triangles (Fig. 2.17). Clearly, this configuration is the most compact that can be formed in two dimensions with identical "hard disks" in contact with each other. Note that the atoms of consecutive **A**-planes form columns parallel to $\vec{c}$.

Applied to the atom located in O', the set of lattice translations generates a second family of atomic-planes (labelled **B**) having identical internal configurations and spacing as **A**, each **B**-plane being at half distance between two **A**-planes. With respect to the columns generated by the **A**-atoms, the **B**-columns are shifted laterally by

$$\vec{w} = \frac{1}{3}(\vec{a_1} + \vec{a_2}) \qquad (2.14)$$

The **B**-atoms lie above or below voids between adjacent **A**-atoms.

The hexagonal close-packing can therefore be described as a stacking **A–B–A–B** ... of planes having each a close-packed atomic configuration in two dimensions, consecutive planes being shifted by $\pm\vec{w}$ with respect to each other.

Relationship between close-packings

In the cubic close-packing, the atomic planes perpendicular to one of the cube-diagonals, and defined by face-diagonals of the cubic cell,

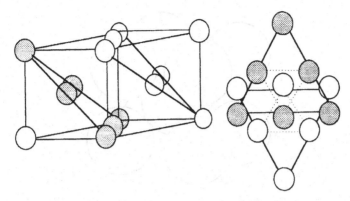

Fig. 2.18.   Decomposition of the cubic close packing. The structure of the most dense planes (defined by face diagonals) A, B, and C are represented on the right.

are the most dense lattice planes of the structure since they correspond to the shortest distances $|\vec{a_i}| = 2\mathbf{r}$ between atoms. It is easy to see that their structure is the close-packed two-dimensional triangular configuration of atoms (Fig. 2.18). Similarly to the hexagonal close packing structure, the cubic close-packed one can be described as a stacking of close-packed planes.

Figure 2.18 shows that the stacking sequence **A–B–C–A–B–C**...
is different from the hexagonal sequence **A–B–A**. Indeed, along a cube-diagonal one finds successively: (a) a plane **(A)** containing the lower corner of the cube; (b) a plane **(B)** containing the diagonal of the lower face of the cube, with atoms shifted by the displacement $\vec{w}$ with respect to **A**; (c) a plane **(C)** containing the diagonal of the upper face, with atoms shifted by $2\vec{w}$ with respect to **A**. The following plane, containing an atom at the upper corner, is an **A**-plane.

Hence, it is the shift between the second and third planes $(\mp \vec{w})$ which distinguishes the structures of the cubic and of the hexagonal close-packings. The similarity of construction of the two structures explains their identical compacity (Fig. 2.18).

Body centered cubic packing

This packing (Fig. 2.19) is obtained by putting an atom at each node of a BCC Bravais lattice (cf. Section 2.3.3). The direct contact between nearest neighbour atoms requires that the radius of an atom equals half the length of the primitive translation of the lattice:

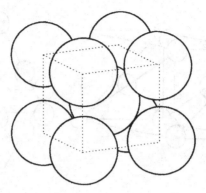

Fig. 2.19.   Body centered cubic packing.

$\mathbf{r} = |\vec{a_i}|/2 = \mathbf{c}\frac{\sqrt{3}}{4}$. The compacity of this packing is

$$\frac{V_{\text{atom}}}{V_{\text{cell}}} = \frac{4\pi r^3/3}{\mathbf{c}^3/2} = 0.68 \qquad (2.15)$$

It is smaller than in the close-packed structures. The shortest distance between atoms being half the length of the cube-diagonal, the most dense planes are defined by two cube-diagonals (or two face-diagonals belonging to opposite faces). The structure of these planes is different from that of the close-packed planes, each atomic sphere being in contact with four atoms only instead of six.

Physical realizations in metals

Part of the Mendeleiev table of chemical elements is reproduced in Fig. 2.20 All the elements of the left-hand side (I) of this table, which are metals in the solid state, give rise to structures which can be considered as examples of the packings described above. Hence, a simple model of hard spheres in contact with each other accounts satisfactorily for the occurrence of a large number of structures actually observed in nature.

However, a more careful examination of the table suggests that the mechanisms behind the observed situation are more complex than the geometrical considerations underlying the construction of the packings. Thus, one can wonder why, in a given chemical element, one of the two close-packed structures is favoured over the other. On the other hand, the high frequency of occurrence of the less-compact

Li bc	Be hc					(I)						B	C	
Na bc	Mg hc											Al fc	Si	P
K bc	Ca fc	Sc hc	Ti hc	V bc	Cr bc	Mn bc	Fe bc	Co hc	Ni fc	Cu fc	Zn	Ga	Ge	S
Rb bc	Sr fc		Zr hc	Nb bc	Mo bc		Ru hc	Rh fc	Pd fc	Ag fc	Cd	In	Sn	Se
Cs bc	Ba bc		Hf hc	Ta bc	W bc		Os hc	Irr fc	Pt fc	Au fc	Hg	Tl	Pb	Te

Fig. 2.20. Portion of the table of chemical elements. Left part (I) mainly observed simple packing structures of metallic elements: bc (body centered packing), fc (cubic close packing), hc (hexagonal close packing). The thick line distinguishes covalent bonded semiconductors.

body centered structure is puzzling in regard of the more "natural" stability of the close-packed structure for a set of mutually attracted hard spheres. The existence, for each element, of several possible structures (e.g. in barium), as a function of temperature or pressure is also a matter of reflexion. Among the refinements which can be addressed to explain the observed situation, two important ones are, in the first place, the *role of conduction electrons* and, on the other hand, the effect of *entropy*.

2.4.2   Structures derived from simple packings

The model-structures built from hard spheres are not only useful to understand the structure of monoatomic metals, they also provide a starting-point for describing more complex structures containing several atoms (identical or different) in their atomic-basis.

An illustrative example is that of sodium chloride, NaCl, in which one has to consider two types of spheres respectively associated to a positive sodium ion (Na^+) and to a negative chlorine ion (Cl^-). In a first approximation, the structure of sodium chloride can be described as a cubic close-packing of chlorine ions. Such a packing has two types of "interstitial" voids between the atoms, characterized by different surroundings. The relevant one in sodium chloride is at the center of the cubic cell, or equivalently at the middle of the edges (Fig. 2.21). This void is surrounded by six nearest-neighbours chlorine ions forming an *octahedron*. In principle, the available space allows to locate in it a sphere of maximum radius 0.41r, in which

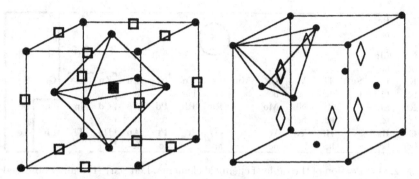

Fig. 2.21. Left: Nodes (filled circles) and octahedral sites (squares) of the cubic close packing. Right: tetrahedral sites (diamonds) of the cubic close packing.

r is the radius of the ion constituting the close-packing. In sodium chloride, the sodium ion is located at these interstitial sites although the radius of the sodium ion is $\sim 0.67r$ ($r_{chlorine} = 1.7\,\text{Å}$). This clearly implies that the configuration of the chlorine ions differs significantly from that of spheres in contact. Nevertheless, the close-packing structure remains a convenient reference for a description of the structure.

A similar example is calcium fluoride CaF_2 which can be approximated as a cubic close-packing of divalent calcium ions Ca^{++}. The fluorine ions F^- occupy a second type of interstitial voids of the packing. These voids are centered on the diagonals of the cubic-cell, at half-distance between the center of the cube and the vertices (Fig. 2.21). Each void is surrounded by four nearest-neighbours calcium ions, forming a *tetrahedron*. One can insert in it a sphere of maximum radius $0.22r$. Here again, the F^- ion is larger than the available space.

As compared to the reference packings, the cohesion of the preceding structures is, nevertheless, very efficiently ensured by the strength of the electrostatic *isotropic* attraction between positive and negative ions.

A similar illustration of an approximate hexagonal close-packing is provided by nickel arsenide NiAs, in which the hexagonal packing is realized by the arsenic atoms, the nickel ones being located in interstitial sites having octahedral surrounding.

2.4.3   Simple covalent structures: Diamond and semiconductors

The chemical elements situated in the right-hand part of the Mendeleiev table do not crystallize as simple atomic packings of hard spheres. Similarly to the case of solid-carbon, this is related to the fact that the cohesion between constituents of the corresponding solids relies on a *covalent* bonding (cf. Chapter 9) having a marked directional character. Hence, carbon (C) tends to be surrounded by four atoms forming a regular tetrahedron. This is also the case for silicon (Si) and for germanium (Ge).

The crystal structure common to the three preceding chemical elements is the so-called *diamond structure*. Its Bravais lattice is of the FCC type (Fig. 2.12). Its basis contains two identical atoms (Fig. 2.22), and the conventional cell therefore contains eight atoms. One is at the origin of the primitive unit cell $(\overrightarrow{a_1}, \overrightarrow{a_2}, \overrightarrow{a_3})$ defined by Eq. (2.10). The second is located at $\frac{1}{4}(\overrightarrow{a_1} + \overrightarrow{a_2} + \overrightarrow{a_3}) = \frac{1}{4}(\overrightarrow{c_1} + \overrightarrow{c_2} + \overrightarrow{c_3})$. Each atom (C, or Si, or Ge) is at the center of a tetrahedron formed by its closest neighbours. If we assume the atoms to be spheres in contact with their neighbours, the relationship between their radius **r** and the cube-edge **c** is $\mathbf{r} = \frac{1}{8}|\overrightarrow{c_1} + \overrightarrow{c_2} + \overrightarrow{c_3}| = \mathbf{c}\frac{\sqrt{3}}{8}$. The compacity

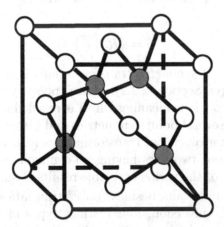

Fig. 2.22. Diamond-type structure. The white atoms are at the nodes of the FCC lattice. The shaded ones in the tetrahedral voids. Each carbon atom is at the center of a regular tetrahedron of atoms.

of the structure can then be evaluated as

$$\frac{8V_{\text{atom}}}{V_{\text{cell}}} = \frac{32\pi\mathbf{r}^3/3}{\mathbf{c}^3} = \frac{\pi\sqrt{3}}{16} = 0.34 \qquad (2.16)$$

It is much smaller than that of the closed-packed structures (0.74). The strong cohesion of the diamond structure, which is reflected, for instance, in the mechanical hardness of the crystals, is not related to the compacity of the atomic configuration but, rather, to the strength of the covalent bonds.

2.5   Classification of Crystal Symmetries

As mentioned in Section 2.1, the space-symmetry of a crystal is the set of geometrical operations which transform the structure of the crystal into itself. We describe briefly in this section the main steps of the argumentation which leads to a classification of the space-symmetries of crystals into a limited number (230) of possible classes.

2.5.1   Symmetry transformations: Space-group

As indicated in Section 2.2.2, any geometrical transformation S belonging to the space-symmetry of the structure can be denoted by:

$$S = \{R_O | \vec{t}\} \qquad (2.17)$$

where R_O is a *point-symmetry* transformation about the origin, and $\vec{t}$ a translation (cf. Section 2.2.2). "Pure" point-symmetry transformations have an obvious meaning which is recalled on Fig. 2.23. The combinations of several point-symmetries and translations can result in fairly complex geometrical transformations (e.g. the succession of two rotations about two axes having no common point). A systematic examination of the various possible results of such combinations shows that one can obtain, besides "pure" translations $\vec{T}$, rotations, reflexions and the inversion, four distinct types of transformations (Fig. 2.23). Two consist in the succession of a rotation and of either an inversion about a point situated on the rotation axis, or a reflexion in a mirror perpendicular to this axis. The two remaining ones are the *screw-rotation*, which is the succession of a rotation and of

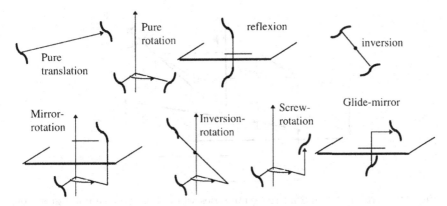

Fig. 2.23. Transformations defining the space-symmetry of a crystal.

a translation parallel to the rotation axis, and the *glide-plane reflexion*, which is the succession of a mirror reflexion and of a translation parallel to the mirror-plane.

Clearly, if two transformations S and S' both have the property of transforming a structure into itself, the succession of S and S', i.e. the product $S'.S$ of these transformations, has the same property. Mathematically, this circumstance is characterized by stating that the set of all the S-transformations (2.17) defining the space-symmetry of the structure form a *group of transformations* **G** which is called the *space-group* of the crystal.

2.5.2 Point-group and Bravais lattice

It can be shown that the succession of two transformations S and S' is

$$S'.S = \{R'_O|\overrightarrow{t'}\}\{R|\overrightarrow{t}\} = \{R'_O.R_O|\overrightarrow{t'} + R'_O.\overrightarrow{t}\} \qquad (2.18)$$

where $R'_O.\overrightarrow{t}$ is the vector obtained by applying to $\overrightarrow{t}$ the rotation, reflexion, or inversion R'_O. From the form of the last member of Eq. (2.18) one infers that the product $R'_O R_O$ is the point symmetry of an element (2.17) of the space group, if R_O and R'_O already have this property. Hence, the set R_O constitutes a group $\widehat{\mathbf{G}}$ which is called the *point-group of the crystal*. $\widehat{\mathbf{G}}$ has an important property: If the origin O is a node of the Bravais lattice, the set of elements of $\widehat{\mathbf{G}}$ transform the Bravais lattice into itself: $\widehat{\mathbf{G}}$ represents a *symmetry of*

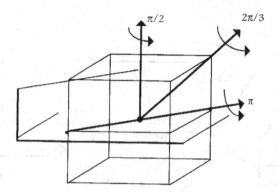

Fig. 2.24. Rotations and mirror symmetries which generate the point-group of a cube. This group has 48 elements.

the lattice.[8] As will be shown in Section 2.5.4, $\widehat{\mathbf{G}}$ is not, in general, the *full* symmetry of this lattice.

2.5.3   Enumeration of point-groups

Since the angles of the rotations can only take the limited set of values listed in Section 2.2.2, the systematic construction of the point-groups of crystals can be achieved by combining certain, or all, of the following elements: rotations with angles restricted by Eq. (2.6), reflexions, and the inversion. It can be shown that 32 distinct point-groups $\widehat{\mathbf{G}}$ can be enumerated. They all possess a relatively small number of elements. A simple example is provided by the set of four rotations of angles $(0, \pi/2, \pi, 3\pi/2)$ around a common axis. The largest, with 48 elements, comprises beside the inversion, all the rotations and the reflexions which leave a cube invariant (Fig. 2.24).

2.5.4   Symmetry of the Bravais lattice

Figure 2.25 shows a crystal structure for which the Bravais lattice is invariant by a point-group larger than $\widehat{\mathbf{G}}$ and containing 16 elements (instead of eight for $\widehat{\mathbf{G}}$). The generalization of this property

---

[8]Note, however, that, in general, R_O is not a symmetry of the crystal structure: it is only the succession of R_O and of its associated translation $\vec{t}$ which has this property.

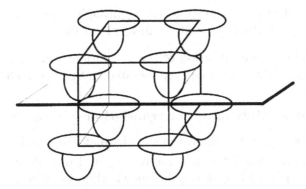

Fig. 2.25. Tetragonal crystal in which the structure is not symmetric with respect to the horizontal plane. The Bravais lattice, where each "bell-shaped" constituent is replaced by a node obviously has this symmetry.

implies that besides being invariant by the point-group $\widehat{\mathbf{G}}$ of the crystal, the Bravais lattice has *additional invariances*, consisting of certain reflexions (or the inversion) which do not belong to $\widehat{\mathbf{G}}$. Hence, while the point-symmetries $\widehat{\mathbf{G}}$ of crystal structures can be any of the 32 groups mentioned hereabove, there are only **seven** possible symmetry-groups of Bravais lattices. These special point-groups, which are part of the 32 point-groups,[9] are associated to the various possible shapes of the Bravais lattices: consistently, seven distinct shapes of conventional cells were listed in Section 2.3.2.

2.5.5   Classification of Bravais lattices

A classification of Bravais lattices was achieved in Section 2.3.3 by using the shape of conventional cells. Alternately, the *equivalence* of two Bravais lattices can be defined by the following conditions:

(a) The two lattices are invariant by the same point-group.
(b) One lattice can be brought to coincide with the other by deforming continuously its generating vectors $(\overrightarrow{a_1}, \overrightarrow{a_2}, \overrightarrow{a_3})$, without, however, lowering the symmetry of the lattice during the process. Hence, it is possible to transform any simple cubic lattice into

---

[9]These groups also involve the restriction on rotations which was already discussed for Bravais lattices in Section 2.2.2.

another lattice of the same type, through a mere dilatation of the edges of the cubic-cell by an identical factor.

The systematic application of the two former conditions leads to the 14 Bravais classes already enumerated in Section 2.3.3.

2.5.6   Constraints on other translations

As shown on Fig. 2.23, translations $\vec{t}$ distinct from the lattice translations $\vec{T}$, are involved in *screw-rotations* and in *glide-plane* reflexions. Using Eq. (2.18), it can be shown that for a screw rotation $\{R_O|\vec{t}\}$ of angle $\theta = \frac{2\pi}{n}$, the component of $\vec{t}_{//}$ parallel to the rotation axis is necessarily of the form $\vec{t}_{//} = \frac{m}{n}\vec{T}$ (m and n integers, with $m < n$), where $\vec{T}$ is a Bravais-lattice translation. Likewise, in a glide-plane reflexion, the mirror-symmetry can only be associated to $\vec{t} = \frac{1}{2}\vec{T}$.

2.5.7   Classification of space-symmetries

The three-fold restriction on: (a) the number of Bravais lattice types; (b) the number of point-groups; (c) the possible translations associated to screw-rotations and glide-planes, makes it possible to classify the symmetries of crystal structures into a finite number of classes. Two structures will belong to the same class if:

- their Bravais lattices are of the same type;
- their point-groups are the same;
- their screw-rotations and their glide planes are respectively associated to the same fractions of lattice translations.

Combining the various possibilities (14 types of lattices, 32 point-groups, a few restricted types of screw-rotations and glide-planes) one finds **230** *possible classes of space-groups.*[10]

---

[10]It is possible to analyze in a more general, though more abstract, manner the classification of crystal space-groups into classes. Such an approach has led to the definition of hypothetical crystals belonging to spaces having more than three dimensions. As often in physics, these considerations, initially unrelated to the experimental reality, have become relevant in interpreting effectively observed situations in complex solids (cf. modulated and quasi-crystals in Section 2.6).

2.6 Complex Translational Orders

The translational order of crystals is associated to a three-fold spatial periodicity (i.e. three non-colinear primitive translations) and to the possibility of paving the three-dimensional space by replication of a *single elementary volume* (a unit cell). Certain systems display similar, though more complex, translational orders as crystals. This is, in particular, the case of *modulated systems* and of *quasi-crystals*. To a certain, extent one can also consider as a complex translational order the situation in systems displaying a magnetic behaviour.

Incommensurably modulated solids

In a first approximation, *modulated crystals* are solids possessing an ordinary crystalline structure, defined by a basis and a set of lattice translations. However, a more careful examination of their structure shows that the atomic groups associated to each node (or to each cell) are all different from each other, and all slightly deformed with respect to the ordinary crystal structure serving as reference. More precisely, parallel to specific directions of the structure (a single one in the simplest case), the amplitude of this deformation is a periodic function of the distance to the origin (Fig. 2.26). In each direction in which such a modulation occurs, the spatial-period is in an *irrational* (*incommensurable*) *ratio* with the period of the crystal-lattice serving as reference.[11] The period of the deformation is generally a few times the value of the lattice period (in the range of tens of angströms). A modulated structure can therefore be considered as the result of the superimposition of a *spatially-periodic deformation-wave* upon an ordinary crystalline order. Such a system has *at least four spatial-periods* instead of three in ordinary crystals. Due to the occurrence

---

[11]Experimentally, this "irrational" character is asserted on the basis of two types of observations: (a) The ratio between the period of the modulation and the period of the lattice is a "large" integer" such as 7 or 15. (b) This ratio appears to change continuously as a function of temperature, and is thus suspected to take irrational values. However, in some cases, theoretical or experimental reasons favour the assignment of "complex rational ratio". This is for instance the case in metallic alloys with "long periods".

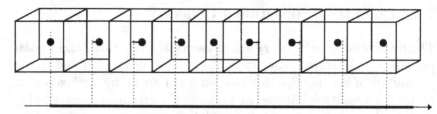

Fig. 2.26.   Incommensurately modulated deformation of a crystal structure along the horizontal axis. Average crystal period (interval between dotted lines). Approximate periods of the modulation (thick line), ≈ 7 average periods, are shown.

of these additional periods, unrelated to the three others, this three-dimensional system appears as *non-periodic* in three dimensions since all its atomic-groups are different and are not mere replications of a basis.

Incommensurately modulated structures have been observed in a large variety of solids: metals, semiconductors, insulators, mineral or organic compounds.[12]

In spite of their lack of periodicity, incommensurately modulated structures are as *perfectly ordered* systems as crystals. Indeed, starting from the group of atoms situated near the origin, one can generate the entire structure on the basis of the knowledge, on the one hand, of the primitive translations of the reference crystalline structure, and, on the other hand, of the amplitudes and periods of the modulation waves. No uncertainty is thus involved in the determination of the positions of the atoms.

The perfect order, and the "multiperiodicity" of modulated structures underly a description of these systems which is a simple extension of the description of ideal crystals. In this view, one uses a mathematical trick consisting in describing their atomic configuration in an abstract space having $(3 + d)$ dimensions, i.e. a number of dimensions equal to the total number of observed spatial-periods of the system (3 primitive translations and d periods of modulation). In this space, the configuration associated to the modulated structure is *an ideal crystal structure having $(3 + d)$ primitive translations.*

---

[12]Prominent examples are the metallic alloy TaS_2, the semi-organic compounds thiourea $SC(NH_2)_2$ and hydrated betaine calcium chloride (see Fig. 2.4), and the insulating compounds K_2SeO_4 and $Ba_2NaNb_5O_{15}$.

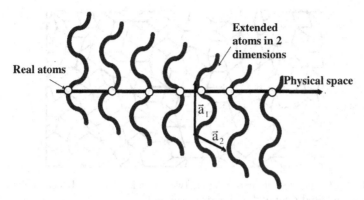

Fig. 2.27. Inbedding of a one-dimensional incommensurately modulated structure in a two-dimensional crystalline structure. Each atom is replaced by a "rib-
bon". The actual structure is the section of these ribbons by the "real axis".

The actual structure in three-dimensions is obtained as the section of this hypothetical structure by the physical three-dimensional
space. Figure 2.27 illustrates such a description for a one-dimensional
modulated structure.

Note that, generally, an incommensurately modulated crystal
phase is stable in a limited temperature-range. On cooling, this phase
transforms into a structure displaying an ordinary crystalline order.

Quasi-crystals

The translational order of crystals implies the possibility of generating a Bravais lattice by means of an infinite set of *adjacent identical cells.* In certain metallic alloys, it has been observed that the
atomic configuration could not be described by a crystalline order.
Nevertheless, the structure has an underlying lattice which can be
constructed by using *two distinct shapes of adjacent cells,* instead of
a single one for ordinary Bravais lattices. The shapes and sizes of
these cells are such as to permit, through definite assembling rules, a
paving of the three-dimensional space by adjacent cells. An illustration, in two-dimensions, is provided by Fig. 2.28. Such a lattice has
no periodicity since any two nodes possess a different surrounding.
However, if attention is given to the immediate surrounding (the
nearest neighbours) of nodes, it is found that an infinity of nodes

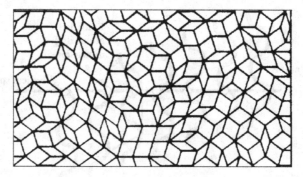

Fig. 2.28. Quasicrystalline paving in two dimensions by adjacent cells of two shapes (elongated diamond and wider diamond).

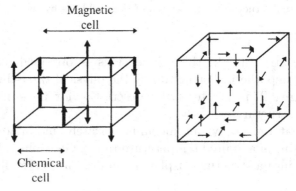

Fig. 2.29. Left: Chemical and magnetic cells of an hypothetical antiferromagnet. Right: Dipole orientation in a complex magnetic structure, nickel iodine boracite.

are equivalent. The similarities with the situation of crystals (paving of space, lattice displaying some regularity) have led to term these systems *quasi-crystals*.[13]

Magnetic order

Magnetic crystals are crystalline solids in which certain atoms of the structure carry magnetic dipoles. The orientations of these dipoles

---

[13]Similarly to incommensurately modulated systems, the structure of these systems can be associated to an ideal crystalline structure in an hypothetical six-dimensional space.

display a definite order, which, though related to the crystalline order is distinct from it. For instance, in a hypothetical *antiferromagnet,* the magnetic atoms can occupy the vertices of a simple cubic lattice, the directions of consecutive dipoles being antiparallel (Fig. 2.29) In such a solid, two translational orders can be distinguished. The *structural order* is relative to the spatial periodicity of the atoms regardless of their magnetic dipoles. This order is associated to a "chemical" Bravais lattice and a unit cell, which in the chosen example, is the cube of edge $\mathbf{c}$. The *magnetic order* is defined by the periodicity of the oriented dipoles, and is associated to a "magnetic" cell. In Fig. 2.29, the resulting magnetic lattice is of the FCC-type with a conventional cell of edge $2\mathbf{c}$.[14]

Bibliography

International Tables for Crystallography, Vol. A, *Edited by the International Union of Crystallography* (2016).

Space-Groups for Solid State Scientists, *M. Glazer and G. Burns*, Academic Press (2013).

Introduction to Crystallography, *Frank Hoffmann*, Springer (2020).

The Landau Theory of Phase Transitions, *J.C. Toledano and P. Toledano*, World Scientific (1987).

Incommensurate Crystals, *T. Janssen and A. Janner*, Advances in Physics Vol. 26, Issue 5 pp. 519–624 (1987).

Useful Quasicrystals, *J.M. Dubois*, World Scientific (2005).

---

[14] A number of examples exist in which the ordering of the dipoles is associated to a period incommensurate with that of the chemical lattice.

Chapter 3

The Reciprocal Space as a Space of Quantum Numbers

Main ideas: Relation between crystal periodicity and Bloch-form of the wavefunctions (Section 3.2). Translational quantum numbers $\vec{k}$ (Section 3.3). Equivalence between quantum numbers; reciprocal lattice and first Brillouin zone (Sections 3.3.2–3.3.4). Born–Von Karman *ansatz* and quantization of the $\vec{k}$-values (Section 3.3.6). Energy-dispersion curves, energy-bands (Section 3.4). Reminder of quantum mechanics (Section 3.5).

3.1 Introduction

We have pointed out, in Chapter 1, that in order to account satisfactorily for the electronic properties of a solid, it was necessary to study these properties in the framework of quantum mechanics. In the present chapter, we show that the quantum states of a crystalline solid have a specific form which is imposed by its three dimensional periodicity. More precisely, the wavefunction $\psi(\vec{r})$ of a particle of the solid, such as an electron, is partly defined by *quantum numbers* whose nature and range of values derive from this periodicity. Such a statement implies that the quantum states of a single electron can be considered independently from the states of the other particles of the solid. We will provisionally accept the validity of this assumption and postpone the discussion of its limitations to Chapter 10. A short

63

reminder of quantum mechanics can be found in the appendix of this chapter.

3.2 Bloch's Theorem

In this section, we determine the functional form, termed the *Bloch functional form*, of the wavefunction $\psi(\vec{r})$ of a single particle in a crystalline solid (e.g. an electron). The derivation of this result requires going through three preliminary steps (Sections 3.2.1–3.2.3).

3.2.1 Quantum operator associated to a translation

Consider a particle of the crystal (e.g. an electron), whose quantum state is described by the wavefunction $\psi(\vec{r})$. This function defines the distribution in space of the *amplitude of probability of presence* of the particle.[1] Such a distribution is schematically represented on Fig. 3.1 by a "cloud" surrounding the "center" of the particle.

If we displace globally the crystal by a translation of vector $\vec{T}$, the cloud representing the amplitude of probability keeps its form and orientation but is displaced by $\vec{T}$. Since the probability of presence of the particle at a given point $\vec{r}$ is modified, so is the quantum state of the particle, and its representation by a wavefunction $\psi'(\vec{r})$.

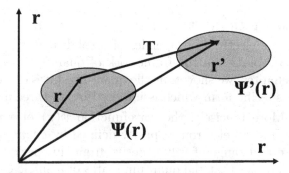

Fig. 3.1. Schematic representation of the spatial distribution of the probability of presence of a particle before and after a translation of the system. The initial wavefunction at $\vec{r}$ is the same as the final one at $\vec{r}' = \vec{r} + \vec{T}$.

---

[1]The probability of finding the particle in the elementary volume $dv = dxdydz = d^3\vec{r}$, is $|\psi(\vec{r})|^2 d^3\vec{r}$, where $|\psi(\vec{r})|^2$ is the density of probability (cf. Section 3.5).

Inspection of Fig. 3.1 shows that the correspondance between the initial and final probability-distributions is expressed by:

$$\psi'(\vec{r} + \vec{T}) = \psi(\vec{r}) \tag{3.1}$$

or equivalently by:

$$\psi'(\vec{r}) = \psi(\vec{r} - \vec{T}) \tag{3.2}$$

Let us define the *quantum operator* τ associated to the translation $\vec{T}$, as the operator which transforms the initial quantum state $\psi(\vec{r})$ into the final quantum state $\psi'(\vec{r})$. This operator is specified by the relation:

$$\psi'(\vec{r}) = \tau\psi(\vec{r}) = \psi(\vec{r} - \vec{T}) \tag{3.3}$$

The total probability (equal to 1) of finding the particle in the crystal, equal to the norm $||\psi||$ of the ψ function, is preserved by the former transformation. Indeed:

$$||\psi|| = \int |\psi(\vec{r})|^2 d^3\vec{r} = \int |\psi(\vec{r} - \vec{T})|^2 d^3\vec{r} = ||\psi'|| \tag{3.4}$$

Mathematically, this preservation expresses the *unitary character* of the τ operator: $\tau^\dagger\tau = \mathbf{I}$.

3.2.2 Eigenvalues and eigenfunctions of translation operators

We now give our attention to the derivation of the eigenvalues and of the functional form of the eigenfunctions of the τ translation-operators. This derivation relies on three properties of these operators, namely: (a) their mutual commutation properties; (b) their unitary character; (c) the expression of the lattice translations as a function of the three primitive translations $\vec{a}_i$ of the crystal (cf. Chapter 2, Sections 2.2–2.3).

Let $(\alpha_1, \alpha_2, \alpha_3)$ be the quantum operators associated, through Eq. (3.3), to the three primitive translations $(\vec{a_1}, \vec{a_2}, \vec{a_3})$. They *commute with each other*. Indeed,

$$\alpha_i[\alpha_j\psi(\vec{r})] = \alpha_i\psi(\vec{r} - \vec{a_j}) = \psi(\vec{r} - \vec{a_i} - \vec{a_j}) = \alpha_j[\alpha_i\psi(\vec{r})] \tag{3.5}$$

One can therefore define, for these unitary operators, a set of *eigenfunctions* $\phi(\vec{r})$ *common* to the three operators α_i. On the other hand, the unitary character also implies that the corresponding eigenvalues are complex numbers of *unit modulus*,[2] hence of the form $\exp(-i\theta)$. We can write, for each eigenfunction $\phi(\vec{r})$:

$$\alpha_1\phi(\vec{r}) = e^{-i\theta_1}\phi(\vec{r}) \quad \alpha_2\phi(\vec{r}) = e^{-i\theta_2}\phi(\vec{r})$$
$$\alpha_3\phi(\vec{r}) = e^{-i\theta_3}\phi(\vec{r}) \tag{3.6}$$

From the commutation properties (3.5), we can deduce that $\phi(\vec{r})$ is also an eigenfunction of the operator τ associated to the Bravais-lattice translation $\vec{T} = n_1\vec{a}_1 + n_2\vec{a}_2 + n_3\vec{a}_3$:

$$\tau\phi(\vec{r}) = (\alpha_1)^{n_1}(\alpha_2)^{n_2}(\alpha_3)^{n_3}\phi(\vec{r})$$
$$= e^{-i(n_1\theta_1+n_2\theta_2+n_3\theta_3)}\phi(\vec{r}) \tag{3.7}$$

Let us put:

$$\theta_1 = \vec{k}.\vec{a}_1 \quad \theta_2 = \vec{k}.\vec{a}_2 \quad \theta_3 = \vec{k}.\vec{a}_3 \tag{3.8}$$

Since the vectors $\vec{a}_i$ are non-coplanar, this system of three equations defines uniquely the components k_i of $\vec{k}$, as functions of the θ_i. Thus, the specification of a vector $\vec{k}$ determines a set of eigenvalues for the operators α_i, and henceforth for all the operators τ associated to the lattice-translations of a crystal:

$$\tau\phi(\vec{r}) = e^{-i\vec{k}.\vec{T}}\phi(\vec{r}) = \phi(\vec{r} - \vec{T}) \tag{3.9}$$

Let us now write the eigenfunction $\phi(\vec{r})$ as

$$\phi(\vec{r}) = e^{i\vec{k}.\vec{r}}.u(\vec{r}) \tag{3.10}$$

and substitute this expression in the eigenvalues equation (3.9). We obtain a relation valid for any lattice-translation $\vec{T}$,

$$u(\vec{r} - \vec{T}) = u(\vec{r}) \tag{3.11}$$

which implies that the unknown function $u(\vec{r})$ has the same three-dimensional periodicity as the crystal structure.

---

[2]This can be easily checked if the operators are expressed in their diagonal form, by expressing the relation $\tau^\dagger\tau = I$.

In conclusion, the eigenfunctions of the lattice-translations operators of a crystal can be written as *the product of a "plane-wave" function* $\exp(i\vec{k}.\vec{r})$, *and of a spatially-periodic function* $u(\vec{r})$ whose basic periods are the primitive translations of the crystal lattice. The corresponding eigenvalue is $\exp(-i\vec{k}.\vec{T})$.

A function having this general functional form is called a *Bloch function*.

3.2.3   Hamiltonian and lattice translations

We will justify in Chapter 10 that the Hamiltonian operator (cf. Section 3.5) relative to a particle of a crystal can be expressed, *within the "single-particle" approximation* as:

$$\widehat{H} = \frac{\widehat{\vec{p}}^2}{2m} + \widehat{V}(\vec{r}) \tag{3.12}$$

In which $\vec{p}$ is the momentum and $\vec{r}$ the coordinate of the particle. The $V(\vec{r})$ potential acting on the single particle represents, within this approximation, the interaction of all the constituents of the crystal with the considered particle.

Since the displacement of a crystal by means of the lattice translations $\vec{T}$ exchanges identical constituents, the potential $V(\vec{r})$ created at a given point $\vec{r}$ is unchanged by the action of the lattice translations (Fig. 3.2). This potential has the same spatial periodicity as

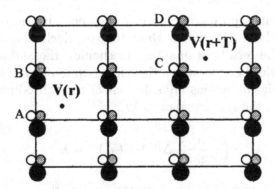

Fig. 3.2.   Fraction of an infinite crystal showing that the environments of two points related by a lattice translation are identical, thus ensuring the lattice-invariance of the potential exerted by the crystal on a particle.

the crystal:

$$V(\vec{r} + \vec{T}) = V(\vec{r}) \tag{3.13}$$

On the other hand, the modulus of the momentum $\vec{p}$ being unmodified by any displacement, the Hamiltonian is invariant by the action of the lattice translations. In order to express this invariance in the framework of quantum mechanics, it can be shown that if an operator τ acts on the wavefunctions of a particle, it also acts on the quantum operators, such as $\widehat{H}$. These operators are transformed as $\widehat{H} \to \widehat{H'}$, with

$$\widehat{H'} = \tau \widehat{H} \tau^{-1} \tag{3.14}$$

The invariance of the Hamiltonian by the lattice translations implies that $\widehat{H'} = \widehat{H}$, with τ defined by Eq. (3.9). Equation (3.14) then takes the form:

$$[\widehat{H}, \tau] = 0 \tag{3.15}$$

In which $[\widehat{H}, \tau] = (\widehat{H}.\tau - \tau\widehat{H})$ is the commutator of the two operators. *The single-particle Hamiltonian commutes with the lattice translation operators.*

3.2.4 Formulation of Bloch's theorem: Band index

We are now in a position to specify the general form of the eigenfunctions of the single-particle Hamiltonian, which describe, in agreement with the results of quantum mechanics, the *stationary* states of the particle, i.e. the states describing *time-independent* properties. These eigenfunctions are solutions of the Schrödinger equation deriving from the Hamiltonian (3.12):

$$\left(-\frac{\hbar^2}{2m}\right)\Delta\psi + V(\vec{r})\psi = E\psi \tag{3.16}$$

The commutation equation (3.15) implies that the ψ functions can be chosen as eigenfunctions of the translation operators τ, and that they are therefore Bloch functions $e^{i\vec{k}.\vec{r}}.u(\vec{r})$. Inserting this functional form into the Schrödinger equation (3.16), yields:

$$\left\{ \left(-\frac{\hbar^2}{2m} \right) (\vec{\nabla} + i\vec{k})^2 + V(\vec{r}) \right\} u = Eu \qquad (3.17)$$

The components of $\vec{k}$ are parameters of this equation. For a given potential $V(\vec{r})$, the possible quantized-values of the energy E, and the form of the u functions will depend on $\vec{k}$.

On the other hand, the solutions $u(\vec{r})$ of Eq. (3.17) are *spatially periodic functions*, with periods equal to the primitive translations of the crystal. On opposite faces of the cell, $u(\vec{r})$ functions will take the same values. The search of the relevant solutions of on Eq. (3.17) can then be restricted to the volume of a single primitive unit-cell. It can be shown that such a condition imposes that, for specified parameters $\vec{k}$ and specified potential $V(\vec{r})$, Eq. (3.17) has a *infinite discrete set of possible energy values* $E_n(\vec{k})$ associated to a discrete set of solutions $u_{\vec{k},n}(\vec{r})$.[3,4]

Hence, the stationary states of a particle in a crystal have the general form:

$$\psi_{\vec{k},n}(\vec{r}) = e^{i\vec{k}\cdot\vec{r}} \cdot u_{\vec{k},n}(\vec{r}) \qquad (3.18)$$

They depend on two types of *quantum numbers*. A first set is constituted by the components (k_x, k_y, k_z) of the $\vec{k}$-vector. These components are *translational quantum numbers* defining the translational properties of the function. Indeed, $u_{\vec{k},n}$ being periodic, the transformation of $\psi_{\vec{k},n}$ when displacing the crystal by a lattice translation $\vec{T}$, is determined by the $\vec{k}$-value. As shown by their definition (Section 3.2.2) the $\vec{k}$-vector components are not constrained to take discrete values. They therefore constitute a set of *continuous quantum numbers*.

---

[3]If the energy $E_n(\vec{k})$ is degenerate there are several functions $u_{\vec{k},n}(\vec{r})$ which should be distinguished by an additional index.

[4]This discrete quantization of the energy and of the periodic part of the wave-function are related to the confinement of the u function in a single cell: as for the bound states of an atom, the non-destructive interference of the wavefunction with itself on the length of the cell imposes phase conditions resulting in the quantization.

The n quantum number, which takes discrete values, is called the *band index*. Its meaning will be further discussed in Section 3.4.

Equation (3.18), which defines the general functional form of the stationary states of a single particle in a crystal, expresses the *Bloch theorem*. Note that the precise form of the $u_{\vec{k},n}$ functions depends on the potential $V(\vec{r})$, which is specific of the crystal-structure considered.[5]

COMMENT. The possible stationary states of crystals defined by the Bloch theorem present similarities with the stationary states of atoms. The operators τ, the plane waves $\exp(i\,\vec{k}\,\vec{r})$ and the translational quantum number $\vec{k}$ play for crystals the same role respectively as the component $\hat{l}_z$ of the angular momentum, its eigenfunctions $\exp(im\phi)$ and the rotational quantum number m, for atoms. Moreover, a Bloch function involves a periodic part $u_{\vec{k},n}$ conform to the symmetry of the Hamiltonian and a non-periodic part, eigenfunction of τ, in the same way as an atomic function involves a spherical part $R(|\vec{r}|)$ conform to the spherical symmetry of the Hamiltonian, and an eigenfunction $Y_{l,m}$ of the angular momentum. The greater simplicity of the translational function $\exp(i\,\vec{k}\,\vec{r})$ as compared to the $Y_{l,m}$ is related to the mutual commutation of the translation operators, while the operators associated to the rotations do not commute.

3.3 Reciprocal Lattice

The space (k_x, k_y, k_z) in which the vectorial quantum numbers are represented is called the *reciprocal space* associated to the crystal.

In this section, we show that a complete repertory of the translational quantum numbers can be obtained by considering a *limited*

---

[5]The Hamiltonian being periodic, with periods equal to the primitive translations, we could expect, on the basis of "common sense" that the solutions $\psi(\vec{r})$ of the Schrödinger equation based on this Hamiltonian possess the same periodicity. This is not the case since $\psi(\vec{r})$ is not reduced to the periodic function $u(\vec{r})$ but also involves a "propagative part" $\exp(i\,\vec{k}.\vec{r})$. One must however bear in mind that $\psi(\vec{r})$ is not directly related to measured physical quantities. The "physical object" is rather $|\psi|^2$ which is equal to $|u(\vec{r})|^2$ and is, in accordance with the "common sense", a periodic function.

region of the reciprocal space. Furthermore, it is shown that the abstract concept of a crystal having infinite extension can be conveniently replaced by that of a *finite Born–Von Karman crystal.* This replacement leads to a *discrete quantization of the* $\overrightarrow{k}$-*coordinates.*

3.3.1 Equivalence between quantum numbers

The eigenvalues of the translation operators τ involve the $\overrightarrow{k}$-vectors as arguments of the function $\exp(-i\,\overrightarrow{k}.\overrightarrow{T})$. Hence, two vectors $\overrightarrow{k}$ and $\overrightarrow{k'}$ define *the same set of eigenvalues* if, for any translation $\overrightarrow{T}$,

$$\overrightarrow{k}.\overrightarrow{T} = \overrightarrow{k'}.\overrightarrow{T} + 2\pi n_T \tag{3.19}$$

where n_T is an integer. In this case, $\overrightarrow{k}$, and $\overrightarrow{k'}$ will be considered as *equivalent.*

3.3.2 Range of the quantum numbers

Let us show that if Eq. (3.19) is satisfied, the difference $(\overrightarrow{k'} - \overrightarrow{k})$ can be expressed as

$$(\overrightarrow{k'} - \overrightarrow{k}) = n_1^* \overrightarrow{a_1^*} + n_2^* \overrightarrow{a_2^*} + n_3^* \overrightarrow{a_3^*} \tag{3.20}$$

where the n_i^* are integers, and where the three vectors $\overrightarrow{a_i^*}$ are specific non-coplanar vectors of the reciprocal space.

In this view, consider a set $(\overrightarrow{a_1}, \overrightarrow{a_2}, \overrightarrow{a_3})$ of primitive translations of the crystal and define a set of three $\overrightarrow{a_i^*}$ through the relations:

$$\overrightarrow{a_i^*}.\overrightarrow{a_j} = 2\pi \delta_{ij} \tag{3.21}$$

The primitive translations $\overrightarrow{a_i}$ being non-coplanar, (3.21) define uniquely the components of the $\overrightarrow{a_i^*}$ vectors. It is easy to check that these equations are satisfied by the set of vectors

$$\overrightarrow{a_1^*} = 2\pi \frac{\overrightarrow{a_2} \times \overrightarrow{a_3}}{\omega} \quad \overrightarrow{a_2^*} = 2\pi \frac{\overrightarrow{a_3} \times \overrightarrow{a_1}}{\omega} \quad \overrightarrow{a_3^*} = 2\pi \frac{\overrightarrow{a_1} \times \overrightarrow{a_2}}{\omega} \tag{3.22}$$

where the mixed product $\omega = (\overrightarrow{a_1}, \overrightarrow{a_2}, \overrightarrow{a_3})$ is the volume of the primitive unit-cell of the crystal.

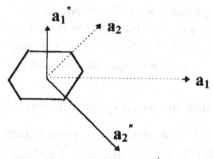

Fig. 3.3. Example of direct $\vec{a_i}$ and reciprocal $\vec{a_i^*}$ lattice vectors in two dimensions, and first Brillouin zone.

Let us now replace in Eq. (3.19) $(\vec{k} - \vec{k}\,')$ by $\sum \beta_i.\vec{a}_i^*$ and $\vec{T}$ by each of the primitive translations $\vec{a}_j$. We find, using Eq. (3.21) that the coefficients β_i are necessarily integers, and that accordingly Eq. (3.19) is satisfied for any $\vec{T}$, thus establishing the validity of Eq. (3.20):

In summary, two vectors $\vec{k}$ and $\vec{k}\,'$ of the reciprocal space define the same translational quantum numbers if

$$\vec{k}\,' = \vec{k} + \vec{G} \quad with \ \vec{G} = \sum n_i^*.\vec{a}_i^* \tag{3.23}$$

Choosing an arbitrary origin Γ in the reciprocal space, the set of points M such as $\overrightarrow{\Gamma M} = \vec{G}$, constitutes a lattice whose generating translations are the $\vec{a}_i^*$. This lattice is called the *reciprocal lattice of the crystal* (cf. Fig. 3.3), the lattice generated by the $\vec{a_i}$ being sometimes specifically termed the *direct* Bravais lattice. Hence $\vec{k}$ and $\vec{k}\,'$ *are equivalent if they differ by a reciprocal lattice vector.*[6]

3.3.3   Properties of the reciprocal lattice

The reciprocal lattices defined by Eq. (3.22) possess all the properties of direct Bravais lattices (cf. Chapter 2, Section 2.2). In particular, each reciprocal lattice belongs to one of the 14 classes of Bravais lattices. It can be shown, using Eqs. (3.19)–(3.21), that the direct

---

[6]If the direct lattice is defined by a different set of primitive translations, an alternate base of the same reciprocal lattice will be defined by Eq. (3.2.2).

and reciprocal lattices of a crystal have the same symmetry. Hence, if the direct lattice is cubic, the reciprocal lattice is also cubic.

However, the two lattices are *not necessarily of the same type*. For instance, if the direct lattice is body centered cubic, the reciprocal lattice is face centered cubic (and conversely). The two lattices will be of the same type only in the case of *simple lattices* i.e. of lattices in which the conventional unit-cell is a primitive unit-cell (cf. Section 2.3.3).

The volume of the primitive unit-cell of the reciprocal lattice is related to the volume ω of the primitive-cell of the direct lattice by

$$\omega^* = (\vec{a_1^*}, \vec{a_2^*}, \vec{a_3^*}) = \frac{8\pi^3}{\omega} \tag{3.24}$$

In Chapter 4, we will point out that a correspondance exists between a node of the reciprocal lattice and a family of lattice planes of the direct lattice.

3.3.4 First Brillouin zone

The equivalence defined by Eq. (3.22) shows that a complete repertory of translational quantum numbers is provided by the $\vec{k}$-vectors belonging to any single unit-cell of the reciprocal lattice. When studying the quantum states of a particle it will be convenient (cf. Chapter 5) to choose a *symmetrical unit-cell* of this lattice, i.e. a Wigner–Seitz cell (cf. Chapter 2, Fig. 2.14). Such a cell, centered at the origin Γ of the reciprocal lattice, is called the *first Brillouin zone of the crystal*.

For a hypothetical one-dimensional crystal with primitive translation a (Fig. 3.4), the reciprocal lattice is the one-dimensional lattice with primitive translation $(\frac{2\pi}{a})$. The method of construction of the Wigner–Seitz cell described in Section 2.3.4 shows that the

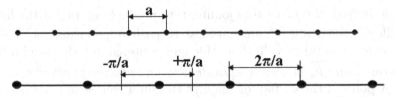

Fig. 3.4.   Direct and reciprocal lattices and first Brillouin zone in one dimension.

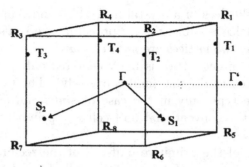

Fig. 3.5.   Equivalent $\vec{k}$-vectors whose endpoints are on the surface of the first Brillouin zone of the simple cubic lattice. For a "general point" S_i, there are two equivalent vectors, while for points of an edge (T_i) or at vertices (R_i) there are respectively four and eight equivalent vectors.

Brillouin zone is the interval $\pm\frac{\pi}{a}$. Likewise, for a simple cubic lattice, the Brillouin zone is a cube whose eight vertices are the points $(\pm\frac{\pi}{a}\pm\frac{\pi}{a}\pm\frac{\pi}{a})$ (Fig. 3.5). In general, the Brillouin zone is a polyhedron whose shape depends, not only of the Bravais class of the direct lattice, but also of the relative magnitude of the primitive translations.

3.3.5   Surface of the first Brillouin zone

The repertory of translational quantum numbers constituted by the first Brillouin zone contains some redundancy. As will be shown in Chapter 5, this redundancy has significant physical consequences on the energy spectrum of the considered particles (the equivalence of vectors of equal moduli will imply a degeneracy of the energy of a particle).

In order to refine this repertory and define the range of *non-equivalent* $\vec{k}$ vectors in the reciprocal space, one must examine carefully the situation of the vectors, denoted $\vec{k_s}$, joining the origin Γ of the reciprocal lattice to points of the *surface* of the Brillouin zone. Indeed, the $\vec{k_i}$ vectors joining Γ to points lying *inside* the first Brillouin zone, are all *unequivalent* to each other, since $(\vec{k_i} + \vec{G})$ belongs to another cell than the one centered on the origin. *In contrast, any $\vec{k_s}$ is always equivalent to one or several other $\vec{k'_s}$.*

A point of the surface of the first Brillouin zone is located on the mediator plane of the segment $\overrightarrow{\Gamma\Gamma'} = \vec{G_s}$ joining the origin Γ of the reciprocal lattice to a "close" node Γ' of this lattice (cf. Section 2.3.4). This plane is defined by

$$2\vec{k}_s.\vec{G}_s = \vec{G}_s^2 \tag{3.25}$$

Consider the *equivalent* vector $\vec{k}_s' = (\vec{k}_s - \vec{G}_s)$. It is easy to deduce from Eq. (3.25) that $\vec{k}_s'$ has the *same modulus* as $\vec{k}_s$, and that it satisfies the equation obtained by replacing in Eq. (3.25) $\vec{G}_s$ by $(-\vec{G}_s)$. Its endpoint, which is in the mediator-plane of the segment $\overline{\Gamma\Gamma''} = -\vec{G}_s$ symmetrical of $\overline{\Gamma\Gamma'}$, *is therefore on the surface of the first Brillouin zone.*

It is of interest to distinguish points which are located on *edges* or at *vertices* of the limiting surface of the first Brillouin zone (Fig. 3.5), from other points located on this surface. If $\vec{k}_s$ ends on a single mediator plane limiting the first Brillouin zone (i.e. neither on an edge nor on a vertex), there a single vector $\vec{G}_s$ satisfying Eq. (3.25), and accordingly $\vec{k}_s$ is equivalent to a single vector. $\vec{k}_s'$ of equal modulus. If $\vec{k}_s$ ends on an edge, intersection of at least two distinct mediator planes, there are, at least, two $\vec{G}_s$ vectors satisfying Eq. (3.25) and, accordingly, two $\vec{k}_s'$ vectors equivalent to $\vec{k}_s$. Likewise, if $\vec{k}_s$ ends on a vertex, it is equivalent to *at least three other vectors* $\vec{k}_s'$, of equal moduli, also ending on a vertex.[7]

3.3.6 Discretization of the quantum numbers

Up to now, the crystal has been considered as an infinitely extended object associated to an infinite set of lattice translations $\vec{T}$. Real crystals being always finite, it is relevant to question the validity of the translational quantum numbers $\vec{k}$ and of their range of possible values.

In this respect, we can note that the constituents of the crystal *interact within distances* which are of the order of the unit-cell

---

[7] Certain points of the surface are on symmetry planes or on symmetry axes of the reciprocal space. As will be seen in Chapter 5, these points play a specific role in the representation of the quantum properties of the crystal. They have been distinguished by letters, for the example considered in Fig. 3.5.

dimensions. The spatial extension of real "macroscopical" crystals being, at least, several hundred cells in any direction, most of the particles of the crystal will not "feel" the presence of the limiting surfaces. Beyond a distance of a few unit cells from the surface, the potential $V(\vec{r})$ (Eq. (3.12)) exerted on a particle is expected to be the same as in an infinite crystal. Besides, as pointed out in Section 3.2.4, the boundary conditions imposed to the periodic $u_{\vec{k},n}(\vec{r})$ solutions of the Schrödinger equation (3.17) can be defined for a single unit-cell within the volume of the crystal. Hence, except in the vicinity of surfaces, the quantum states of real crystals can be assimilated to the Bloch-functions determined for an infinite crystal, and specified by the vectorial quantum numbers $\vec{k}$.

However, in spite of its physical validity, this strategy raises a *formal difficulty* relative to the *normalization* of the Bloch functions. Indeed, the norm of ψ is equal to the total probability of presence of the particle in the crystal. This can be expressed as

$$p = \sum_{cells} \int_{unit-cell} |u_{\vec{k},n}|^2 d^3\vec{r} \qquad (3.26)$$

since $|\psi|^2 = |u_{\vec{k},n}|^2$. Owing to the periodicity of $|u_{\vec{k},n}|^2$, the probability of presence is the same in all the unit-cells. *Imposing a finite total probability ($p = 1$) in an infinitely extended crystal implies the non-physical condition of a vanishingly small probability in each cell.* Conversely, a finite probability in each cell determines an infinite probability in the entire crystal.

Born–Von Karman crystal

In order to remove the former difficulty, the considered particle is assumed to belong to a hypothetical crystal, termed *the Born–Von Karman crystal*, which is identical to the real finite crystal except in the vicinity of the surfaces. The Born–Von Karman crystal has both a *global periodicity* which ensures the formal correctness of the Bloch functional-form, and a *finite character* which allows a normalization of the wavefunction.

Consider a finite crystal of primitive translations $(\vec{a_1}, \vec{a_2}, \vec{a_3})$ comprising N_i cells in the $\vec{a_i}$ direction and a total number of

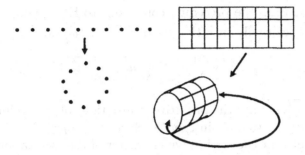

Fig. 3.6. Illustration for one- and two-dimensional crystals of the abstract "folding" of a crystal which ensures its global periodicity in spite of its finite spatial extension.

$N = N_1 N_2 N_3$ cells.[8] Making this crystal infinitely periodic consists in considering as identical the points $\vec{r}$ and $\vec{r} + N_i \vec{a_i}$. One realizes in this manner a *folding* of the crystal along the three directions $\vec{a_i}$ by bringing in coincidence the pairs of faces limiting the crystal in each direction $\vec{a_i}$. Figure 3.6 illustrates the nature of this abstract operation in the case of one and two-dimensional crystals.

Discrete values of the translational quantum numbers

The preceding folding ensures the strict periodicity of the Born–Von Karman crystal. However, it has the effect of replacing the infinite set of lattice translations by a finite set. Indeed, a translation $\vec{T} = N_i \vec{a_i}$ must be considered as identical to the translational of null vector, since it transforms $\vec{r}$ into $(\vec{r} + N_i \vec{a_i}) = \vec{r}$. Hence, the set of lattice translations comprises $N = N_1 N_2 N_3$ distinct translations only.

Note that the condition $\vec{T} = N_i \vec{a_i} = 0$, implies that the operator τ associated to $\vec{T}$ is the identity operator $\tau(\vec{T}) = \mathbf{I}$. Hence its eigenvalues satisfy the condition:

$$\tau(N_i \vec{a_i}) = e^{i \vec{k} \cdot (N_i \vec{a_i})} = 1 \qquad (3.27)$$

leading to:

$$\vec{k} \cdot (N_i \vec{a_i}) = (N_i \vec{k}) \vec{a_i} = 2h\pi \qquad (3.28)$$

---

[8]This crystal is not a rectangular parallelepiped if the primitive translations $\vec{a_i}$ are not mutually perpendicular.

By comparing the preceding conditions to Eqs. (3.20)–(3.21) we find that $\vec{k}$ can only take the *discrete set of values*:

$$\vec{k} = \frac{n_1}{N_1}\vec{a_1^*} + \frac{n_2}{N_2}\vec{a_2^*} + \frac{n_3}{N_3}\vec{a_3^*} \qquad (3.29)$$

where $(\vec{a_1^*}, \vec{a_2^*}, \vec{a_3^*})$ are the primitive translations of the reciprocal lattice and (n_1, n_2, n_3) are integers such as $1 \leq n_i \leq N_i$. It is worth pointing out that the *continuum* character of the translational quantum numbers, underlined in Section 3.2.4, was related to the *infinite extension* of the Bravais lattice.

Density of k-quantum numbers in reciprocal space

The endpoints of the $\vec{k}$-vectors in Eq. (3.28) form a regular lattice of points *uniformly distributed in the reciprocal space*. This lattice is much more dense than the reciprocal lattice since its periods are $\vec{a_i^*}/N_i$ instead of $\vec{a_i^*}$. Scaled to the dimensions of the first Brillouin zone, this lattice forms *a quasi-continuum set of points*, comprising $N = N_1 N_2 N_3$ points. Note that this number is equal to the number of (direct) unit-cells in the Born–Von Karman crystal. Taking into account the volume ω^* of the reciprocal unit-cell, determined by Eq. (3.24), the density of quantum numbers $\vec{k}$ in the reciprocal space is

$$g(\vec{k}) = \frac{N}{\omega^*} = \frac{N\omega}{8\pi^3} = \frac{\Omega}{8\pi^3} \qquad (3.30)$$

where Ω is the volume of the Born–Von Karman crystal.

Figure 3.7 specifies the field of values of the translational quantum number $\vec{k}$, in the case of a crystal with simple cubic Bravais lattice, by summarizing for this case the results of Sections 3.3.4 and 3.3.6.

3.4   Energy Bands

We have seen in Section 3.2.2 that the energy spectrum $E_n(\vec{k})$ of a particle in a crystalline solid could be indexed as functions of the quasi-continuous translational quantum number $\vec{k}$ and of

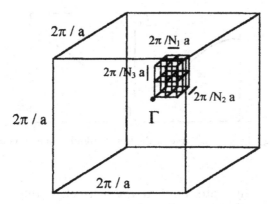

Fig. 3.7.   First Brillouin zone of a simple cubic lattice (large cube) and portion of the fine mesh (not to scale) determined by the Born–Von Karman periodic conditions.

the band index n. Referring to Eq. (3.16), in which the components of $\vec{k}$ are parameters, we can expect that small variations of these components will induce small variations of the spectrum $E_n(\vec{k})$. When $\vec{k}$ takes the quasi-continuous set of values specified by Eq. (3.28), for each value of the band index n, the function $E_n(\vec{k})$ takes a quasi-continuous set of energy values. The correspondance between the variations of $\vec{k}$ and those of $E_n(\vec{k})$ is an *energy dispersion relation* in the reciprocal space. Figure 3.8 illustrates such a correspondance, represented by *an energy dispersion curve*, in the case of a one-dimensional crystal. We can see that the energy $E_n(\vec{k})$ varies between two finite limits. Within these limits it takes a quasi-continuous set of values forming an energy *band*, which, in accordance with the adopted terminology, is in correspondence with the band index n.[9]

Note that if, for given $\vec{k}$, the quantized energy value $E_n(\vec{k})$ is associated to a single solution $u_{\vec{k},n}(\vec{r})$ of Eq. (3.16), *the corresponding band is not degenerate*. In that case, each vector $\vec{k}$ defines a single quantum state $u_{\vec{k},n}(\vec{r})$ of the particle. The entire energy band will

---

[9]As will be shown in Chapter 5, the lack of discontinuity is only true within the first Brillouin zone: points of the surface can be associated to discontinuities in a band, which are of crucial importance to understand the properties of a solid.

Physics of Electrons in Solids

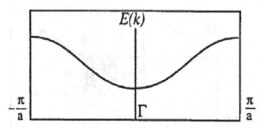

Fig. 3.8. Energy dispersion curve in the first Brillouin zone of a one-dimensional crystal.

"contain" as many states as there are distinct $\vec{k}$-values. As pointed out in Section 3.3.6, this number is equal to the number of primitive unit-cells N of the crystal.

3.4.1 *N*th Brillouin zone: Reduced and extended zone schemes

Let $\psi_{\vec{k}+\vec{G},n}$ be a Bloch wavefunction solution of Eq. (3.17) for the translational quantum number $(\vec{k} + \vec{G})$. We can note that:

$$\psi_{\vec{k}+\vec{G},n} = e^{i(\vec{k}+\vec{G}).\vec{r}} . u_{\vec{k}+\vec{G},n}(\vec{r}) = e^{i\vec{k}.\vec{r}} [e^{i\vec{G}.\vec{r}} u_{\vec{k}+\vec{G},n}(\vec{r})]$$

(3.31)

The function between brackets in the last member of Eq. (3.31) is periodic with the same basic periods as $u_{\vec{k}+\vec{G},n}(\vec{r})$. Accordingly, the expression of $\psi_{\vec{k}+\vec{G},n}$ in the last member of Eq. (3.31) has the form of a Bloch function relative to the vector $\vec{k}$, while its expression in the second member is that of a Bloch function relative to the vector $(\vec{k} + \vec{G})$.

Hence, the (infinite) set of solutions of Eq. (3.17) corresponding to $\vec{k}$ is identical to the set determined for $(\vec{k} + \vec{G})$. This implies that the set of energies is also the same: *the energy spectrum $E(\vec{k})$ is identical to the spectrum $E(\vec{k} + \vec{G})$.*

However, Eq. (3.17) is clearly *not invariant* under the replacement of $\vec{k}$ by $(\vec{k} + \vec{G})$. The meaning of this situation is the following: Setting the value of the index n and following the band $E_n(\vec{k})$

by varying continuously the translational quantum number from $\vec{k}$
to $(\vec{k} + \vec{G})$, we see that the energy of the band at $(\vec{k} + \vec{G})$ is
$E_n(\vec{k} + \vec{G}) \neq E_n(\vec{k})$. Likewise, the wavefunction $\psi_{\vec{k}+\vec{G},n}$ is distinct
from $\psi_{\vec{k},n}$.

Hence, although the entire spectrum is the same for translation-
ally equivalent quantum numbers $\vec{k}$ and $(\vec{k} + \vec{G})$, there is no continuity between the bands having the same position in the spectrum.

This situation is at the origin of two possible representations of
the energy spectrum of a particle.

The first representation is the *reduced zone scheme*, in which one
restricts the representation of the energy bands $E_n(\vec{k})$ of a particle to the $\vec{k}$-vectors of the first Brillouin zone. To each $\vec{k}$ vector
corresponds an infinite discrete set of energy levels. In this scheme,
the knowledge of the spectrum $E_n(\vec{k}')$ corresponding to any vector
$\vec{k}'$ of the reciprocal space is provided by the spectrum $E_n(\vec{k})$ corresponding to $\vec{k} = \vec{k}' - \vec{G}$ with $\vec{k}$ belonging to the first Brillouin
zone.

The second representation is the *extended zone scheme*, in which
one represents the bands in the entire reciprocal space.

In order to specify this representation, it is necessary to define the
concept of an *nth-Brillouin zone*. In this view, we consider the set of
all (infinitely extended) mediator planes of the segments $\Gamma\Gamma'$ joining
the origin of the reciprocal space to the nodes of the reciprocal lattice.
The *nth-Brillouin zone* is the portion of reciprocal space comprising
all the points Λ such as $\Gamma\Lambda$ cuts $(n - 1)$ mediator planes. This definition clearly includes that of the first Brillouin zone, which is the
portion of space enclosed by the planes closest to the origin. The
nth Brillouin zone is generally *multiply connected*, i.e. constituted by
regions of space not communicating with each other.

It is possible to show that there is a one-to-one correspondance
between the quantized $\vec{k}$-vectors of the nth-Brillouin zone and the
$N\vec{k}$-vectors of the first Brillouin zone. On the basis of this correspondance, the extended zone scheme assigns *one band of the energy
spectrum to each Brillouin zone*. For instance, we can consider the
infinite spectrum at $\vec{k} = 0$, and assign the successive band-indices n
to increasing energy values. The lowest energy-band $(n = 1)$ will be

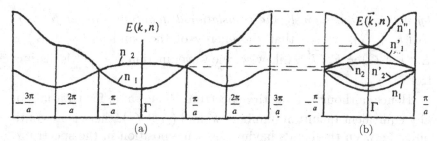

Fig. 3.9. Representation of the same dispersion curve in the extended zone scheme (a) and the reduced zone scheme (b).

represented in the first Brillouin zone, the second band in the second Brillouin zone, etc.

Figure 3.9 illustrates, for a one-dimensional crystal, the two preceding representations of the energy spectrum.

3.5   Appendix: Reminder of Quantum Mechanics

In this appendix, after summarizing some basic ideas underlying quantum mechanics, we recall briefly the main notions used in the book. Additional indications will be provided in some paragraphs of the following chapters whenever needed (e.g. perturbation method, Ehrenfest equations, etc.).

3.5.1   A few ideas

When dealing with microscopic objects such as atoms, electrons, nuclei, subatomic particles, the theory which accounts for their properties is quantum mechanics. The size of these objects imposes specific characteristics to this theory.

A probabilistic theory

Consider first measuring the position and the velocity of an object. To achieve this measurement one can think of sending light (a photon) on the object, and from the geometrical location of the reflected light deduce the required position. Likewise, from measurement of the Doppler effect induced by the velocity of the object, we infer the value and direction of this velocity. This method can indeed work if the object is a macroscopic object (e.g. a billiard ball moving on its trajectory), because the tiny photon will not perturb the motion of the ball, and hence the result obtained will correspond to the "true" location and velocity.

However, if the object is an electron or any other microscopic object, the incident photon will have an energy similar to that of the measured object. The measurement (impact of the photon) will modify the position and/or the velocity to be measured. There is therefore an uncertainty on the values of the considered quantities (position and velocity) *before* the measurement. This uncertainty will exist, for the reason stated (interaction between the measurement and the measured object), for any microscopic object.

Quantum mechanics is a theory which incorporates this essential uncertainty, and governs the relationship of matter with experiment and not a hypothetical objective reality. Instead of providing predictive results of the properties of microscopic objects (position, velocity etc.), it provides *probable* results.

A theory which deals with "dual" objects

In classical physics, *particles* and *waves* are distinct types of objects. Particles are localized in a small volume of space while waves have an extended localization. This is not the case in quantum mechanics. The signature of waves is the possibility to produce interferences. Figure 3.10 summarizes the result of an interference experiment achieved with particles, for example, electrons. An electron beam incident on a screen pierced by two slits generates by diffraction two beams which interfere and produce interference figures. This type of experiment has led to the conclusion that a particle (which has a trajectory $\vec{r}(t)$ and a velocity $\vec{v}$) has also the characteristics of a wave (an amplitude A and a wavelength λ). The wavelength of the particle is related to its momentum $p = m\mathbf{v}$ by the De Broglie relation $\lambda = h/p$ where h is the Planck constant, and p the momentum of the particle.

Wave nature and discrete results of measurements

In classical mechanics, the results obtained in the measurement of a physical quantity can be any value in a continuous set of values. The

Physics of Electrons in Solids

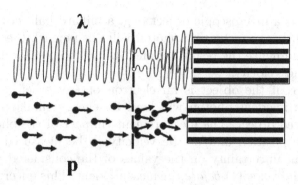

Fig. 3.10.   Upper: Principle of an interference experiment with a coherent wave
λ. The first screen is pierced with two horizontal slits, which diffract the incoming
wave and generates two cylindrical beams which interfere and determine alternate
clear and dark fringes on the second screen. Lower: Same experiment performed
with mono-kinetic electrons having the momentum p.

situation is different in quantum mechanics for "closed systems",
i.e. systems which are confined in a small region of space (e.g. an
electron in an atom). Indeed, the wave associated to the particle,
will circulate in the closed region of space and will tend to interfere
with itself. If the length of the path of the particle is not a multiple
of its wavelength $\lambda = h/p$, this interference, will be destructive, i.e.
the related particle is not in a stable state. Hence, particles with
definite values only of wavelengths can be stable. The wavelengths
are "quantized": they can only have certain discrete values. As the
energy of the wave associated to the particle is $E = h\nu = hc/\lambda$,
the energy of the wave/particle is also quantized. It can only take a
series of discrete values E_n. This is not the case for systems which
are "open" (e.g. a particle diffused to infinity by a potential) because
the wavelength of the particle does not interfere with itself. In that
case, no quantization of the energy or other related quantities occurs.

Wavefunction and description of the state of a system

In interference measurements the result observed, e.g. on a screen,
corresponds to the intensity of the interfering waves. The amplitude
of the wave (e.g. $A\exp(-i.\overrightarrow{k}.\overrightarrow{r})$) has to be considered for predict-
ing the result of the interference experiment, but the observed result
is proportional to the square $|A|^2$ of the amplitude. Similarly, the
physical state of a particle, which is related to the results of experiments, has to be described by a complex amplitude $\psi(\overrightarrow{r})$. However,

the (probabilistic) observed results of the experiments will depend on the square of the modulus of this function. $\psi(\overrightarrow{r})$ is the *wavefunction* which describes the state of a system, similar to the amplitude of a wave. It is an *amplitude of probability describing the state of the system and allowing to predict the properties.* The probabilistic results of the experiments will be related to $|\psi(\overrightarrow{r})|^2$, which is a *probability density.*

Stationary states

Consider an atom in which, in a "classical" description, an electron has an orbital trajectory around the nucleus, and is therefore a system in constant motion. In another description, considering the uncertainty, mentioned above, on the positions and velocity of the electron, the density of probability of the electron is described, if the system is stable, by a function $|\psi(\overrightarrow{r}, t)|^2 = |\psi(\overrightarrow{r})|^2$ independent of time. In the latter description the electron is represented by a spatial "static" distribution of charge within the atom. This dual picture of a system in motion which however is a stable system at rest defines a *stationary quantum state.* To a certain extent, this duality also exists in classical mechanics: a planet can have a stable orbit (a stationary state) although it is permanently in motion.

In Chapters 7 and 8 examples of non-stationary states will also be considered.

Interacting particles

In classical and quantum mechanics when several particles are interacting, studying separately their motions is not possible strictly speaking (methods of approximations exist, however, to perform such studies see Chapter 10). In the essential chapters of this book, we will consider non-interacting particles whose state can therefore be studied separately.

3.5.2 Formalism and determination of the stationary states

The Schrödinger equation

In classical mechanics, the motion of a particle $\overrightarrow{r}(t)$ is determined by the Newton equation $\overrightarrow{F} = m\overrightarrow{\gamma}$ and initial conditions.

Equivalently, this motion can be derived in the framework of the so-called Hamiltonian formalism (differential equations involving the position and momentum of the particle) in which the Hamiltonian H is the total energy of the particle expressed as function of the position and the momentum: $H = (p^2/2m) + V(\overrightarrow{r})$.[10] The first term of the second member is the kinetic energy and the second term is the potential energy of the particle.

In the wavefunction formulation of quantum mechanics, the stationary states of the particle are determined by the Schrödinger equation:

$$-\frac{\hbar^2}{2m}\Delta\psi(\overrightarrow{r}) + V(\overrightarrow{r}).\psi(\overrightarrow{r}) = E.\psi(\overrightarrow{r})$$ (3.32)

where Δ is the Laplacian of the function, and E is the energy of the stationary state. This equation can also be written as

$$\left[-\frac{\hbar^2}{2m}\Delta + V(\overrightarrow{r})\right]\psi(\overrightarrow{r}) = E.\psi(\overrightarrow{r})$$ (3.33)

In which the term between brackets $[-\frac{\hbar^2}{2m}\Delta + V(\overrightarrow{r})]$ is a differential _operator_, associated to the Hamiltonian H, and acting on the wavefunction $\psi(\overrightarrow{r})$.

Since $|\psi(\overrightarrow{r})|^2$ is a probability density, and that the total probability in the entire space is equal to 1, we must have:

$$\int_{\text{space}} |\psi(\overrightarrow{r})|^2 d\overrightarrow{r} = 1$$ (3.34)

Besides, conditions of regularity of the probability impose other conditions such as the continuity of the derivative of ψ.

Interpretation of the equation

Note that the Schrödinger equation is a linear differential equation. If $\psi_1(\overrightarrow{r})$ and $\psi_2(\overrightarrow{r})$ are two solutions of the Schrödinger equation, their sum is also a solution. Likewise $\lambda\psi(\overrightarrow{r})$ is also a solution. The solutions of the Schrödinger equation can be considered to belong

---

[10]The case of a Hamiltonian in presence of a magnetic field will be treated in Chapters 7 and 8.

to a *vector space* of functions $\psi(\vec{r})$. As obvious from the expression of the classical Hamiltonian H one can associate, in the Schrödinger equation, the kinetic energy to the *differential operator* $-\frac{\hbar^2}{2m}\Delta$ with ($\hbar = h/2\pi$), and the potential energy to the *multiplicative operator* $V(\vec{r})$. The sum of the kinetic and potential energy being equal to the total energy of the particle the constant E in the second member is, as expected, the value of the energy of the particle in its stationary state. Equation (3.33) can then be interpreted as the *eigenvalue equation* of the differential operator associated to H, the energy E being a (real) eigenvalue of this operator.

Note that a more general correspondance can be established between each physical quantity such as H and an operator. Hence, the following correspondance holds for the three components of the momentum $\vec{p}$.

$$(p_x, p_y, p_z) \rightarrow \left(\frac{\hbar}{i}\frac{\partial}{\partial x}, \frac{\hbar}{i}\frac{\partial}{\partial y}, \frac{\hbar}{i}\frac{\partial}{\partial z} \right) \tag{3.35}$$

The differential operator associated to the kinetic energy in the Hamiltonian H is accordingly constructed formally as

$$\frac{p_x^2 + p_y^2 + p_z^2}{2m} \rightarrow -\frac{\hbar^2}{2m}\left(\frac{\partial^2}{\partial x^2} + \frac{\partial^2}{\partial x^2} + \frac{\partial^2}{\partial x^2} \right) = -\frac{\hbar^2}{2m}\Delta \tag{3.36}$$

Likewise any function of the components of $\vec{r}$ will be a *multiplicative* operator. As an example the first component $(yp_z - zp_y)$ of the orbital momentum $\vec{L} = \vec{r} \times \vec{p}$ will be associated to the operator:

$$(yp_z - zp_y) \rightarrow \frac{\hbar}{i}\left(y.\frac{\partial}{\partial z} - z\frac{\partial}{\partial y} \right) \tag{3.37}$$

Considering the form of the operators, the vector space of the wavefunctions is defined on the set of complex numbers, i.e. one can consider combinations of the type $\psi_1 + \lambda\psi_2$ with λ a complex number. However, it is important to note that since the eigenvalues of the various operators are values of physical quantities, they have to be real. Thus, the operators associated to physical quantities (which are also termed "observables") which necessary have real eigenvalues, possess the mathematical property of being *Hermitians operators*.

Vectorial formulation and operator formalism

Since the physical states of a quantum system belong to a vector space, we can denote the states as vectors. They are denoted as $|\psi\rangle$ and the Hermitian operators A associated to the physical quantities as $\widehat{A}$. Hence, Eq. (3.32) which is the eigenequation of the Hamiltonian can be written as

$$\widehat{H}|\psi\rangle = \left[\frac{\widehat{p}^2}{2m} + \widehat{V(\vec{r})}\right]|\psi\rangle = E|\psi\rangle \qquad (3.38)$$

One could ask why it is necessary to replace the wavefunction $\psi(\vec{r})$ description of states, solution of a differential equation, by the more abstract vector $|\psi\rangle$ description. The reason is the discovery of a large number of physical quantities which were unknown to classical physics. They are called quantities *without classical analog*. They characterize either atomic properties such as *the spin* or particles belonging to the nucleus, or other "elementary" particles. In that case, one cannot write a function of the coordinates $\psi(\vec{r})$ since the position of the particle has no clear meaning. One has then to rely on the vector description. In vector spaces, an operator can be defined by its action on its eigenvectors. Hence, the operators $\widehat{A}$, representing the "non-classical" quantities, will be entirely defined by their action on their eigenvectors. The different states are defined by their linear combinations as function of the these eigenvectors. In this book, the only such "non-classical" quantity considered will be the spin of the electron.

Relation between $|\psi\rangle$ and $\psi(\vec{r})$

As other physical quantities, the momentum $\vec{p}$ and the position $\vec{r}$ are associated to operators. These observables $\widehat{\vec{p}}$ and $\widehat{\vec{r}}$ have *continuous (not quantized) sets of eigenvectors and of eigenvalues.* For instance, the eigenvectors of the operator $\widehat{x}$, associated to the first coordinate of the position can be denoted $|x\rangle$ where x is a continuous variable. The eigenvalue of $\widehat{x}$ corresponding to the eigenvector $|x\rangle$ is the coordinate x which, as already stated, can take any value. We can write $\widehat{x}|x\rangle = x|x\rangle$. The wavefunction $\psi(x)$ defined above is then the *projection* of the state $|\psi\rangle$ on the eigenvector $|x\rangle$ of the position operator $\widehat{x}$. The wavefunction $\psi(x)$ is the projection of $|\psi\rangle$ on $|x\rangle$, thus, $\psi(x) = \langle x|\psi \rangle$ where $\langle x|\psi \rangle$ is the scalar product of the state $|\psi \rangle$ and of the eigenvector $|x \rangle$.

As shown by Eq. (3.37), the stable states, solutions of the Schrödinger equation, are also eigenvectors of the Hamiltonian operator. The eigenvalues E being the corresponding values of the energy of the system. For a closed system, these values are necessarily quantized and have discrete values. This means that, in this case, $\widehat{H}$, contrary to the case of the observables $\widehat{p}$ and $\widehat{x}$, has a discrete series of eigenvalues. This is also the case of other quantities such as the observables associated to the orbital moment, or to the spin.

Results of the physical measurements

As shown by Eqs. (3.37) or (3.31), the determination of a stationary state $|\psi_\lambda \rangle$ will also provide the corresponding value of the energy E_λ which is an eigenvalue of $\widehat{H}$. We can therefore infer that the only results of the measurement of the energy of the considered particle is an eigenvalue of $\widehat{H}$. This holds for other physical quantities: the possible results of the measurement of a physical property A are the eigenvalues of the operator (observable) $\widehat{A}$ associated to this property.

Commutation relations and quantum numbers

Consider now two operators $\widehat{A}$ and $\widehat{B}$ which commute with each other: $\widehat{A}.\widehat{B} = \widehat{B}.\widehat{A}$. They possess a set of common eigenvectors $|\psi_\lambda \rangle$ which span the entire vector space. For a given common eigenvector, the respective eigenvalues are a_λ and b_λ. If the system considered is in the stationary state $|\psi_\lambda \rangle$ common eigenvector of $\widehat{A}$ and $\widehat{B}$, the result of the measurement of property A will be a_λ and the result of the measurement of B will be b_λ. *Hence, it is possible to measure simultaneously two physical quantities whose operators commute.* For instance, it can be shown that in an isolated atom, the operators associated to the total energy (e.g. the Hamiltonian), the operator associated to the square $\overrightarrow{L}^2$ of the orbital momentum and the operator associated to the component L_z of this momentum, all commute. Therefore, a stationary state $|\psi_\lambda \rangle$ of the atom will be characterized by the three eigenvalues of these quantities, namely the energy E_λ (of the form $-R_H/n^2$), an eigenvalue (of the form $l(l + 1\hbar^2)$) of $\overrightarrow{L}^2$ and an eigenvalue of L_z (of the form $m\hbar$). The numbers n, l and

m related to the eigenvalues of the three commuting operators are *quantum numbers*.

It can be shown that, in general, a stationary state can be uniquely defined as common eigenvector of a so-called *"complete set"* of operators which commute with each other. Hence, the simultaneous measurement of the eigenvalues of this set will allow to determine without ambiguity the stationary state of the system, which can be denoted by the set of quantum numbers relative to the set of eigenvalues.

Bibliography

The Theory of Metals, *A.H. Wilson*, Cambridge University Press (2011).
Quantum Mechanics, *A. Messiah*, Dover Books on Physics (2014).

Chapter 4

The Reciprocal Space as a Space
of Diffraction Patterns

Main ideas: Diffraction equations expressed as functions of reciprocal space vectors (Section 4.3). Existence of a diffraction only for definite orientations of the incident beam with respect to a crystal (Section 4.3.4). Transfer vector equal to vectors of the reciprocal lattice (Section 4.3.1). Correspondence between reciprocal lattice vectors and lattice planes (Section 4.4). Relationship between diffracted intensity and atomic structure (Section 4.3.2). Experimental determination of crystal structures (Section 4.5).

4.1 Introduction

The diffraction, by a crystal, of X-rays or of particle-beams (electrons, neutrons) is a subject which has a dual function in the study of the properties of a solid.

In the first place, diffraction is the main experimental method giving access to the spatial configuration of the constituting atoms. It therefore provides the indispensable structural substrate for the interpretation of the physical properties of the solid.

Besides, as shown in the next chapter, the properties of an electron in a crystal can be seen as resulting from a "quantum diffraction" of the electron wavefunction by the periodic potential exerted by the crystal. This diffraction is *formally similar* to the diffraction of a beam by a periodic crystal structure. The considerations developed in the present chapter can therefore be usefully transposed to the study of the quantum states of electrons in a solid.

4.2 Atomic Scattering Process

Underlying a diffraction experiment, there is an *elastic scattering* process consisting in the interaction of a single atom with an incident electromagnetic beam or a particle beam. The interaction determines the emission of a beam having the *same wavelength*.[1,2] In the schematic description of such a process, one considers an incident *scalar plane wave* $A_q \exp(-i\vec{q}.\vec{r})$ with $|\vec{q}| = 2\pi/\lambda$.

The interaction of this wave with an atom produces a scattering in every direction of space (Fig. 4.1).[3] In a given direction, the scattered plane wave is $A_{q'} \exp(-i\vec{q}'.\vec{r})$. Its amplitude can be written as

$$A_{q'} = f(\lambda, \theta).A_q \tag{4.1}$$

The function f is the *atomic scattering factor* which determines the angular dependence of the amplitude of the scattered wave.

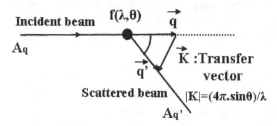

Fig. 4.1. Schematic representation of the elastic scattering of an electromagnetic beam or of a particle beam.

---

[1] The De Broglie wavelength $\lambda = h/p$ in the case of a beam of particles with momentum p.

[2] The elastic character of the scattering implies the conservation of the energy of the particles (photons, neutrons or electrons), and therefore the conservation of their wavelength. Consideration of non-elastic scattering processes will be briefly discussed in Section 4.5.3.

[3] The emission in every direction of space, and not only in the incident direction, is the result of the mutual bonding of the particles constituting the atom. Hence, a particle is submitted not only to the forces exerted by the incident beam, but also by those exerted by the other particles of the atom.

Table 4.1.   Characteristics of three usual types of beams.

	X-rays	Thermal neutrons	Electrons
Wavelength	1 Å	1 Å	10^{-2} Å
Particle energy	10 keV	10^{-2} eV	100 keV
Particle flux by square cm	10^{12}–10^{18} cm^{-2}s^{-1}	10^7 cm^{-2}s^{-1}	10^{19} cm^{-2}s^{-1}
Source	Vacuum tube; synchrotron	Research nuclear reactor	Electron microscope

Owing to the fact that the scattering by an atom is itself a complex process resulting from the interaction of the beam with the constituents of the atom (electrons, nucleus), the f-factor depends on the nature of the incident beam (X-rays, neutrons, electrons, etc.) and of the atom considered (Table 4.1).

The *X-ray* wave scattered by an atom is the coherent addition of the waves scattered by the *electrons*.[4] The magnitude of the atomic scattering factor f is thus proportional to the atomic number: "heavy atoms" will diffract more efficiently than "light" ones. Besides, there is a marked angular dependence of $f(\lambda, \theta)$ owing to the fact that electrons are located at mutual distances of the order of 1 Å, a distance comparable to the X-rays wavelength. Thus, the phase shifts between the waves scattered by the different electrons vary pronouncedly as a function of the scattering direction and of the wavelength (Fig. 4.2).

The resulting typical dependence of the X-ray atomic scattering factor as function of $(\sin\theta/\lambda)$ is shown on Fig. 4.3. Neutrons interact with the atomic nuclei. This "target" is very small as compared to the incident wavelength, and accordingly the neutron atomic scattering power is almost independent of θ. Also in contrast with the case of X-rays, the magnitude of the scattering does not vary regularly as a function of the atomic number. Consecutive elements of the Mendeleev table, or isotopes of the same element, can possess very different scattering factors (cf. Table 4.2).

---

[4]The oscillating electric field of the wave acts both on the electrons and on the protons. However, electrons are considerably less bound and their motion induced by the field has a larger amplitude, and consequently emits a wave with a larger amplitude.

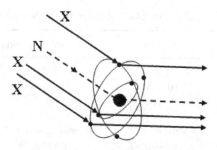

Fig. 4.2. Atomic mechanism of scattering for X-rays and neutrons. The angular-dependent phase shifts between scattered beams are large for X-rays, which are scattered by electrons, since their distances are of the order of the beam wavelength. These distances are almost vanishing for neutrons which are scattered by the nucleus.

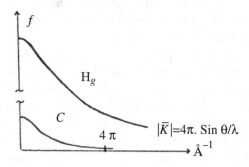

Fig. 4.3. Variations as a function of the transfer vector of the X-ray scattering factors f of carbon ($Z = 6$) and of mercury ($Z = 80$). f is proportional to the atomic number Z. The scale for K is usefully compared to the shortest distance between reciprocal lattice nodes ($2\ \text{Å}^{-1}$) for a cubic crystal of unit cell edge 3 Å.

Table 4.2.   Scattering cross-sections S for neutrons in Barns (10^{-24}cm^2).

Element	H	D	Li	Be	Na	N	O	Co	Ni	Cu	Au	Pb
Atomic number	1	1	3	4	11	7	8	27	28	29	79	82
S		1.7	5.6	0.6	7.6	1.7	11	4.2	0.8	13	7.5	7.3	11.3

Electrons interact with both the electrons and the nuclei of the atoms through the electric field generated by these atomic constituents. The f scattering factor has a similar angular and atomic-number dependence as for X-rays. Owing to its electrostatic origin, the efficiency of the scattering by electrons is much higher than that of X-rays or neutrons. As a consequence, an electron beam incoming on a crystal will be significantly scattered by each atom, and its intensity will therefore decrease rapidly on penetrating the crystal.

4.3   Diffraction by a Crystal

Let us now consider the *diffraction* by a crystal of a monochromatic plane wave of wavevector $\vec{q}$, i.e. its *collective scattering* by all the atoms of the structure in the direction defined by the wavevector q'. The basic result of interest derived in this section concerns the intensity $I(\vec{q}, \vec{q'})$ of the diffracted wave, which will be shown to be the product of two terms:

$$I(\vec{q}, \vec{q'}) = I(\vec{K}) = I_1(\vec{K}).I_2(\vec{K}) \qquad (4.2)$$

where $\vec{K} = (\vec{q'} - \vec{q})$ is the *transfer vector*. Since the elastic character of the scattering implies that $|\vec{q}| = |q'|$ (Fig. 4.1), we have:

$$|\vec{K}| = 4\pi \frac{\sin\theta}{\lambda} \qquad (4.3)$$

The term I_1 only depends of the geometrical characteristics of the Bravais lattice of the crystal, while I_2 is determined by the specific configuration of the atoms in the primitive unit cell.

The derivation of Eq. (4.2) involves simplifying assumptions. In the first place, it is based on the hypothesis that each atom scatters a *negligible fraction* of the incident beam intensity into other directions than $\vec{q}$, and that, accordingly, all the atoms of the structure are submitted to *the same incident beam intensity*. The latter situation also implies that *the crystal is finite* because, otherwise, however small the scattering, the incident beam would eventually be weakened after its interactions with a very large number of atoms. On the other hand, the weakness of the scattering warrants that secondary scattering of a beam primarily scattered by another atom is negligible. These assumptions, which are generally valid for X-ray diffraction, are infirmed in the case of electron-diffraction, owing to the large magnitude of the atomic scattering factors for electrons (cf. Section 4.2).

4.3.1   Diffraction by a crystal with a monoatomic-basis

Phase difference between two scattered waves

Consider two atoms whose centers are located at points O and M, both submitted to the incident beam $(\vec{q}, A_q)$ and reemitting scattered waves with wavevector $\vec{q}'$. There is a phase difference $\Delta\phi$ between the scattered waves (Fig. 4.4):

$$\Delta\phi_{OM}(\vec{q}, \vec{q}') = \frac{2\pi}{\lambda}(BM - OA) = (\vec{q} - \vec{q}') \cdot \overrightarrow{OM}$$

$$= -\vec{K} \cdot \overrightarrow{OM} \qquad (4.4)$$

Here $\vec{K}$ being the transfer vector $(\vec{q}' - \vec{q})$.

Intensity of the diffracted wave

Consider now a monoatomic crystal whose atoms, all identical, are located at the nodes $\overrightarrow{OM} = (\sum n_j \vec{a}_j)$ of a Bravais lattice, with $0 \le n_j \le (N_j - 1)$. The crystal is thus a parallelepiped with edges along the primitive translations, and it contains a total number of atoms $N = N_1 N_2 N_3$.

The waves *diffracted* by the crystal are the results of the *interference* of the waves scattered by the N identical atoms of the crystal. In the $\vec{q}'$ direction, each atom scatters a wave $A_{q'} \exp(-i\vec{q}'.\vec{r})$, whose amplitude is of the form $A_{q'} = f A_q$ (Eq. 4.1), where f is the atomic scattering factor (cf. Section 4.2). Its phase shift, with respect to the wave scattered by the atom located at the origin O, is provided by Eq. (4.4). The *complex amplitude* of the diffracted wave can be

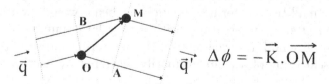

Fig. 4.4.   Phase shift $\Delta\phi$ between waves scattered by two atoms of the crystal, as a function of the transfer vector and of the relative positions of the atoms.

written as

$$A(\overrightarrow{K}) = A_q . f(\overrightarrow{K}) \left\{ \sum_{n_1 n_2 n_3} e^{-i\overrightarrow{K}.(\sum n_j . \overrightarrow{a}_j)} \right\}$$

$$= A_q . f(\overrightarrow{K}) \prod_{j=1}^{3} \frac{1 - e^{-iN_j . \overrightarrow{K} . \overrightarrow{a}_j}}{1 - e^{-i\overrightarrow{K} . \overrightarrow{a}_j}} \qquad (4.5)$$

and its intensity is therefore:

$$I(\overrightarrow{K}) = |A|^2 = N^2 |A_q|^2 |f(\overrightarrow{K})|^2 \prod_{j=1}^{3} \frac{\sin^2(N_j \overrightarrow{K} . \overrightarrow{a}_j / 2)}{N^2 \sin^2(\overrightarrow{K} . \overrightarrow{a}_j / 2)} \qquad (4.6)$$

For large values of N, the function $Y = \sin^2(Nx)/N^2 \sin^2(x)$ takes very small values (of order $(1/N^2)$ except at values $x = h\pi$ (Fig. 4.5) for which $Y = 1$. Even for small crystallites, a few microns in each dimension, N^2 is of the order of 10^{20}. Hence, the intensity equation (4.6) is only non-zero for transfer vector values strictly satisfying[5] the set of conditions:

$$\overrightarrow{K} . \overrightarrow{a_1} = 2h_1 \pi \quad \overrightarrow{K} . \overrightarrow{a_2} = 2h_2 \pi \quad \overrightarrow{K} . \overrightarrow{a_3} = 2h_3 \pi \qquad (4.7)$$

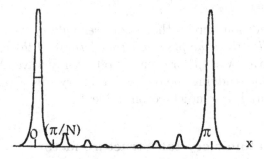

Fig. 4.5. Form of the function $\sin^2(Nx)/N^2 \sin^2 x$ for large values of N. The width of the main peaks is of the order of π/N.

---

[5]For very small crystals (e.g. corresponding to values of the $N_i \lesssim 10$), a diffraction will be observable in directions corresponding to transfer vectors located in a significant interval of values around $\overrightarrow{K}$. In particular, if a crystal is a thin wafer, the intervals will be "rods" perpendicular to the wafer.

Referring to Eq. (3.19) in Chapter 3, we can see that a $\vec{K}$ vector complying with Eq. (4.7) is necessarily equal to a vector $\vec{G}$ of the reciprocal lattice:

$$\vec{K} = \vec{G} \qquad (4.8)$$

The intensity diffracted by a crystal is non-zero only if the incident and scattered beam directions are such that $(\vec{q} - \vec{q'}) = \vec{K}$ *is a vector of the reciprocal lattice.* Equation (4.8) expresses the diffraction *Laue condition.*

For $\vec{K} = \vec{G}$, the diffracted intensity is equal[6] to $A_q N^2 |f(\vec{G})|^2$. It therefore varies, for the various orientations, as $|f(\vec{G})|^2$ which decreases smoothly with the magnitude of $|\vec{G}|$ (Fig. 4.3).

The diffraction equation is simply formulated as a function of vectors of the reciprocal space. *This "natural" representation of the geometry of the diffraction in the reciprocal space will be further illustrated by the considerations in Sections 4.3 and 4.3.5.*

Note that $|\vec{q}|$ is of the same order of magnitude as the modulus $|\vec{G}|$ of the reciprocal lattice vector involved in the Laue equation. For the *shortest* $\vec{G}$ vectors, $|\vec{G}| \approx 2\pi/|\vec{a_i}|$ ($\vec{a_i}$ being the primitive translations). Consequently:

$$\lambda = \frac{2\pi}{|\vec{q}|} \approx \frac{2\pi}{|\vec{G}|} \lesssim |\vec{a_i}| \qquad (4.9)$$

The Laue condition implies that *the wavelength of the incident beam must be smaller than the primitive translations of the crystal lattice.* An optical wave ($\lambda \approx 1\mu$) or even a "soft" X-ray wave ($\lambda \approx 100$ Å) is not suitable for studying the diffraction in crystals. Commonly used X-ray wavelength are indicated on Table 4.3.

4.3.2   Polyatomic crystal: Structure factor

Let us now consider a crystal comprising M atoms in its basis. The expression of the amplitude of the diffracted beam will differ from

---

[6]The proportionality to N^2, which would imply, even for small values of the structure factor, that the intensity grows to infinity for an infinite crystal is clearly induced by the assumption, unrealistic for large crystals, of an identical incident amplitude on every atom.

Table 4.3. Wavelengths of main X-ray sources.

Source (vacuum tube)	Copper anticathode	Silver anticathode	Iron anticathode
Wavelength	≈1.55 Å	≈0.55 Å	≈1.93 Å

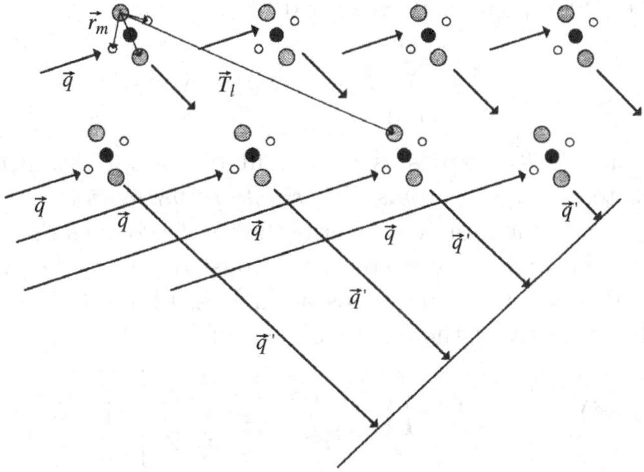

Fig. 4.6.   Diffraction by a polyatomic crystal. The phase shifts between the waves scattered by the atoms of the structure are function, on the one hand, of the Bravais lattice vectors T_l and of the positions r_m of the M atoms composing an elementary unit cell.

Eq. (4.5) in two aspects. In the first place, if several types of atoms (denoted m) are present, the corresponding scattering factors f_m must be taken into account. On the other hand, the positions of the atoms of the structure cannot be mapped on the nodes of the Bravais lattice. Taking as origin a point O within one of the primitive unit cells, the positions of the atoms of the structure are (Fig. 4.6):

$$\overrightarrow{OM} = \overrightarrow{T_l} + \overrightarrow{r_m} = \left[\sum_{j=1}^{3} n_j^l \overrightarrow{a_j} \right] + [\alpha_m \overrightarrow{a_1} + \beta_m \overrightarrow{a_2} + \gamma_m \overrightarrow{a_3}] \quad (4.10)$$

Here $\overrightarrow{T_l}$ being the lattice translation defining the lth unit cell, and $\overrightarrow{r_m}$ being the position of the mth atom within the basis (or the primitive unit cell) surrounding the origin. Using Eq. (4.4), the expression of the diffracted amplitude, which substitutes Eq. (4.5), can be written as

$$A(\vec{K}) = A_q \cdot F(\vec{K}) \left[\sum_{n_1,n_2,n_3} e^{-i\vec{K}\cdot(n_1\vec{a_1}+n_2\vec{a_2}+n_3\vec{a_3})} \right] \qquad (4.11)$$

where the *structure factor* $F(\vec{K})$ substitutes the atomic scattering factor $f(\vec{K})$ in Eq. (4.5). It is defined by

$$F(\vec{K}) = \sum_{m=1}^{M} f_m(\vec{K}) \exp(-i\vec{K}\cdot\vec{r_m}) \qquad (4.12)$$

Note that $F(\vec{K})$ expresses the result of the interference of the waves scattered by the atoms of a *single primitive unit cell*, in the same manner as the atomic scattering factor f provides the coherent addition of the waves scattered by the constituents of the atom.

Going through the same steps as in Eqs. (4.5)–(4.6), we finally obtain the intensity of the diffracted beam:

$$I(\vec{K}) = \left[N^2|A_q|^2 \prod_{j=1}^{3} \frac{\sin^2(N_j\vec{K}\cdot\vec{a_j}/2)}{N^2\sin^2(\vec{K}\cdot\vec{a_j}/2)} \right] \cdot |F(\vec{K})|^2$$

$$= I_1(\vec{K}) \cdot I_2(\vec{K}) \qquad (4.13)$$

The factor between brackets, similar to (4.6) is only non-zero if $\vec{K} = \vec{G}$ is a vector of the reciprocal lattice. Its value $I_1(\vec{G})$ *is the same for all the nodes of the reciprocal lattice*. The second factor $I_2(\vec{G})$ depends of the nature and of the configuration of the M atoms composing the primitive unit cell of the structure. If we express $\vec{G}$ as a function of the primitive translations $\vec{a_i^*}$ of the reciprocal lattice (cf. Eq. (3.21)):

$$\vec{G} = h.\vec{a_1^*} + k.\vec{a_2^*} + l.\vec{a_3^*} \qquad (4.14)$$

where (h, k, l) are integers, and use the coordinates $\vec{r_m}$ of the atoms (Eq. (4.10)), the structure factor F takes the form:

$$F(h,k,l) = \sum_{m=1}^{M} f_m(h,k,l) \exp[-2\pi i(h.\alpha + k.\beta + l.\gamma] \qquad (4.15)$$

For X-rays, the scattering by the various atoms is due to the *electrons*, the number and spatial distribution of which can be characterized, in the unit cell, by a *continuous function* which is the electronic density $\rho(\vec{r})$. On this basis, instead of expressing F as a *discrete sum* over the positions $\vec{r_m}$ of the atoms of a unit cell, one can use a *continuous formulation* using the electron density $\rho(\vec{r})$ and the individual scattering factor f of the electron:

$$F(\vec{K}) = f \iiint_{\text{unit cells}} \rho(\vec{r}) \, e^{-i\vec{K}.\vec{r}} \, d^3\vec{r} \qquad (4.16)$$

In this form, the structure factor F appears proportional to the *Fourier transform* of the periodic function $\rho(\vec{r})$.

4.3.3 Effect of the thermal vibrations of atoms

As mentioned in Chapter 1, at any temperature, the atoms of a crystal *oscillate* about their average position in the structure. Consequently, the scattering centers do not possess the exact periodic configuration assumed in the preceding calculations. There is an "instantaneous disorder" which has the effect of determining in any direction a non-zero intensity of the scattered radiation, constituting a *continuous background* $I(\vec{K})$ in the reciprocal space. Part of the intensity diffracted in the directions defined by the Laue condition (4.8) is diverted to this background scattering. As a consequence, the diffracted intensity is reduced, and a corrective multiplying factor, the *Debye–Waller factor*, smaller than one, must be introduced in Eq. (4.15):

$$I(\vec{G}) = e^{-2W}.I_1(\vec{G}).I_2(\vec{G}) \qquad (4.17)$$

This factor is related to the thermal vibration of atoms through the statistical average $W \propto \langle (\vec{G}.\vec{Q})^2 \rangle$, where $\vec{Q}$ is the amplitude of the atomic oscillations. Its value decreases on raising the temperature, owing to the increase of the amplitude of oscillation of the atoms (and correlatively to the more pronounced deviation from a strict periodicity). The expression of W shows that it also decreases as function of the magnitude of $\vec{G}$. As will be pointed out in Section 4.4, a reciprocal vector $\vec{G}$ defines a family of lattice planes of interplane spacing $d = 2\pi/|\vec{G}|$. Clearly, a thermal vibration of given amplitude Δr will perturb mostly the periodicity of lattice planes having the smallest spacing ($\Delta r/d$ being larger). We can therefore understand that the intensity-reduction effect will be larger for the diffraction directions corresponding to the large values of the transfer vector $\vec{G}$. At room temperature, this reduction is of the order of 10% for small $|\vec{G}|$ values, but it can be much larger for directions corresponding to more distant nodes of the reciprocal lattice.

4.3.4   Bragg equation

In a diffraction setup, the wavelength λ of the incident beam is specified. Since the elastic character of the scattering imposes the equality $|\vec{q}| = |\vec{q'}| = (2\pi/\lambda)$, the Laue condition $\vec{K} = \vec{G}$ determines, *for each value of* $\vec{G}$, the three edges of the triangle $(\vec{q}, \vec{q'}, \vec{K} = \vec{G})$. This implies that diffraction will occur only for certain orientations of the incident (resp. diffracted) beam with respect to the crystal.

The Bragg equation specifies explicitly the constraint imposed to these orientations, for each "order" of diffraction associated to a given reciprocal lattice vector $\vec{G}$ (Fig. 4.7). It is straightforwardly obtained by substituting $\vec{K} = (\vec{q'} - \vec{q})$, in the Laue condition (4.8), and taking into account the equality $|\vec{q'}| = |\vec{q}|$:

$$2\vec{q'}.\vec{G} = \vec{G}^2; \quad 2\vec{q}.\vec{G} = -\vec{G}^2 \qquad (4.18)$$

The first equation is identical to Eq. (3.25) which defines, in particular, the vectors joining the origin Γ of the reciprocal lattice to a point of the surface of the first Brillouin zone. More generally,

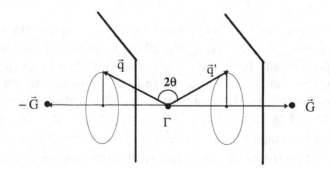

Fig. 4.7.   Geometrical representation of the Bragg diffraction equation: the diffraction wavevector whose origin is taken on a node of the reciprocal lattice has its extremity on a Brillouin plane, mediator of a reciprocal lattice vector. For given λ and $\vec{G}$, the incident and diffracted wavevectors are on definite cones.

Eq. (4.18) defines *the mediator planes* of the vectors $\overrightarrow{\Gamma M} = \pm \vec{G}$. These *reciprocal-space* planes, whose relevance has already been pointed out in the representation of the quantum energy levels of a solid (Section 3.3.5), are called *Brillouin planes* (if one refers to the fact that they define the borders between the successive Brillouin zones) or *Bragg planes* (if one refers to their importance in the geometry of the diffraction).

Note that, for given $\vec{G}$, the Bragg equation only determines the value of the projection of $\vec{q}$ and $\vec{q'}$ on $\vec{G}$. The additional condition set by the value of λ constrains these two vectors to be on cones of axis $\vec{G}$.

4.3.5   The Ewald construction

While the Laue and Bragg conditions emphasize the condition imposed either to the transfer vector $\vec{K}$, or to the incident or diffracted wavevector, the *Ewald construction* takes into account *simultaneously* the two constraints $\vec{K} = \vec{G}$ and $|\vec{q'}| = |\vec{q}|$, in order to specify the orientations of the incident and diffracted beams with respect to the reciprocal lattice (and hence, to the crystal lattice). Figure 4.8 shows that the two former conditions imply that two nodes **A** and **B** of the reciprocal lattice (such as $\overrightarrow{AB} = \vec{G}$) are located on a sphere of radius $|\vec{q}|$, called the *Ewald sphere*. The directions of the incident and diffracted beams are respectively oriented along the radii pointing towards **A** and **B**.

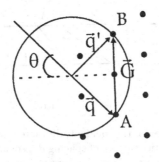

Fig. 4.8.   Ewald construction permitting the determination of the relative orientations of the incident and diffracted beam with respect to the reciprocal lattice. The dashed line marks the direction of lattice planes associated to **G**.

In practice, the "diffraction plane" defined by $\vec{q}$, $\vec{q'}$ and $\vec{G}$ is imposed by the material configuration of the diffraction instrument (see Fig. 4.11). For a given type of diffraction (specified by the vector $\vec{G} = \overrightarrow{AB}$), the Ewald circle in this plane is defined by its radius $|\vec{q}| = 2\pi/\lambda$, and its center located on the mediator of $\overrightarrow{AB}$. Placing this circle, one can then deduce the relative orientations of the incident and diffracted beams.

4.4   Diffraction and Lattice Planes

Let us show that an alternate derivation of the Bragg equation can be obtained by considering that diffraction, instead of resulting from the coherent interference of waves *scattered by the individual atoms*, results from the coherent interference of waves *reflected by a family of lattice planes*. This equation is then given its more "elementary" expression:

$$2d\sin\theta = h\lambda \qquad (4.19)$$

where d is the spacing of a family of regularly spaced lattice planes of the crystal and θ is the angle of incidence of the incoming beam on the planes (Fig. 4.9).

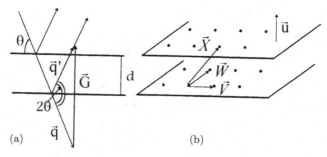

Fig. 4.9.   (a) Configuration of the incident, diffracted, and transfer vectors, with respect to the family of lattice planes, of spacing d associated to **G**. The diffraction is the constructive interference between waves *reflected on the lattice planes*. (b) Set of primitive translations of the crystal adapted to a family of lattice planes.

4.4.1 Lattice planes and reciprocal vectors: Miller indices

In Chapter 2 (Fig. 2.8), we have seen that one could define an infinity of families of equally spaced parallel lattice-planes passing through all the nodes of the Bravais lattice. Consider such a family defined by its normal unit-vector $\vec{u}$ and its spacing d (Fig. 4.9). As shown in Section 2.2.2, a set of primitive translations of the crystal can be constituted by two vectors $\vec{V}$ and $\vec{W}$ lying within one of the lattice planes, and a third vector $\vec{X}$ joining the nodes of two consecutive planes. The set $(\vec{V}, \vec{W}, \vec{X})$ defines a set of primitive translations $(\vec{V^*}, \vec{W^*}, \vec{G_0})$ of the reciprocal lattice. On the basis of the relations $\vec{V}.\vec{u} = 0$, $\vec{W}.\vec{u} = 0$, and $\vec{X}.\vec{u} = d$, it is easy to show that

$$\vec{G_0}(\vec{u}, d) = 2\pi \frac{\vec{V} \times \vec{W}}{(\vec{V}, \vec{W}, \vec{X})} = \left(\frac{2\pi}{d}\right) \vec{u} \qquad (4.20)$$

Hence, each family of lattice planes $(\vec{u}, d)$ can be associated to a vector $\vec{G_0}(\vec{u}, d)$ of the reciprocal lattice.[7]

From this property derives the definition of *Miller indices* which specify a direction of lattice planes. These integer indices (h, k, l) are the components of the vector $\vec{G_0} = (2\pi/d)\vec{u}$ in the basis $(\vec{a_1^*}, \vec{a_2^*}, \vec{a_3^*})$ of the reciprocal lattice. These indices have a simple geometrical interpretation.

Thus, if we refer the *direct lattice* to the primitive translations $(\vec{a_1}, \vec{a_2}, \vec{a_3})$, the set of *equally spaced* planes of the considered family will necessarily contain a plane A passing through the origin, and also parallel planes passing through the endpoints of $(\vec{a_1}, \vec{a_2}, \vec{a_3})$. In this family of lattice planes, one, at a distance d of the origin, is *closest to the node taken as origin*. It intersects the three primitive translations at $(1/m_1, 1/m_2, 1/m_3)$, with (m_1, m_2, m_3) integers (Fig. 4.10): Indeed this simple fractional character of the coordinates of the intersections derives from the equal distances between lattice planes. Accordingly, the set of planes will divide each $\vec{a_i}$ into equal parts.

---

[7]The converse is not true. In the set of all $\vec{G}$ vectors, parallel to a given direction $\vec{u}$ only the shortest is associated to the spacing d through Eq. (4.18).

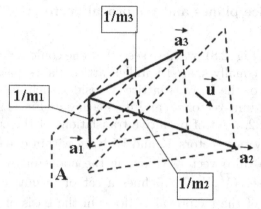

Fig. 4.10. Miller indices (m_1, m_2, m_3) of a family of lattice planes. d is the spacing between consecutive planes in the direction $\vec{u}$. The case represented corresponds to planes of Miller indices m_i (2, 3, 2).

Let us show that the indices (m_1, m_2, m_3) are equal[8] to the components (h, k, l) of $\overrightarrow{G_0}$. We have (Fig. 4.10):

$$\frac{\overrightarrow{a_i}}{m_i} . \vec{u} = d \qquad (4.21)$$

Taking into account Eq. (4.18) and the expression of $\overrightarrow{G_0}$ as a function of the reciprocal lattice vectors $\vec{a_i^*}$, we can write:

$$\frac{\overrightarrow{a_i}}{m_i} . (h\overrightarrow{a_1^*} + k\overrightarrow{a_2^*} + l\overrightarrow{a_3^*}) = 2\pi \implies (m_1 = h; \quad m_2 = k; \quad m_3 = l)$$

$$(4.22)$$

4.4.2 Bragg equation for lattice planes

If we consider any multiple $\overrightarrow{G} = (2h\pi/d)\vec{u}$ of the reciprocal vector $\overrightarrow{G_0}$ (Eq. (4.18)) associated to a family of lattice planes, its compliance with the Bragg equation (4.19), where $\vec{q} . \vec{u} = (2\pi/\lambda) \sin \theta$,

---

[8]This correspondence is not valid if the Miller indices are defined as intersections with the primitive translations of the conventional cell. In this case, (h, k, l) is the set of smallest integers proportional to (m_1, m_2, m_3).

implies

$$\frac{4h\pi}{d}\vec{q}\cdot\vec{u} + \frac{4h^2\pi^2}{d^2}\vec{u}^2 = \frac{4h\pi}{d}\left(\frac{2\pi}{\lambda}\sin\theta + \frac{h\pi}{d}\right) = 0$$

which is equivalent to Eq. (4.19).

Hence any reciprocal lattice vector $\vec{G}$ *satisfying Eq. (4.18) can be associated to a diffraction on a definite family of lattice planes of the crystal.*

4.5 The Determination of Crystal Structures

Equation (4.15), which separates the contributions of the Bravais lattice and that of the atomic configuration of the basis, to the diffraction spectrum, underlies the method of studying a crystal structure. A first step of this study consists in determining the *geometry of the various diffracting configurations* ($\vec{q}, \vec{q}', \vec{G}$) in order to specify the characteristics of the Bravais lattice (type of lattice, directions and magnitudes of the primitive translations, primitive and conventional unit cells). In a second step, the measurement of the relative *intensities* of the diffracted beams associated to each node of the reciprocal lattice will lead to the specification of the atomic configuration.

4.5.1 Specification of the Bravais lattice

A description of the different experimental diffraction methods used to determine the geometrical characteristics of a crystal lattice is beyond the scope of this course.[9] Let us only underline here the principles of this determination and its difficulties, dwelling on the case of X-ray diffraction, which, for reasons which will be put forward in Section 4.5.3, has a central place among these methods.

In its principle, a standard X-ray diffraction experiment consists in exposing a crystal to a monochromatic beam having a given wavevector $\vec{q}$. The orientation of the crystal is adjusted in a manner to obtain the emission of a diffracted wave. In accordance with

---

[9]Such methods are described in a number of specialized textbooks such as the references indicated at the end of the chapter.

the Ewald construction, the determination of the angle between $\overrightarrow{q}$ and $\overrightarrow{q'}$, and the prior knowledge of their common modulus $(2\pi/\lambda)$, permit the specification of the modulus of a vector $\overrightarrow{K} = \overrightarrow{G}$ of the reciprocal lattice, as well as of its orientation in a reference frame linked to the instrument (and to the external surfaces of the crystalline sample). A systematic search of "diffracting configurations" of the preceding type, will allow to infer the primitive translations of the reciprocal lattice, through specification of a sufficient number of $\overrightarrow{G}$ vectors. The equations of correspondence (3.22) eventually determine the primitive translations of the *direct lattice*.

The implementation of this principle can be achieved by using various experimental techniques.

For instance, it can proceed through a *goniometric technique* (Fig. 4.11) in which the crystal and an X-ray detector can be rotated around a common axis perpendicular to the "diffraction plane" $(\overrightarrow{q}, \overrightarrow{q'})$. A reciprocal lattice plane of the crystal has been brought to coincide with this plane.[10] As apparent on Fig. (4.7), the angle of

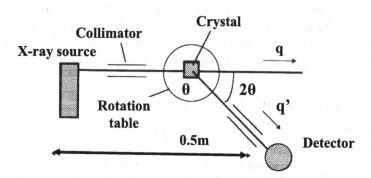

Fig. 4.11.   Schematic X-ray goniometer setup. The diffraction plane (defined by the incident and diffraction wavevectors) is materially set by the configuration of the source and the detector and is perpendicular to the common rotation axis of the crystal and of the detector. To obtain a diffraction, a reciprocal lattice plane must be brought in coincidence with this plane.

---

[10]As apparent in the Ewald configuration, for given angle between the incident beam and the direction of the detector, there is not necessarily an orientation of the crystal permitting a diffraction. Hence, a prior step to the described goniometric experiment consists in a trial and error procedure of adjustment aiming

incidence of the incoming beam being half the angle between $\overrightarrow{q}$ and $\overrightarrow{q'}$, a sequence of diffractions associated to the considered diffraction plane is obtained for discrete values of θ, by rotating synchronously the crystal by the angle θ and the detector by an angle 2θ. Each value corresponds to a definite reciprocal lattice vector, or, equivalently to a definite spacing between lattice planes.

Another method consists in recording, for each diffraction setting, the spot-trace left by the diffracted beam on a photographic film placed at some distance from the crystal. When the crystal is rotated to reach another diffracting orientation, the film is displaced in a manner correlated to the motion of the crystal (various types of motions of the film are possible and correspond to various photographic "precession" X-rays methods). After completion of the experiment the revealed film will display a set of "diffraction spots" which are in a one-to-one correspondence with nodes of the reciprocal lattice. Figure 4.12 illustrates the result of a precession experiment in

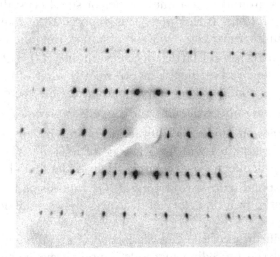

Fig. 4.12.   Image of a plane of the reciprocal lattice of an organic crystal obtained by means of an X-ray precession camera (courtesy of Françoise Denoyer CNRS). In this method, a photographic film has a precession motion around the crystal synchronously with the crystal rotation, and records the diffracted beams for each diffracting position of the crystal in a fixed incident beam.

---

at bringing a reciprocal lattice plane in coincidence with the diffraction plane (through a "tilting" of the crystal, and setting independently θ_{crystal} and θ_{detector}).

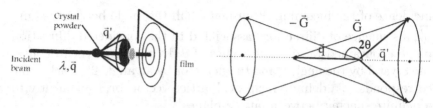

Fig. 4.13.   Principle of the Debye–Scherrer method which relies on the use of a powder of randomly oriented small crystallites. For specified incident wavevector, and specified modulus of a reciprocal lattice vector, the figure on the right shows that the diffraction plane being arbitrary, the Bragg condition is satisfied by a set of diffracted wavevectors forming a cone.

which one observes on the film a regular configuration of spots homothetical to the configuration of a fraction of the *nodes of a reciprocal lattice plane.*

In a third type of method, the Debye–Scherrer method (Fig. 4.13), the crystal is ground into a fine powder of small crystallites having *random orientations.* If this powder is exposed to a monochromatic beam of definite wavevector $\vec{q}$, the assumed randomness warrants that, for each reciprocal lattice vector verifying $|\vec{G}| < 2|\vec{q}|$ (a condition equivalent to $|\sin\theta| < 1$ in Eq. (4.19)) a set of crystallites is in a diffracting configuration corresponding to this vector. Since the Bragg condition only constrains the angle θ between $\vec{q}$ and $\vec{q}'$, but not the diffraction plane, the diffraction beams $\vec{q}'$ corresponding to the various crystallites will form a cone of axis $\vec{q}$, whose trace on a film perpendicular to $\vec{q}$ is a circle. Hence, the film shows a set of circles corresponding to a repertory of $|\vec{G}|$ values smaller than $2|\vec{q}|$, or, equivalently, to a repertory of distances between lattice planes of the structure.

The specification of the Bravais lattice of an *unknown crystal* by any single of the preceding methods is not always straightforward. Indeed, initially, the considered crystal will be arbitrarily oriented with respect to the incident beam. A search of the diffracting configurations by means of a goniometric technique would require exploring with a high angular precision all the possible orientations of the crystal, and, for each such orientation, observe the possible existence of a diffraction, through a systematic scan of the detector angle.[11] Likewise, the Debye–Scherrer technique, which provides directly a set of lattice spacings, can be difficult to interpret, except in the case of cubic lattices.

In practice, the identification of the Bravais lattice of an unknown crystal is performed through successive approximations using several of the preceding techniques, the goniometric technique being used for the final refinement of the description of the lattice.

4.5.2   Determination of the atomic configuration

The Bravais lattice having been determined as explained in the preceding paragraph, the configuration of the atoms in the primitive unit cell remains to be specified. In principle, this configuration can be obtained by measurement of the diffracted intensities at each node of the reciprocal lattice through Eqs. (4.13)–(4.16) which provide the expression of the structure factor. However, such a strategy meets an important obstacle.

To clarify the nature of the difficulty, it is convenient to rely on the continuous expression (4.16) of the structure factor. Indeed, the measurement of $F(\overrightarrow{K})$ at all the nodes of the reciprocal lattice should allow a determination of the electron density $\rho(\overrightarrow{r})$ in the crystal through the inversion of the Fourier correspondence between $\rho(\overrightarrow{r})$ and $F(\overrightarrow{K})$. Such an inversion requires the knowledge of *both the modulus and the phase of the complex number* $F(\overrightarrow{K})$.

X-rays detectors are only sensitive to the intensity of the diffracted beam, hence to the value of the modulus of $F(\overrightarrow{K})$. In itself this is not an absolute obstacle. In principle, it is possible to deduce the phase of a wave, as is currently performed in optics, through an interference experiment, by superimposing the investigated beam to a reference beam. However, in the case of X-rays, exploiting such a method

---

[11] In the last decades, spatially extended linear or circular detectors constituted by a large number of detectors of small dimension have facilitated this search by allowing to detect simultaneously the possible existence of a diffracted beam in a diffraction plane.

is made impossible by the poor phase coherence of the incident beam (of the order of a few wavelength) which will blur the interference patterns on the distance separating the crystal from the detector.[12] In consequence, the loss of information concerning the phase of the structure factor prevents a straightforward determination of the atomic configuration.

In practice, this determination is achieved by a "trial and error" procedure. The intensities $I(\vec{K})$ expected for a given model-structure are compared to the measured intensities. Corrections are brought to the model until a satisfactory agreement is reached. The successful application of such a strategy to complex structures requires the (automatized) measurement of the intensity corresponding to several thousands nodes of the reciprocal lattice, the use of statistical methods for treating the results and comparing them to the atomic model, and also, in order to elaborate a suitable initial model, a thorough knowledge of the standard configurations adopted by the atomic subunits of the considered structure.

The need to analyze biological structures involving thousands of atoms has stimulated the development of "direct" methods, which aim at obtaining partial information on the phases of the structure factors. These methods rely on the fact that the investigated *electron density* has two remarkable features. It is a *non-negative* function of the space coordinates. Also, its spatial variations are not smooth, but rather involve *sharp peaks of concentration* around the positions of the atoms. It can be shown that these criteria impose significant restrictions to the relative values of the phases of the different structure factors.

4.5.3 Comparison of the use of X-rays, neutrons and electrons

(a) Neutrons and X-rays

A beam of thermal neutrons possesses a De Broglie wavelength of the order of 1 Å. It can be used, in principle, in the determination

---

[12]The phase undetermination could be lifted if coherent X-ray sources became available. Significant progress have been currently made in the development of such sources (X-ray lasers).

of the lattice and of the structure of a crystal in the same manner as X-rays. However, the available neutron beams have weak intensity, making the detection of the diffraction more difficult for given illuminated volume of the sample. As recalled by Table 4.1, a typical flux is 10^7 neutrons/s, while an X-ray beam will correspond to a flux of 10^{12}–10^{18} photons/s (the last figure corresponding to the flux available in a synchrotron).

In spite of this drawback, neutron diffraction plays an important role in the *refinement* of structures containing both heavy atoms and light ones. In X-ray diffraction, the dominant contribution will come from scattering by the heavy atoms. The atomic configuration derived from this diffraction will therefore be unaccurate regarding the positions of light atoms. In contrast, the neutron atomic scattering factor can be comparable for light and heavy atoms (cf. Table 4.2), and accordingly, a refinement of the structure by means of neutron diffraction, will improve this accuracy.

Two other prominent uses of neutron scattering are worth mentioning.

The first one pertains to the *inelastic scattering* of neutrons by a crystal, for which the incident and scattered beams have *different energies* ($|\vec{q}| \neq |\vec{q'}|$). The difference in energy is exchanged with the crystal, and is equal to the energy-spacing between two quantum states of the crystal. Such a scattering provides, for instance, a means of studying the energy spectrum of the *thermal oscillation of atoms* around their equilibrium positions. Other types of spectra can also be examined (e.g. the energy levels of electrons). The measured energy spacings are generally in the range 10^{-3}–10^{-7} eV. The specific role of neutrons in these studies has a dual basis. In the first place, as shown on Table 4.1, the energy of thermal neutrons is of the order of 10^{-2} eV instead of 10^4 eV for X-ray-photons. The *relative variation* of energy between the incident and scattered beams is therefore much larger for neutrons ($\sim 10^{-1}$) and henceforth easier to measure.

Another important use of thermal neutrons is the determination of *magnetic structures* (cf. Section 2.6). The translational and orientational ordering of magnetic dipoles in these structures cannot be probed by using X-ray beams owing to the lack of interaction of X-ray photons with the magnetic field of the dipoles. In order to study these structures, one can use *polarized neutron beams*, i.e. neutrons whose magnetic moment (related to their spin) has a definite orientation. The amplitudes of the scattered neutron beams are functions of the relative orientations of the neutrons and of the magnetic dipoles of the crystal. The characteristics of the diffraction will reflect both the "chemical" configuration of the crystal and the magnetic one.

(b) Electrons and X-rays

The advantage of using the diffraction of electrons within a *transmission electron microscope* (Fig. 4.14) resides in the possibility to obtain, in the focal plane of a magnetic focusing lens of the microscope, a planar pattern of diffraction providing an undistorted, and magnified, "image" of the nodes $\vec{G}$ of a reciprocal lattice plane of the crystal. In principle, the relative intensities of the spots reproduce the intensities $I(\vec{G})$ of the corresponding diffraction beams.

The observation of such diffraction patterns is related, on the one hand, to the *short wavelength* $(10^{-2}$ Å) of the electrons, and, on the other hand, to the required use of *thin samples* (a few hundred unit cells) in the direction of the incident beam (cf. Fig. 4.15).

The short wavelength implies that, at the scale of the distances between neighboring nodes of the reciprocal lattice, *the Ewald sphere*, whose radius is $|\vec{q}|$, *can be assimilated to a plane.* The need to use a thin crystal sample results from the strong absorption of the electron beam (related to the high value of the atomic scattering factor) which would not be transmitted across a thick sample. In practice, samples are, at most, ~ 1000 Å, thick. For such samples, as mentioned in footnote 6, transfer vectors are associated, in the reciprocal space, to rod-shaped nodes perpendicular to the crystal-wafer examined. The

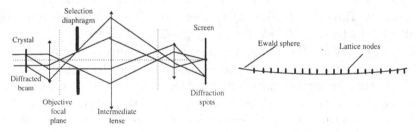

Fig. 4.14. Schematic representation of the formation of the diffraction pattern in an electron microscope. Due to the thinness of the crystal along the electron beam, the row of nodes of the reciprocal lattice, represented on the right, is rod-shaped. At the scale of the internode distance, the Ewald sphere is almost planar (owing to the short electron-wavelength) and intersects many nodes.

Fig. 4.15.   Electron diffraction by a nickel crystal (courtesy of Alain Barbu Ecole Polytechnique). The almost flat Ewald sphere (drawing above) intersects a large number of reciprocal lattice nodes which are rod-shaped, owing to the thinness of the crystal in the direction of the beam.

conjunction of the flatness of the Ewald sphere and of the rod-shaped ranges of permissible transfer vectors have the consequence that, for a suitable orientation of the crystal with respect to the incident beam, the Ewald sphere will intersect many rods.[13] In other terms, it is possible to orient the crystal in a manner of satisfying *simultaneously the Bragg condition, for many nodes of a reciprocal lattice plane.*

Note, however, that electron diffraction is not, in general, a suitable method for determining the configuration of atoms in the structure. Indeed, owing to the high values of the atomic scattering factors, the intensity of the scattered beams cannot be neglected with respect to that of the incident beam (which is also significantly weakened by the strong absorption). Hence, Eqs. (4.13)–(4.15) are not valid since they do not incorporate the effects of the multiple scattering.

4.5.4   Diffraction by partly or fully disordered solids

In the case of solids characterized by a non-crystalline type of ordering, or by the complete absence of such an ordering, diffraction

---

[13]The Fourier correspondence implies that the length ΔK_z of the rods in reciprocal space is of the order of $(1/\Delta z)$ where Δz is the thickness of the sample. For a 500 Å thick sample, $\Delta K_z \sim 2 \times 10^{-3}$ Å^{-1}. As the Ewald sphere has a radius of ~ 1000 Å^{-1}, it will intersect the rods on a section of ~ 10 Å^{-2}, corresponding to several tens of reciprocal lattice nodes.

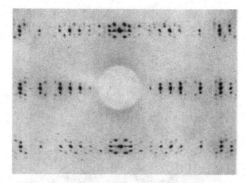

Fig. 4.16. X-ray diffraction spots in a modulated crystal of barium sodium nio-bate (courtesy of Jacques Schneck). Each intense spot is surrounded, in the ver-tical direction, by two satellite spots at a distance corresponding to a "fourth primitive translation" of the system.

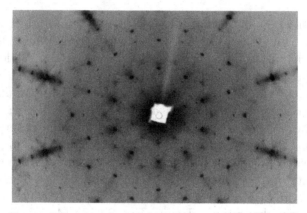

Fig. 4.17. X-ray diffraction by a decagonal quasi-crystal. The pattern is unchanged by rotations of angle $(2\pi/10)$ a value not allowed in a strictly periodic crystal.

is nevertheless a useful means of access to the atomic configu-ration. Figures 4.16–4.18 will provide illustrations of the results obtained from the diffraction of X-rays or of electrons for such systems.

Modulated crystals (cf. Section 2.6) display diffraction spots sim-ilar to those of ordinary crystals. Their non-crystalline character is revealed by the fact that the observed spots are not all situated at the nodes of a three-dimensional Bravais lattice. For instance, in the case represented on Fig. 4.16, the vectors whose endpoints coincide with

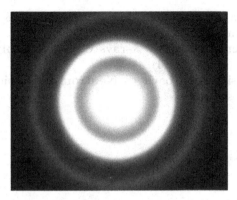

Fig. 4.18. Electron microdiffraction of an amorphous metal (courtesy of Alain Barbu Ecole Polytechnique). The diffraction intensity is non-zero in the entire reciprocal space. The maxima disclose the occurrence of a "local" order of the atomic configuration.

these spots are of the form $(h\vec{a_1^*} + k\vec{a_2^*} + l\vec{a_3^*} + m\,\vec{b})$. Their positions involve a "reciprocal lattice" generated by the four independent basis vectors $(\vec{a_1^*}, \vec{a_2^*}, \vec{a_3^*}, \vec{b})$, in agreement with the occurrence of four spatial periodicities in the direct space (cf. Section 2.6).[14].

A similar situation prevails in the case of quasi-crystals (the spots are nodes of a lattice generated by six independent vectors) (cf. Note in Section 2.6). An additional remarkable feature of these systems is the fact that the set of diffraction spots is invariant by "forbidden rotations", which in the example shown on Fig. 4.17 are rotations by multiples of $(2\pi/10)$.

Bibliography

Neutron Diffraction, *G.E. Bacon*, Pergamon Press (1966).
Electron Diffraction and High Resolution Electron Microscopy, *V.A. Drits*, Springer (1987).
X-ray Diffraction, *B.E. Warren*, Dover Edition (1990).

---

[14]In principle, a lattice generated by four independent vectors generates, in the three-dimensional space, a quasi-continuous set of spots. However, it can be shown that the intensity of these spots decreases rapidly with increasing values of the fourth index m. Hence one only observes a discrete set of spots corresponding to small values of the index m.

X-ray Diffraction Crystallography, *Y. Waseda, E. Matsubara and K. Shinoda*, Springer (2011).

Structure Determination by X-rays Crystallography, *M. Ladd and R. Palmer*, Springer (2013).

A Comprehensive Approach to Neutron Diffraction, *N. Pitt*, NY Research Press (2015).

Chapter 5

Quantum States of an Electron in a Crystal

Main ideas: The quantum energy levels of an electron form quasi-continuous bands separated by "empty gaps" (Section 5.2.2). A weak periodic potential induces the opening of narrow empty gaps in the continuous energy band of a free-electron. In the tight binding approximation, each discrete energy level of electrons tightly-bound to the atomic nuclei gives rise to a narrow energy band of the solid (Section 5.3). Different and complementary representations of the quantum spectrum are of interest (dispersion curves, equal-energy surfaces, density of states).

5.1 Introduction

In this chapter, devoted to the central topic of the book, we examine the quantum states and the energy spectrum of a single electron in a crystal. The next three chapters will rely on the results thus obtained, in order to account for the physical properties of crystalline solids, considered as systems of N non-interacting electrons.

As stated in Chapter 3, we provisionally accept that each electron of the solid is an independent entity submitted to a *"crystal potential"* $\widehat{V}(\overrightarrow{r})$. If $\overrightarrow{r}$, $\overrightarrow{p}$, and m are respectively the coordinate, the momentum and the mass of the electron, the single-electron Hamitonian quantum operator is then:

$$\widehat{H} = \frac{\overrightarrow{p}^2}{2m} + \widehat{V}(\overrightarrow{r}) \qquad (5.1)$$

119

where $\widehat{V}(\vec{r})$ is spatially periodic, with periods equal to the primitive translations of the crystal. As shown in Chapter 3, Eq. (3.17), the quantum states of an electron, eigenstates of $\widehat{H}$, are *Bloch functions* $\exp(i\,\vec{k}.\vec{r}).u_{\vec{k},n}(\vec{r})$, indexed by a reciprocal-space vector $\vec{k}$ and a band index n. The corresponding electron eigenenergy values $E(\vec{k},n)$ (the electron energy spectrum) is composed of quasi-continuous *energy bands*.

To determine, beyond this general functional form, the possible states of an electron, a precise knowledge of the potential $\widehat{V}(\vec{r})$ is necessary. This potential replaces the complex interactions existing between an electron and the other constituents of the solid. These are atomic nuclei of various species distributed periodically in space and electrons. From a qualitative standpoint, the *electrostatic* crystal potential $V(\vec{r})$ is *attractive*, in accordance with the fact that electrons have a *binding energy* which confines them within the volume of the solid. It has a dominant part resulting from the attraction of the positive atomic nuclei forming the structure of the crystal, and a repulsive part due to the set of electrons. The methods for its determination will be considered in Chapter 10. Figure 5.1 shows the expected schematic shape of $V(x)$ in the case of a speculative one-dimensional crystal, with potential "wells" at the sites of the positive nuclei. In such a potential, an electron can occupy different states corresponding to its total energy. The lowest energy levels (e.g. E_1 on Fig. 5.1) correspond to the most strongly bound electron states in which an electron is essentially confined in the vicinity of an atomic nucleus. These states are expected to be similar to those of

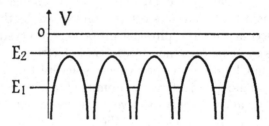

Fig. 5.1.　Schematic shape of the periodic binding potential exerted by the constituents of a one-dimensional crystal on a single electron. E_1 and E_2 are electron-energy levels corresponding respectively to strongly bound (core) electrons and to "almost-free" electrons of the crystal.

electrons in an isolated atom. Higher energy levels (e.g. E_2), in the range of an almost spatially uniform potential, correspond to states in which an electron has a probability of presence more uniformly distributed through the solid.

From this standpoint, let us examine two limiting situations underlying two approximation methods used to study the electronic states. These are respectively described in Sections 5.2 and 5.3. In Section 5.4, the relevance of each of these methods to the *band structure of real crystal is* discussed, and a brief mention is made of more elaborate methods permitting the determination of the band structure of solids.

The first method, adapted to energy levels such as E_2, consists in considering an electron submitted to a *quasi-uniform potential* having a small periodic component. The resulting energy-spectrum will be shown to consist of *wide energy-bands*, whose states resemble the quantum states of a free-electron. Two consecutive bands are separated by a "forbidden band" which is *a narrow interval, empty from quantum states.*

The second method concerns the case of an electron submitted to the potential of almost isolated atoms/ions weakly interacting with each other. In this case, the spectrum is formed of *narrow energy bands*, in a one-to-one correspondance with the discrete energy levels of the atoms/ions of the crystal. These bands are separated by *wide forbidden bands.*

5.2 Almost-Free Electron Approximation

In this section, we proceed in two steps. We first consider an electron maintained in the volume of the crystal by a *spatially uniform potential* V_0. This simple situation[1] constitutes the quantum-mechanical background of the *Sommerfeld model* of metals, whose physical implications are analyzed in Chapter 6. In a second step, we study the effect of a *small spatially periodic* potential $V(\overrightarrow{r})$ added to V_0.

---

[1]This is the situation of a "free-electron in a box" considered in elementary treatises of quantum physics.

5.2.1   Uniform potential: Free electron

Form and energy of the quantum states

It is possible to determine directly the eigenfunctions and eigenvalues of the Hamiltonian equation (5.1) for a uniform potential confining the electron in the crystal ($V_0 < 0$). Using an alternate method, let us rather apply the general results of Chapter 3 by considering that the uniform potential V_0 is a periodic potential corresponding to *infinitesimal periods* in the three directions of space. V_0 is invariant by translations $\overrightarrow{\delta}$ of any direction and amplitude. Referring to the solution of Eq. (3.16) in Chapter 3, the eigenstates of an electron are Bloch functions whose periodic part satisfy the equality $u_{\overrightarrow{k},n}(\overrightarrow{r} + \overrightarrow{\delta}) = u_{\overrightarrow{k},n}(\overrightarrow{r})$ for any $\overrightarrow{\delta}$. These functions are, necessarily, mere constants, and the eigenstates are thus reduced to plane waves

$$\phi_{\overrightarrow{k}}(\overrightarrow{r}) = A.e^{i\overrightarrow{k}.\overrightarrow{r}} \qquad (5.2)$$

indexed by the vectorial quantum number $\overrightarrow{k}$. The density of probability of presence of the electron is $|\phi_{\overrightarrow{k}}(\overrightarrow{r})|^2$ in the entire volume of the crystal.

Substituting Eq. (5.2) into the Schrödinger equations (3.16)–(3.17), and choosing the origin of the energy scale as to make $V_0 = 0$, we obtain

$$E(\overrightarrow{k}) = \frac{\overrightarrow{p}^2}{2m} = \frac{\hbar^2 \overrightarrow{k}^2}{2m} \qquad (5.3)$$

The translational quantum number $\overrightarrow{k}$ is proportional to the momentum of the electron: $\overrightarrow{p} = \hbar\overrightarrow{k}$. Hence, *in the eigenstate $\phi_{\overrightarrow{k}}$, the electron is a free-particle circulating at a constant velocity* $(\hbar\overrightarrow{k}/m)$.

Note that a limitation to the first Brillouin zone (Section 3.3.4) of the set of unequivalent $\overrightarrow{k}$-values is irrelevant here. Indeed, the infinitesimal character of the periods of the potential implies that the unit cell of the *direct* lattice is itself infinitesimal. Consequently, the Brillouin zone occupies the entire reciprocal space. For instance, taking a cubic direct cell of edges δ_i, the first Brillouin zone is a cube of edges $(2\pi/\delta_i)$, with $\delta_i \to 0$. *All the vectors of the reciprocal space define unequivalent quantum numbers.*

In contrast, the discretization of the $\overrightarrow{k}$-values imposed by the Born–Von Karman folding condition (Section 3.3.6) keeps entirely its relevance. However, for a parallelepipedic sample of edges (L_1, L_2, L_3), Eq. (3.29) providing the discrete set of $\overrightarrow{k}$-vectors loses its meaning and must be replaced by

$$\overrightarrow{k} = 2\pi \frac{n_1}{L_1} \overrightarrow{i_1} + 2\pi \frac{n_2}{L_2} \overrightarrow{i_2} + 2\pi \frac{n_3}{L_3} \overrightarrow{i_3} \qquad (5.4)$$

In agreement with (3.30), each quantum state (defined by its $\overrightarrow{k}$-vector), occupies the reciprocal volume $(8\pi^3/\Omega)$, where $\Omega = L_1 L_2 L_3$ is the volume of the sample. Thus, in the reciprocal space, the quantum states are uniformly distributed, and their uniform density is $(\Omega/8\pi^3)$.

It is worth pointing out that the Bloch functions, as well as their quantum numbers $\overrightarrow{k}$, define the *orbital state* of the electron. *To each orbital state one must associate the two possible spin-states* of the electron. These states have the same energy since the Hamiltonian (5.1) does not depend of the spin. Hence, *when the spin is taken into account, the density of quantum states in the reciprocal space is:*

$$\rho(\overrightarrow{k}) = (\Omega/4\pi^3) \qquad (5.5)$$

Representations of the energy spectrum: Density of states

The graphic representation of the energy spectrum of a quantum system whose states depend of a single discrete quantum number is simple. Thus, in the case of an atom or a molecule, the allowed energy values are displayed on a vertical scale, each discrete energy level being associated to a reduced set of quantum numbers (analogous to the n, l, m quantum numbers of the isolated atom). Such a representation provides a global overview of the characteristics of the spectrum (number of states in the different energy ranges, magnitude of the intervals between energy levels).

In the case of a solid, the representation of $E(\overrightarrow{k}, n)$ is not as trivial, owing to the fact that the full display of each band (n), requires the use of a four-dimensional system of coordinates (E, k_x, k_y, k_z).

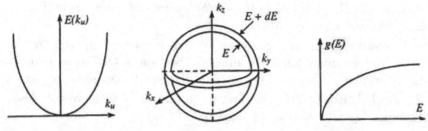

Fig. 5.2. Three partial representations of the electron energy spectrum in a uniform potential. Left: Dispersion relation in a given direction $\vec{u}$ of the reciprocal space. Center: Spherical surface corresponding to a given energy. Right: Density of states $g(E)$.

This difficulty explains that *several partial representations* are necessary, in order to visualize the various relevant characteristics of the spectrum. Namely, one uses three types of diagrams (Fig. 5.2)

(a) A plot of the variations $E(k_u, n)$ of the energy of the different bands in a given direction k_u of physical interest. Such a plot has the advantage of distinguishing the contributions of the different bands, but the possible singular shape of certain bands in specific directions of the reciprocal space can be overlooked.

(b) The *equal-energy surfaces* in the reciprocal space: $E(k_x, k_y, k_z, n) = E_0$.[2]

(c) The plot of the *energy-density of states* $g(E)$, which provides, *for a quasi-continuum of quantum states*, the number of quantum states $g(E).dE$ having their energy in the interval $(E, E + dE)$. This representation derives in a natural manner from the simple energy-level diagram of an isolated atom recalled above. However, the set of quantum numbers associated to each energy interval will not appear on such a representation.

In the case of a free electron, the two first representations are straightforwardly obtained. There is a single band $E(k_u)$ (n being irrelevant) which has the same form in all directions k_u, since the parabolic dispersion relation $E(\vec{k}) = (\hbar^2 \vec{k}^2/2m)$ is *isotropic*.

---

[2]A particularly important surface of this type is the Fermi surface which corresponds to the Fermi energy $E_0 = E_F$ (cf. Chapter 6).

The equal-energy surfaces $E(\vec{k}) = E_0$ are spheres of radius $|\vec{k}| = \sqrt{2mE_0/\hbar^2}$. Hence, *the orbital degeneracy of the energy E is equal to the number of $\vec{k}$-states having the same modulus.*

The density of states is easily derived.[3] The quantum states whose energy is in the interval $(E, E + dE)$ are represented by $\vec{k}$-vectors whose endpoints are comprised between the equal-energy spheres corresponding to E and $(E + dE)$ (Fig. 5.2) of respective radius $|\vec{k}|$ and $(|\vec{k}| + dk)$, with

$$|\vec{k}| = \sqrt{2mE/\hbar^2} \quad dk = \sqrt{\frac{m}{2\hbar^2}} \cdot \frac{dE}{\sqrt{E}} \tag{5.6}$$

The volume between the spheres is $4\pi|\vec{k}|^2.dk$. The quantum states are *uniformly* distributed in the reciprocal space with the density $(\Omega/4\pi^3)$ (*spin states included*) (cf. Eq. (5.5)), the density of states $g(E)$ is

$$g(E).dE = \frac{\Omega}{4\pi^3} 4\pi|\vec{k}|^2.dk = \frac{4\pi\Omega}{h^3}(2m)^{3/2}\sqrt{E}.dE \tag{5.7}$$

This density increases proportionally to $\sqrt{E}$. The quantum states are more densely packed for higher energies. In an apparent paradox, Fig. 5.2 shows that in a given direction k_u, the density of quantum states has an opposite behaviour and decreases for higher energies. Actually, the growth of the density of states is due to the fact that the number of states *integrated over the different directions* of the reciprocal space, grows faster with E than the spacing between the states in each direction.

5.2.2 Periodic potential: Qualitative results

In the preceding paragraph, we have considered that the particles of the crystal act on a single electron by exerting a spatially-uniform

---

[3]It is consistent to calculate a density of states for free electrons since their quantum states form a quasi-continuum, at the scale of the average energy of an electron of the solid. Indeed, electrons in a solid have an energy of the order of $1\,eV$. The relevant k-scale deduced from this value is $\sim 1\,\text{Å}^{-1}$. As shown in Chapter 3, the lattice of quantum states is, at least, 10^3 more dense in each direction.

binding potential. We now take into account the spatially-periodic variations of the potential which reflect the periodicity of the crystal structure. We assume that these variations consist in a *weak* modulation $V(\vec{r})$ around its average uniform value ($\langle V \rangle = V_0 = 0$). This weakness is scaled by comparison to the magnitude of the average *kinetic energy* ($\vec{p}^2/2m$) of the electron:

$$|V(\vec{r})| \ll \left\langle \frac{\hbar^2 \vec{k}^2}{2m} \right\rangle \tag{5.8}$$

Let us first examine qualitatively the effects of such a periodic potential in the case of a *one-dimensional Born–Von Karman crystal*, i.e. of a crystal folded into a *ring* by the Born–Von Karman procedure (Fig. 5.3).

The electron is assumed to be initially "prepared" in a quantum state represented by a plane wave $\exp(-ikx)$ circulating clockwise. As pointed out in footnote 3, the wavelength $\lambda = (2\pi/k)$ determined by the magnitude of the kinetic energy is of the order of 1 Å. If the potential is spatially uniform ($V = 0$), the wave propagates without meeting obstacles in this ring-shaped system. In accordance with the Born–Von Karman condition (Section 3.3.6) the value of k is equal to $(2n\pi/L)$ in order to ensure a phase-shift of $2n\pi$ on the length L of the crystal. The fulfilment of this condition ensures the indefinite preservation of the amplitude of the wave during its propagation. One can therefore understand that, for $V = 0$, such a wave constitutes a *stationary* quantum state of the electron.

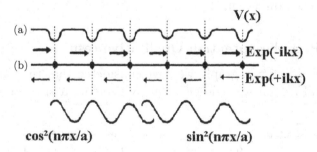

Fig. 5.3.   Perturbation of a plane wave by a one-dimensional periodic potential of period a. (a) Shape of the weakly modulated potential. (b) Reflections of the incident electron encountering the obstacles constituted by the attracting potential of the atoms located at the nodes of the lattice.

Let us now take into account the spatially periodic component $V(x)$ of the potential whose periodicity is that of the lattice of period $a = L/N$. To understand intuitively the effect of $V(x)$, we can consider that each atom located at a node of the lattice will interact with the wave and partly reflect it backwards. The wavelength being of the same order of magnitude as the lattice spacing, interferences will occur between the incident wave and the various reflected waves. Besides, $V(x)$ being "small", each reflection is a small fraction of the incident waveplane. Two consequences stem from this situation. On the one hand, the atoms are all approximately submitted to the same incident wave. Also, *multiple reflections* of a reflected wave can be neglected *in the first place* (i.e. before the steady state regime, complying with the Bragg condition, is established).

The situation is thus similar to that described in Chapter 4 for the diffraction of an X-ray or particle beam by a crystal structure. This analogy leads to an important result: *the incident plane wave will remain a stationary state of the system, transmitted without alteration, except for the specific k-values defined by the Bragg diffraction equation.*

Quantitatively, the phase-shift between two waves reflected by consecutive atoms is $(2a).(2\pi/\lambda) = 2ak$. If this phase-shift *differs* from $2n\pi$, the sum of a large number of reflected waves *all having the same amplitude* is equal to zero (cf. Fig. 5.4). The wavefunction of the electron is not modified by the reflections, and the stationary state remains equal to the incident wave $\exp(-ikx)$ with an energy $\hbar^2 k^2/2m$.

If the phase shift is equal to $2n\pi$ (i.e. $k = n\pi/a$), the reflected waves add-up, all in phase. Although their individual amplitude is

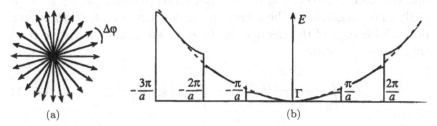

(a)   (b)

Fig. 5.4.   (a) The sum of a large number of plane waves having the same amplitude and having an arbitrary phase shift is equal to zero. (b) Modification of the energy dispersion relation by the periodic crystal potential.

small, their number, equal to the number of cells in the crystal, is large enough to determine a global reflected amplitude *comparable to the amplitude of the incident wave*, and, as it propagates counterclockwise, has the form $\exp(+ikx)$. In this situation, neither the existence of multiple reflections, nor the variations of the amplitude of the incident wave can be neglected. Thus, the reflected wave will generate, by interacting with the atoms, forward-propagating waves $\exp(-ikx)$ which add-up to the incident wave. From the symmetry of the lattice, we can infer that the final result of the diffraction of the waves $\exp(-ikx)$ and $\exp(+ikx)$ will determine for the electron a stationary state constituted by the combination of the two waves, propagating in opposite directions, and having *equal amplitudes*:

$$\psi_k \propto [\exp(ikx) \pm \exp(-ikx)] \quad \text{with } k = (n\pi/a) \qquad (5.9)$$

Let us calculate the energies of the two former stationary waves (normalized on the length of a single cell), $\psi_+ = \sqrt{2/a}\cos kx$ and $\psi_- = \sqrt{2/a}\sin kx$. Their kinetic energy is the same as that of the initial plane-wave:

$$\frac{2}{a}\int_0^a \cos kx \left(-\frac{\hbar^2}{2m}\Delta\right)\cos kx.dx$$

$$= \frac{2}{a}\int_0^a \sin kx \left(-\frac{\hbar^2}{2m}\Delta\right)\sin kx.dx = \frac{\hbar^2 k^2}{2m} \qquad (5.10)$$

In contrast, *their potential energy is different*. Indeed, we note on Fig. 5.3 that the probability density $(\cos kx)^2$ is maximum at the nodes of the lattice, while the density $(\sin kx)^2$ is maximum at mid-distance between the nodes. Since the potential $V(x)$ is obviously more negative at the attracting-sites located on the atoms, the potential energy of the electron in the ψ_+ state is more negative than that of the ψ_- state:

$$\int \cos^2 kx.V(x).dx < \int \sin^2 kx.V(x).dx \qquad (5.11)$$

Hence, for $k = \pm(n\pi/a)$, the two possible quantum states (ψ_+, ψ_-) of an electron have different energies, in the presence of the periodic potential $V(x)$, while in a uniform potential $V = 0$

these two states would have the same energy (reduced to its kinetic part).

Figure 5.4(b) summarizes the results of the above discussion. The dispersion curve $E(k)$ is not modified by the periodic potential except for $k = \pm n\pi/a$, where "energy gaps" are induced by the potential. These gaps are the result of the *lifting of the energy degeneracy* between the pairs of quantum states $\sin kx$ and $\cos kx$ by the small periodic potential.

Note that for a one-dimensional crystal, the reciprocal lattice vectors are $G = (2n\pi/a)$ and that the Bragg-equation $2\vec{k}.\vec{G} = \pm\vec{G}^2$, thus reduces to $k = \pm n\pi/a$, in agreement with the effect pointed out above.

Figure 5.4(b) shows that the opening of a gap at $k = \pm n\pi/a$ is anticipated, in the vicinity of these k-values by a progressive deviation of $E(k)$ from the free-electron dispersion curve. As will be shown in Section 5.2.3, this is a second-order effect which will only emerge from a quantitative study of the considered situation.

5.2.3 Periodic potential: First-order perturbation study

The periodic potential $\widehat{V}(\vec{r})$ being small as compared to the kinetic energy can be considered as a *perturbative component* of the Hamiltonian (3.12). Studied *at the lowest order,* the effects of a perturbation on the energy depend, as resulting from quantum mechanics, on the fact that the considered energy level is *non-degenerate* or *degenerate.*

In the former case, in agreement with the results of perturbation theory, the wavefunction is identical to the non-perturbed function $\phi(\vec{k})$, and the energy $E_0(\vec{k})$ is shifted by the quantity $\langle \phi(\vec{k})|\widehat{V}|\phi(\vec{k})\rangle$.

If the non-perturbed level is degenerate, i.e. if several states $\phi(\vec{k_m})$ possess the same kinetic energy $E_0(\vec{k}_m)$, the degeneracy is partly or completely lifted by the potential. The resulting energy values and wavefunctions of the "perturbed" electron are respectively the eigenvalues and the eigenfunctions of the "secular matrix"

$$M_{mn} = \langle \phi(\vec{k_m})|\widehat{H}|\phi(\vec{k_n})\rangle \tag{5.12}$$

The quantum states of the electron are therefore linear combinations of the plane waves $\phi(\vec{k}_m)$ which possess the same kinetic energy $E_0(\vec{k})$.

The application of this general method is considerably simplified by the specific properties of the matrix elements of the kinetic energy and of the periodic potential $\widehat{V}(\vec{r})$.

Diagonal matrix elements of the perturbed Hamiltonian

The diagonal matrix elements of the potential $V(\vec{r})$ are equal to zero. Indeed:

$$\langle\phi(\vec{k_m})|\widehat{V}|\phi(\vec{k_m})\rangle = \int e^{i\vec{k_m}\cdot\vec{r}}\cdot V(\vec{r})\cdot e^{-i\vec{k_m}\cdot\vec{r}}\cdot d^3\vec{r}$$

$$= \int V(\vec{r})d^3\vec{r} = 0 \tag{5.13}$$

since the space-average of the potential is taken as origin of the energy scale (cf. Section 5.2.2). Hence, the diagonal elements M_{mm} of the secular matrix are reduced to their kinetic part. The plane waves $\phi(\vec{k_m})$ being orthonormal eigenfunctions of the kinetic energy, we can write:

$$M_{mm} = E_0(\vec{k}_m) = (\hbar^2\vec{k}^2/2m) \tag{5.14}$$

Vanishing of off-diagonal elements for non-equivalent k-vectors

The kinetic energy has no off-diagonal contribution, and the $M_{mn}(m \neq n)$ are reduced to the potential part $\langle\phi(\vec{k_m})|\widehat{V}|\phi(\vec{k_n})\rangle$. Expressing, on the one hand, the translational invariance of the $\widehat{V}$ potential $(\tau\widehat{V}\tau^\dagger = \widehat{V})$ by means of the translation quantum operators τ introduced in Section 3.2, and, on the other hand, the fact that the plane waves $\phi(\vec{k})$ are eigenfunctions of the τ-operators corresponding to the eigenvalue $\exp(-i\vec{k}\cdot\vec{T})$ (cf. Eq. (3.9)) we can write:

$$\langle\phi(\vec{k_m})|\widehat{V}|\phi(\vec{k_n})\rangle = \langle\phi(\vec{k_m})|\tau^\dagger\tau\widehat{V}\tau^\dagger\tau|\phi(\vec{k_n})\rangle$$

$$= e^{-i(\vec{k_n}-\vec{k_m})\cdot\vec{T}}\langle\phi(\vec{k_m})|\widehat{V}|\phi(\vec{k_n})\rangle \tag{5.15}$$

Comparing the two end-members of this equation shows that the considered matrix element vanishes unless the exponential is equal to unity, which implies:

$$(\vec{k_m} - \vec{k}_n)\vec{T} = 2h\pi \tag{5.16}$$

a condition shown in Eqs. (3.20)–(3.23) to be equivalent to $(\vec{k_m} - \vec{k_n}) = \vec{G}$. Hence, the off-diagonal matrix elements of the potential vanish if $\vec{k}_m$ and $\vec{k}_n$ are unequivalent vectors, i.e. vectors which do not differ by a reciprocal lattice vector: $\vec{k_m} \neq \vec{k_n} + \vec{G}$.

Another derivation relies on the fact that an eigenfunction ψ of the matrix M_{mn}, linear combination of the $\phi(\vec{k_m})$ is also, necessarily, a Bloch function:

$$\psi = e^{i\vec{k}\cdot\vec{r}} \sum_{\vec{k_m}} C_{\vec{k_m}} \cdot e^{i(\vec{k_m} - \vec{k})\cdot\vec{r}} = e^{i\vec{k}\cdot\vec{r}} \cdot u_{\vec{k},n}(\vec{r}) \tag{5.17}$$

The $u_{\vec{k},n}$ function being spatially-periodic, each exponential in the second member must also be periodic with respect to the lattice translations $\vec{T}$ of the crystal, thus implying condition (5.17).

Relevant and irrelevant degeneracies

The energy of the states of the unperturbed system is reduced to the kinetic energy $(\hbar^2 \vec{k}^2/2m)$. Hence, all the states defined by $\vec{k}$-vectors of equal modulus are degenerate. There is an infinite quasi-continuous set of such states represented, in the reciprocal space, by points of the sphere of radius $|\vec{k}|$. However, only a small number among these states are involved in the determination of the effects of the periodic potential $\hat{V}(\vec{r})$. Indeed, only the states which contribute to the values of the matrix elements M_{mn} are relevant. As shown in the preceding paragraph, if $\phi(\vec{k})$ is the state under consideration, the only *relevant states* are those indexed by *equivalent vectors*. Though the degeneracy of each state of the unperturbed system is infinite the *relevant degeneracy* of an energy-value is defined by the set of conditions $|\vec{k_m} = \vec{k} \pm G| = |\vec{k}|$, or equivalently by:

$$2\vec{k}\cdot\vec{G} \pm \vec{G}^2 = 0 \tag{5.18}$$

identical to the Bragg equation (4.18).

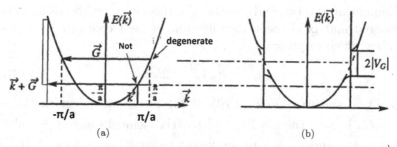

Fig. 5.5. (a) Location, for a one-dimensional crystal of the vectors $\vec{k}$ and $(\vec{k} + \vec{G})$. (b) Dispersion relation determined by the periodic potential.

Study of the non-degenerate states

Let us first examine the case (Fig. 5.5(a)) in which there is no vector $(\vec{k} + \vec{G})$ of modulus equal to, or close to $|\vec{k}|$. In this situation, converse to the one defined by Eq. (5.18), the endpoint of $\vec{k}$ is away from Brillouin planes. The quantum state $\phi(\vec{k})$ must then be considered as non-degenerate. The first-order correction to its energy, induced by the potential $\widehat{V}(\vec{r})$, is $\langle \phi(\vec{k})|\widehat{V}|\phi(\vec{k})\rangle$. As established by Eq. (5.13), this "diagonal" matrix element is *equal to zero*.

An electron state associated to a $\vec{k}$-vector sufficiently distant from the planes separating Brillouin-zones, is not modified, to first order, by the periodic potential. In particular, its energy remains equal to $\hbar^2 \vec{k}^2/2m$.

Second-order correction to the non-degenerate energies

Based on the results of perturbation theory, it can be shown that, the second-order correction to a non-degenerate energy is negative and is provided by the expansion:

$$E(\vec{k}) = E_0(\vec{k}) - \sum_{\vec{G}} \frac{|V_G|^2}{E_0(\vec{k} + \vec{G}) - E_0(\vec{k})} \qquad (5.19)$$

in which Each $\vec{G}$ vector of the reciprocal lattice contributes to the modification of the energy. [4]

---

[4]The correction diverges for $E_0(\vec{k} + \vec{G})$ approaching $E_0(\vec{k})$, in agreement with the loss of validity of the calculation which assumes that $|E_0(\vec{k}) - E_0(\vec{k} + \vec{G})| \ll |V_G|$.

Study of the degenerate states: Energy gaps

Let us now examine an electron state $\phi(\vec{k})$ complying with Eq. (5.18). Several states $\phi(\vec{k} + \vec{G}), \phi(\vec{k} + \vec{G'}), \ldots$ then have the same energy $E_0(\vec{k})$. In order to evaluate the effects of a periodic potential, let us perform the diagonalization defined by the secular equation (5.12).

In the first place, the p relevant states $\phi(\vec{k_m})$ involved in the diagonalization must be enumerated. In Chapter 3 (Section 3.3.5), the enumeration of equivalent $\vec{k}$-vectors of equal modulus, ending on the surface of the *first Brillouin zone* was discussed. A distinction was made between planes, edges, or intersection of edges of this surface. The same type of geometrical distinction holds for the surfaces limiting the nth Brillouin zone.

For instance, if the endpoint of $\vec{k}$ is on a single Brillouin plane, at sufficient distance from the intersection between two or more planes, there is a single vector $(\vec{k} + \vec{G})$ of modulus equal to $|\vec{k}|$. The effective degeneracy of $\phi(\vec{k})$ is $p = 2$. If the endpoint of $\vec{k}$ is on the intersection of several Brillouin planes, the degeneracy will be, at least, equal to three.

Let us place ourselves in the simplest degenerate case corresponding to $p = 2$. The secular matrix which determines the states of the electron is a 2×2 matrix. Putting $V_G = \langle \phi(\vec{k}) | \widehat{V} | \phi(\vec{k}) \rangle$, and applying (5.12) the form of the secular equation is

$$\begin{bmatrix} E_0(\vec{k}) - E & V_G \\ \overline{V}_G & E_0(\vec{k}) - E \end{bmatrix} = 0 \tag{5.20}$$

Its solutions are

$$E(\vec{k}) = E_0(\vec{k}) \pm |V_G| \tag{5.21}$$

Hence, the degeneracy of the energy level $E_0(\vec{k})$ is lifted. This effect creates a "gap" in the dispersion curves $E(\vec{k})$. To first order of the perturbation $\widehat{V}(\vec{r})$, the energy is continuous $(E(\vec{k}) = E_0(\vec{k}))$ in the entire reciprocal space *except on the Brillouin planes* (Fig. 5.5(b)).

Assuming that V_G is real, and substituting in Eq. (5.20) the energy values (5.21), one finds straightforwardly that the eigenfunctions of the secular matrix corresponding respectively to $(E_0(\vec{k}) \pm |V_G|)$ are:

$$\psi_+ \propto \phi(\vec{k}) + \phi(\vec{k} + \vec{G}); \quad \psi_- \propto \phi(\vec{k}) - \phi(\vec{k} + \vec{G}) \tag{5.22}$$

This form generalizes Eq. (5.9). Hence, the analysis performed for a one-dimensional crystal holds for a three-dimensional crystal.

The notation $V_G = \langle \phi(\vec{k}) | \hat{V} | \phi(\vec{k}) \rangle$ in which $\vec{k}$ does not appear explicitly, is justified by the fact that V_G is the coefficient, relative to $\vec{G}$, of the expansion of the potential $V(\vec{r})$ into a (triple) Fourier series, as a function of the plane waves $\phi(G)$:

$$V_G = \langle \phi(\vec{k}) | \hat{V} | \phi(\vec{k}) \rangle \propto \int e^{-i\vec{k}.\vec{r}} V(\vec{r}) e^{i(\vec{k}+\vec{G}).\vec{r}} d^3\vec{r}$$

$$= \int e^{i\vec{G}.\vec{r}} V(\vec{r}) d^3\vec{r} \tag{5.23}$$

V_G is thus common to all the sets of vectors $(\vec{k}, \vec{k} + \vec{G})$. Using the converse of Eq. (5.23) provides an expression of the potential as function of the V_G:

$$V(\vec{r}) \propto \sum_{\vec{G}} V_G . e^{-i\vec{G}.\vec{r}} \tag{5.24}$$

Note that the value of the gap (5.21), which occurs on a Brillouin plane, only depends of V_G, and is thus the same for all the $\vec{k}$-vectors ending on this plane. However, as will be clarified in Section 5.2.5, this property is only valid if the endpoint of $\vec{k}$ is sufficiently distant from the intersection of Brillouin planes.

In the case of a degeneracy higher than 2, the secular matrix will determine a partial or complete lifting of the degeneracy which depends on the specific geometrical situation of the various $\vec{G}$-vectors involved. The expression of the resulting energies $E(\vec{k})$ can be more complicated than (5.21).

5.2.4   Vicinity of the gaps: Quasi-degenerate states

The first-order perturbation study can be completed in order to obtain the behaviour of $E(\vec{k})$ in the vicinity of the discontinuities occurring at the Brillouin planes. In this range of $\vec{k}$-vectors, the unperturbed energies $E_0(\vec{k})$ and $E_0(\vec{k}+\vec{G})$ are not equal but are close to each other (Fig. 5.5). From the standpoint of perturbation theory, if

$$|E_0(\vec{k}) - E_0(\vec{k}+\vec{G})| \lesssim |V_G| \tag{5.25}$$

one has a situation of *quasi-degeneracy* in which the modified energy $E(\vec{k})$ has to be determined by diagonalizing the secular matrix associated to the set of quasi-degenerate states $(\phi(\vec{k}), \phi(\vec{k}+\vec{G}), \phi(\vec{k}+\vec{G}'), \ldots)$. In the simplest case, the secular equation is

$$\begin{bmatrix} E_0(\vec{k}) - E & V_G \\ \overline{V_G} & E_0(\vec{k}+\vec{G}) - E \end{bmatrix} = 0 \tag{5.26}$$

Assuming, for instance, that $E_0(\vec{k}+\vec{G}) > E_0(\vec{k})$, the solution of (5.26) is

$$E(\vec{k}) = \frac{E_0(\vec{k}) + E_0(\vec{k}+\vec{G})}{2}$$
$$\pm \frac{1}{2}\sqrt{\{E_0(\vec{k}+\vec{G}) - E_0(\vec{k})\}^2 + 4|V_G|^2} \tag{5.27}$$

Taking into account the condition $|E_0(\vec{k}) - E_0(\vec{k}+\vec{G})| \ll |V_G|$ provides modified values for $E_0(\vec{k})$ and for $E_0(\vec{k}+\vec{G})$. By continuity with the result (5.19), we can infer that the correction to the former energy is negative, while it is positive for the latter:

$$E(\vec{k}) - E_0(\vec{k}) = -|V_G| + \Delta_k \quad E(\vec{k}+\vec{G}) - E_0(\vec{k}+\vec{G})$$
$$= +|V_G| - \Delta_k \tag{5.28}$$

with

$$\Delta_k = \frac{E_0(\vec{k}+\vec{G}) - E_0(\vec{k})}{2} - \frac{\{E_0(\vec{k}+\vec{G}) - E_0(\vec{k})\}^2}{8|V_G|} \tag{5.29}$$

The resulting variations are sketched on Fig. 5.5(b). The secular equations (5.20) and (5.26) also determine the modified wavefunctions. On the Brillouin planes, these are proportional to $[\phi(\overrightarrow{k} + \overrightarrow{G}) \pm \phi(\overrightarrow{k})]$ in agreement with Eq. (5.22).

5.2.5 Representations of the energy-spectrum

We have seen in Section 5.2.1, that a *free-electron* has an isotropic energy-spectrum associated to a single energy-band involving no discontinuities. The periodic potential induces the occurrence of gaps at the Brillouin planes. This circumstance destroys the isotropy of the energy spectrum through the conjunction of two effects.

On the one hand, the distances between the origin of the reciprocal space and the Brillouin planes vary with the orientation of $\overrightarrow{k}$. As a consequence in the various $\overrightarrow{u}$ directions the gaps in $E(k_u)$ will not occur for the same values of k_u. On the other hand, the Fourier components V_G of the crystal potential, which determine the amplitudes of the gaps depend on the direction of $\overrightarrow{G}$.

This *anisotropy* of $E(\overrightarrow{k})$ is the source of a greater complexity of the various representations of the spectrum, as compared to the situation depicted in Fig. 5.2.

Spectrum in a given reciprocal direction

Whe have seen in Chapter 3 (Section 3.4) that one can represent the spectrum $E(k_u)$ in a given direction of the reciprocal space either within an *extended zone scheme*, in which one reproduces the variations of a single energy-band in the entire reciprocal space, or within a *reduced zone scheme*, in which there are several bands restricted to the first Brillouin zone. Figure 5.6 shows the schematic variations in the framework of each scheme for a simple two-dimensional situation, in a specific direction k_u (e.g. along the edge of a simple reciprocal cubic cell).

The gaps induced by the periodic potential appear on the two representations. The reduced scheme shows clearly that the spectrum consists of "allowed energy bands" composed of the possible quantum states of the electron, and of gaps in which no electron state has its energy.

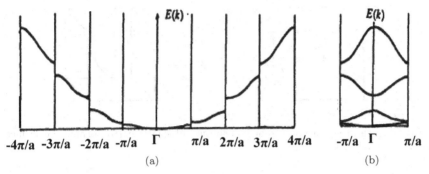

Fig. 5.6.   Variations of $E(k)$ (a) in an extended zone scheme and (b) in a reduced zone scheme.

Such an example does not show all the complexity of the energy spectrum. It can happen that the allowed energy bands in certain directions "overlap" with part, or all, of the energy ranges occupied by gaps in other directions (cf. for instance Fig. 5.17, Fig. 6.7, and exercises in Appendices A and B). As a consequence, any energy value E is associated to one or several quantum state. Though the spectrum $E(k_u)$ in certain directions has gaps, the entire spectrum $E(\vec{k})$ has no gaps in the considered energy range. In other cases, the overlap of the energy-bands is not complete and gaps are present in $E(\vec{k})$.

Equal-energy surfaces

Owing to the absence of isotropy of $E(\vec{k})$, the equal-energy surfaces are not spheres. Let us consider the example of a simple cubic crystal. The Brillouin planes closest to the origin Γ of the reciprocal lattice are the faces of the reciprocal cubic cell (Fig. 5.7), *all equivalent by symmetry*. The energy gap is $2V_G$ with $\vec{G} = \vec{a}_1^* = (2\pi/a)\,\vec{i}$. Let us assume that the Fourier component corresponding to $\vec{G}$ in the crystal potential is small, and examine the shape of the surface $E(\vec{k}) = U$ for U values belonging to two distinct ranges.

Consider first the situation $U \ll E_0(\vec{G}/2) = \hbar^2\vec{G}^2/8m$. This condition implies that we consider quantum states $\phi(\vec{k})$ of the free-electron with $\vec{k}$ within the first Brillouin zone, distant from Brillouin planes. For such states, the periodic potential does not modify the

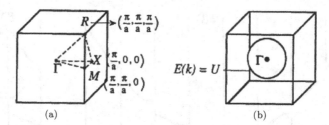

Fig. 5.7.   (a) First Brillouin zone in a simple cubic lattice. (b) Form of the surface $U = cst$ for U.

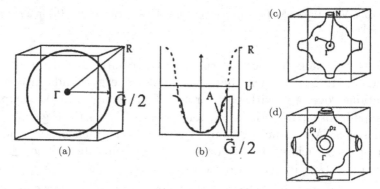

Fig. 5.8.   (a) Equal energy surface, for zero value of the potential. (b) Modification of $E(k)$ in different directions close to $\vec{G}$. (c) Shape of the equal energy surface in presence of a periodic potential. (d) Shape corresponding to a higher value of the energy.

energy (apart from a small second-order correction). Hence, the isotropy of $E(\vec{k})$ is approximately preserved, and the surface $E = U$ remains spherical (Fig. 5.7(b)). The second-order energy correction has the effect of slightly increasing the sphere radius with respect to the "free-electron" situation.

Consider now the electron-energy U complying with the conditions $[E_0(\vec{G}/2) - |V_G|] < U < E_0(\vec{G}/2)$ (Fig. 5.8(a)). The sphere of radius $r = |\vec{k_0}|$ constituting the equal energy surface $E_0 = U$ of the free-electron is still inside the first Brillouin zone but *almost tangent to the Brillouin plane*, mediator of $\vec{G}$. On this sphere, one can distinguish two categories of $\vec{k}$-vectors.

For the vectors whose orientation deviates significantly from $\vec{G}$ (e.g. vectors along the diagonal of the cube), the distance between their endpoint and the Brillouin planes is much larger than V_G. According to Eq. (5.19), the second-order correction induced by the periodic potential is negligible. Since $E(\vec{k}) \simeq E_0(\vec{k})$, the equal-energy surface remains spherical in these directions. Its radius is slightly increased in comparison with $|\vec{k}_0|$.

For the $\vec{k}$-vectors whose direction is close to that of $\vec{G}$, the situation is the same as for the A-point in Fig. 5.8(b). The potential lowers significantly the energy (by a quantity of the order of V_G) and thus, the energy value U falls into the range of the gap. There are no quantum states in these directions: the equal-energy surface $E = U$ has a circular "hole". Besides, for more inclined $\vec{k}$-directions, the condition $E(\vec{k}) < E_0(\vec{k})$ implies that the electron states with energy U correspond to a larger $\vec{k}$-vector than that of the free-electron having the same energy: $|\vec{k}| > |\vec{k}_0|$. Hence, in this range of directions, the surface $E = U$, is "sucked-out" of the sphere $r = |\vec{k}_0|$ towards the neighbouring Brillouin plane. The resulting shape of the equal-energy surface is sketched on (Fig. 5.8(c)).

The radius ρ of the hole corresponds to the $\vec{k}$ directions for which the *perturbed energies* is equal to U, and for which the equal-energy surface is in contact with the Brillouin plane:

$$E_0(\overrightarrow{\Gamma N}) - |V_G| = U \tag{5.30}$$

with

$$E_0(\overrightarrow{\Gamma N}) = \frac{\hbar^2}{2m}(\overrightarrow{\Gamma N})^2; \quad (\overrightarrow{\Gamma N})^2 = \left(\frac{\vec{G}}{2}\right)^2 + \rho^2 \tag{5.31}$$

Hence

$$\rho = \sqrt{\frac{2m}{\hbar^2}} \cdot \sqrt{U - \left[\frac{\hbar^2}{2m}\left(\frac{\vec{G}}{2}\right)^2 - |V_G|\right]} \tag{5.32}$$

For larger values of U the surface can have a more complicated form, with a circular ring of radii (ρ_1, ρ_2) deprived of quantum states and a "cap" corresponding to $\vec{k}$-directions surrounding $\vec{G}$ (Fig. 5.8(d)).

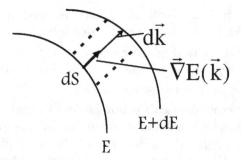

Fig. 5.9.   Relationship between density of states and equal energy surfaces.

Density of states $g(E)$

To calculate the density of states in a given energy range, we proceed as in the case of the uniform potential, by evaluating the reciprocal volume comprised between two neighbouring equal-energy surfaces corresponding to $U = E$ and $U = E + dE$. In general, this evaluation is not simple owing to the non-spherical form of the surfaces $E = U$.

The elementary volume dw^* comprised between two portions of the considered surfaces is equal to the product $dS.dk$ of the elementary surface dS belonging to $U = E$, and of the distance dk between the two neighbouring surfaces in the direction perpendicular to dS (Fig. 5.9).

Since

$$dE = \{\vec{\nabla}_k E(\vec{k}).d\vec{k}\} = |\vec{\nabla}_k E|.dk \qquad (5.33)$$

we can write, taking into account the two spin-states of the electron[5]:

$$dw^* = dE \int_{surface\, U = E} \frac{dS(\vec{k})}{|\vec{\nabla}_k E(\vec{k})|} \qquad g(E).dE = \frac{\Omega}{4\pi^3}dw^* \qquad (5.34)$$

The conclusions obtained in the case of the equal-energy surfaces imply that for values of $E \ll E_0(\vec{G}/2)$, the density of states has the same shape as for the free-electron. In contrast, for E close to the lower limit of the gap which occurs in the $\vec{G}$-direction, $g(E)$ is larger

---

[5]The actual computation of $g(E)$ using the derived expression requires a convenient parametrization of the equal-energy surface.

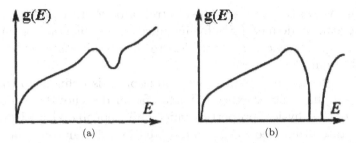

Fig. 5.10. Possible shapes of the density of states in the case of an almost uniform potential. Such forms will be analyzed in Chapter 6, Section 6.4.2.

than the density of states of the free-electron. Indeed, $|\vec{\nabla}_k E(\vec{k})|$ becomes small, and the integral in Eq. (5.34) is increased. For still larger values of E, various forms of $g(E)$ can arise, depending on the value of V_G. Hence, for two different situations the density of states has the variations qualitatively represented on Figs. 5.10(a) and 5.10(b).

5.3 Tight Binding Approximation

In the preceding section, we have considered the limiting case in which the periodic potential exerted on the electron induces small corrections with respect to the quantum state of a free-electron in a uniform potential. In this section, we examine quantum states for which it is relevant to use as reference the strongly bound states of the electron in the constituent atoms or ions. In this framework, we will express the electron states in the solid as linear combinations of atomic wavefunctions centered on the nuclei of the various constituents. Such a formulation generalizes the so-called LCAO-procedure[6] used to determine the electronic quantum states of molecules. It implies considering a crystal as a giant molecule comprising N-atoms.

5.3.1 Qualitative origin of the energy-bands

Let us consider a one-dimensional crystal formed by N identical atoms situated at the nodes of a lattice of period a. We assume

---

[6]LCAO: Linear Combination of Atomic Orbitals.

that in each of the atoms, a one-electron atomic stationary quantum state φ can be defined (φ is for instance an atomic state of type 1s, 2s, 2p, ...). Let $\varphi(x - x_i) = \varphi_i$ denote the atomic state centered on the ith-atom.

If an atom is isolated, its wavefunction φ is defined up to a an arbitrary phase-factor $\exp(i\alpha)$. Such a factor does not determine the values of the physical properties which only depend on $|\varphi|^2$. However, if the state of an electron is represented by a linear combination of several $\varphi(x - x_i)$ whose *spatial extension overlap, the relative phases* of the different φ_i will influence the energy of the state.

Atoms far apart from each other: N-degeneracy

Figure 5.11(a) represents schematically the amplitude of probability, $\varphi(x - x_i)$ associated to each atom, as well as the potential $V_i = V(x - x_i)$ exerted by this atom on an electron. Both functions decrease exponentially with the distance $|x - x_i|$ to the nucleus (they are negligible at a distance of a few angströms).

Assume first that the period of the crystal is large compared to the spatial extension of φ and of V. The electron in the state φ_i only feels the potential exerted by the ith atom. Its energy is then equal to the energy E_0 associated to φ in the isolated atom, and it does not depend on the phase α_i of the wavefunction (the potential energy being obtained by integrating $V.|\varphi|^2$).

Since each of the φ_i is, equivalently, a possible quantum-state of an electron of the solid, and that obviously, the N functions φ_i are

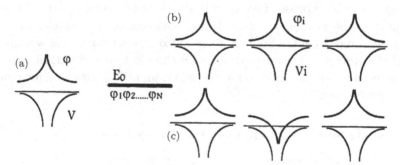

Fig. 5.11. Representation of the wave function of a crystal by a set of identical atomic functions, not overlapping with each other, due to the large distance between atoms.

linearly independent, *their common energy-level E_0 must be considered as possessing a degeneracy equal to N.*

Lifting of the N-degeneracy: Energy band

We now assume that the period a of the crystal is such that the wavefunctions φ_i and the potentials V_i *of neighbouring atoms* overlap (Fig. 5.12). In the intermediate region between atoms i and $i+1$, the states φ_i and φ_{i+1} both have significant amplitudes. In this range, the probability of presence of an electron is determined by a linear combination of the two states. More generally, this probability is associated to a linear combination ψ of all the φ_i. Owing to the identity of the wavefunctions considered on the N identical atoms, the coefficients in the linear combination have the same modulus ρ, and this combination can thus be defined by a distribution of phase-factors $\exp(i\alpha_i)$:

$$\psi = \rho \sum_{i=1}^{N} e^{i\alpha_n}.\varphi_i \qquad (5.35)$$

Let φ be a *positive wavefunction* (e.g. an atomic s-function), and examine the effect of a relative phase shift between consecutive atoms in the lattice. Compare, for instance, the situation in which all the phases α_n are identical (Fig. 5.12(a)), and the situation in which adjacent wavefunctions are phase-shifted by π (Fig. 5.12(b)). In the first case, the amplitude of probability ψ_+ of the electron is positive in the entire crystal, while in the second case ψ_- changes sign in consecutive atoms and vanishes at half-distance between the nuclei.

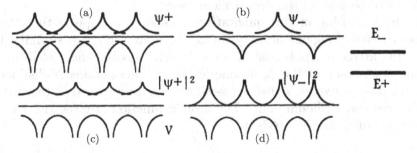

Fig. 5.12. Representation of an electron state when the atomic wave functions of neighbouring atoms overlap.

The corresponding densities of probability $|\psi_\pm|^2$ determine different potential energies for the electron. Indeed, in the state ψ_+ the sum of the attracting potentials of two neighbouring atoms at half-distance between the nuclei determines a significant (negative) contribution $V|\psi_+|^2$ to the energy E_+ of the electron, while such a contribution vanishes for ψ_-. We can infer that $E_+ < E_-$.

More generally, we can construct other electron states by assigning sets of positive and negative signs to the N functions φ_i. The resulting probability densities will vanish in certain intervals separating the nuclei, and add-up positively in other intervals. As shown above, the different assignments will generally correspond to different values of the energy. Clearly, the lowest energy corresponds to ψ_+ and the highest one to ψ_-.

Note that an arbitrary distribution of signs is not physically acceptable. Indeed, the Bloch-functional form of the ψ-functions implies a constant phase-shift between two consecutive functions φ_i (the phase factor $\exp(i\alpha_i)$ can be considered as the discrete realization of the plane-wave factor $\exp(ik.r)$). Besides, the Born–Von Karman condition requires a global phase-shift between the two ends of the crystal equal to a multiple of 2π. These two requirements leave the possibility of only N periodic sign-distributions over the φ_i functions. Each distribution consists in the repetition of a sequence of N/p consecutive positive functions and N/p negative ones $(p = N, N/2, N/4, \ldots, 1)$.

Through association to the Bloch-theorem, one can attach a discrete-value of the wavevector $k = (2\pi/pa)$ to each of the phase-distributions. Thus ψ_+ corresponds to $k = 2\pi/a$ or equivalently to $k = 0$ (center of the first Brillouin-zone) while ψ_- corresponds to $k = \pi/a$ (border of the first Brillouin zone).

In conclusion of this qualitative discussion, it appears that the occurrence of regions of a crystal in which an electron is submitted to the attracting potential of several nuclei has the effect of lifting the degeneracy of the N atomic levels φ_i. One obtains N-distinct levels associated to N distinct k-values regularly spaced in the one-dimensional Brillouin zone. The atomic function φ gives rise to a quasi-continuous energy-band.[7]

---

[7]This qualitative analysis is not strictly correct. Indeed, owing to the normalization of the wavefunctions, the increase of $|\psi|^2$ in an interval implies its decrease in

The width of the band depends of the overlap between consecutive functions φ_i, and also of the magnitude of the potential in the region of the overlap. For given distance a between nuclei, the atomic functions φ whose spatial extension is small (e.g. s-functions) will generate narrow bands, while more extended functions (generally those corresponding to higher values of the principal quantum number) will generate wider bands.

An alternate explanation of the formation of bands through the degeneracy lifting of N identical atomic levels relies on the observation that whenever two atoms are close to each other there is a possibility for the electrons of one atom of shifting to the other atom *through quantum tunneling*. Two effects contribute to this tunneling: (a) decrease of the width of the barrier separating two atoms by reduction of their distance; (b) lowering of the barrier height through the addition of the attracting potential of the two atoms. A simple quantum mechanical calculation shows that the tunneling effect lifts the degeneracy between the quantum states of the electron in each atom. Such a description is equivalent to the preceding qualitative analysis. Indeed, considering the tunneling effect consists in taking into account, on the site of a given atom, the residual amplitude of the electron wavefunction of a neighbouring atom: we have considered this residual amplitude through its overlap on the neighbouring wavefunctions. On the other hand, the lowering of the potential barrier is replaced by taking into account of the conjugate effect of the potential of the two consecutive atoms in the interval separating consecutive atoms.

5.3.2   Principle of the band calculation

Bloch functions built from atomic orbitals

In order to formalize the preceding qualitative model in the simplest situation, let us consider N identical atomic orbitals $\varphi(\vec{r} - \vec{T_j})$ centered at the nodes $\vec{T_j}$ of a three-dimensional lattice. By assigning

---

other intervals. Thus, the probability density associated to ψ_- must increase on the sites of the nuclei. However, as shown by the quantitative analysis performed in Section 5.3.2, the effects predicted on the basis of the above arguments are real.

a phase factor $\exp(i\vec{k}.\vec{T_j})$ to each orbital, we can construct the function:[8]

$$\psi_k(\vec{r}) = \frac{1}{\sqrt{N}} \left\{ \sum_{j=1}^{N} e^{i\vec{k}.\vec{T_j}}.\varphi(\vec{r} - \vec{T_j}) \right\}$$

$$= \frac{1}{\sqrt{N}} e^{i\vec{k}.\vec{r}} \left\{ \sum_{j=1}^{N} e^{i\vec{k}.(\vec{r}-\vec{T_j})}.\varphi(\vec{r} - \vec{T_j}) \right\} \quad (5.36)$$

The second member has the Bloch-functional form, since the expression between brackets is clearly periodic, with periods $\vec{T_i}$ (the transformation $\vec{r} \rightarrow \vec{r} + \vec{T_i}$ consists in a relabelling of the summation index in Eq. (5.36).

Energy of the tight-binding Bloch functions

Considering the rapid spatial-decrease of the φ functions it is reasonable to assume that the *overlap integrals* $S = \langle \varphi(\vec{r} - \vec{T_l}) | \varphi(\vec{r} - \vec{T_m}) \rangle$ vanish if the nodes located at $\vec{T}_l$ and $\vec{T_m}$ are not *nearest-neighbours lattice nodes*. Besides, the φ functions being all identical, the non-vanishing overlap integrals S are all equal. We then deduce from (5.36):

$$\langle \psi_k | \psi_k \rangle = \frac{1}{N} \sum_{l,m} e^{i\vec{k}.(\vec{T_l}-\vec{T_m})} \langle \varphi_l | \varphi_m \rangle = 1 + S. \left[\sum_{\mu=1}^{z} e^{i\vec{k}.\vec{T_\mu}} \right]$$

$$(5.37)$$

in which $\vec{T_\mu}$ is one of the z lattice-vectors joining a node to its z nearest neighbours. S is the overlap integral $\langle \phi_l | \phi_m \rangle$.

If we denote $w(\vec{r})$ the potential exerted on the electron in an isolated atom (in the framework of the one-electron approximation

---

[8]In spite of the $1/\sqrt{N}$ factor the defined function is not necessarily normalized, owing to the fact that, since the φ functions overlap, they are not orthogonal.

already mentioned), the one-electron Hamiltonian can be written as

$$\hat{H} = \frac{\widehat{\vec{p}}^2}{2m} + \sum_{n=1}^{N} w(\vec{r} - \vec{T_n}) \tag{5.38}$$

The electron energy corresponding to the ψ_k state is then:

$$E_k = \frac{\langle \psi_k | \hat{H} | \psi_k \rangle}{\langle \psi_k | \psi_k \rangle} \tag{5.39}$$

It can be calculated from the expressions of ψ_k and $\hat{H}$ in Eqs. (5.38) and (5.39). Let us first evaluate $\langle \psi_k | \hat{H} | \psi_k \rangle$:

$$\langle \psi_k | \hat{H} | \psi_k \rangle = \frac{1}{N} \sum_{j,j'} e^{i\vec{k}.(\vec{T_{j'}}-\vec{T_j})}. \left[A_{j,j'} + B_{j,j'} \right] \tag{5.40}$$

in which two contributions have been distinguished:

$$A_{j,j'} = \langle \varphi_j | \frac{\widehat{\vec{p}}^2}{2m} + w_j | \varphi_{j'} \rangle \quad B_{jj'} = \sum_{l \neq j} \langle \varphi_j | w_l | \varphi_{j'} \rangle \tag{5.41}$$

Putting $\varphi_j = \varphi(\vec{r} - \vec{T_j})$ and $w_j = w(\vec{r} - \vec{T_j})$, we have:

$$A_{jj} = E_0 \quad A_{j,j'}(j \neq j') = E_0 S \langle jj' \rangle \tag{5.42}$$

where $\langle jj' \rangle$ is equal to unity if the indices j and $j\prime$ refer to nearest-neighbour nodes and is equal to zero in the other cases. The results in Eq. (5.42) derive from the fact that the φ_j are eigenfunctions of the Hamiltonian $[(\widehat{\vec{p}}^2/2m) + w_j]$ of the isolated atom corresponding to the energy E_0. The contribution of the $A_{j_{j'}}$ to (5.42) then takes the form:

$$E_0[1 + S\sigma(\vec{k})] \tag{5.43}$$

where the $\vec{k}$-dependent sum $\sigma(\vec{k})$ is equal to:

$$\sigma(\vec{k}) = \sum_{\mu=1}^{z} \exp(i\vec{k}.\vec{T_\mu}) \tag{5.44}$$

In order to calculate the contribution of $B_{jj'}$, the rapid spatial-decrease of $\varphi(\vec{r})$ and of $w(\vec{r})$ can be invoked to justify the following simplifications:

$$\langle \varphi_j | w_l | \varphi_j \rangle = -\alpha \langle lj \rangle \quad \langle \varphi_j | w_l | \varphi_{j'} \rangle = -\gamma \langle jj' \rangle \delta_{lj'} \qquad (5.45)$$

where the symbol $\langle lj \rangle$ has the same meaning as in Eq. (5.42). On this basis, we obtain

$$\frac{1}{N} \sum_{j,j'} B_{jj'} \exp\left[i \vec{k}.(\vec{T_{j'}} - \vec{T_j}) \right] = -z\alpha - \gamma\sigma(\vec{k}) \qquad (5.46)$$

Finally, Eqs. (5.43), (5.44), (5.46) yield:

$$E_k = E_0 - \frac{z\alpha + \gamma\sigma(\vec{k})}{1 + S\sigma(\vec{k})} \qquad (5.47)$$

If the overlap integral S is small, this expression reduces to

$$E_k = (E_0 - z\alpha - \gamma\sigma(\vec{k})) \qquad (5.48)$$

$(-\alpha) < 0$ expresses the attraction of an electron in the state φ_j by the potential of a neighbouring atom. Equation (5.48) shows that, in the first approximation (S small), this attraction lowers *uniformly* the energies of the states composing the energy band. In contrast, the $\vec{k}$-dependent term ($-\gamma\sigma(\vec{k})$), which arises from the overlap between neighbouring wavefunctions, induces a dispersion of the spectrum as a function of $\vec{k}$. In the case of a φ-wavefunction having a positive sign for any $\vec{r}$ (e.g. an s-atomic orbital) the product $\varphi_j\varphi_{j'}$ is positive, and in this case $\gamma > 0$ (cf. Eq. (5.45)).

As indicated by Eq. (5.45) α and γ are small if the overlaps between neighbouring wavefunctions φ_j and potentials w_n are small. In that case, the energy band will be narrow (as compared to the energy spacings in the isolated atom). In contrast, if relatively large values of these coefficients are induced, for instance, by reducing the distance between constituents a shift and a widening of the bands will result. Figure 5.13 shows, for a crystal formed by a single type of atom, the location and width of bands associated to different atomic levels.

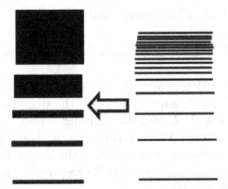

Fig. 5.13. Correspondance between the atomic levels (right) and bands (left). The lowest atomic levels do not overlap and remain narrow while the upper levels are widened and give rise to quasi continuous bands.

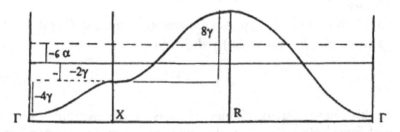

Fig. 5.14. Dispersion curves $E(\vec{k})$ deduced from Eq. (5.49) along different directions of the first Brillouin zone specified on Fig. 5.7(a).

Example: Energy spectrum of a monoatomic cubic crystal

In the case of a *one-dimensional* monoatomic crystal each atom has two neighbours ($z = 2$) at distances $\pm a$, and therefore $\sigma(k) = 2\cos ka$. Making the approximation $S = 0$, the dispersion relationship is

$$E_k = E_0 - 2\alpha - 2\gamma \cos ka \tag{5.49}$$

The dispersion curve representing $E(k)$ has the form in Fig. 5.14.

Let us examine the three-dimensional example of a crystal constituted by identical atoms located at the vertices of a simple cubic lattice. Each atom has six neighbours ($z = 6$) defined by the primitive translations of the crystal : $\vec{T_\mu} = \pm \vec{a_j}$. We then deduce (again assuming $S = 0$) from Eqs. (5.47) and (5.48):

$$E(\vec{k}) = E_0 - 6\alpha - 2\gamma \left[\cos k_x a + \cos k_y a + \cos k_z a\right] \qquad (5.50)$$

This expression determines the electronic spectrum associated to the atomic level E_0. Figure 5.14 shows the dispersion curves along three segments $\Gamma X, XR$, and $R\Gamma$, joining *high-symmetry points* of the first Brillouin zone. The crystal potential being invariant by the symmetries of the atomic configuration, the dispersion curves will be identical along the similar segments corresponding to the other faces of the cubic Brillouin zone.

Let us take as origin of the energies $E = (E_0 - 6\alpha - 6\gamma)$. The spectrum (5.50) is comprised between $E = 0$ and $E = 12\gamma$. The *equal energy surfaces* as well as the *density of states* are defined in this interval.

At the bottom of the considered energy band, for $E \approx 0$, i.e. for small values of $|\vec{k}|$, Eq. (5.50) reduces to:

$$E = \gamma a^2 \vec{k}^2 \qquad (5.51)$$

In this range of energy values, the spectrum is isotropic as in the case of free-electrons. Hence, the equal-energy surfaces are spheres. Note, however, that the coefficient of k^2 is γa^2 instead of $(\hbar^2/2m)$ for free electrons. Likewise, the density of states is $g(E) = K\sqrt{E}$, the curvature of the parabola being different from that of free-electrons, and smaller for a narrow band (γ small).

The upper part of the band ($E \approx 12\gamma$) corresponds to $\vec{k}$-vectors which are along the diagonal ΓR of the first Brillouin zone. In this energy range, the equal-energy surfaces reduce (Fig. 5.15) to triangular sectors nearby the points $\vec{k_0} = (\pm\pi/a, \pm\pi/a, \pm\pi/a)$ (Fig. 5.15(b)). In the vicinity of these points, the energy (5.50) can be expressed as a function of the small parameter $\vec{\delta} = (\vec{k} - \vec{k_0})$ as:

$$\Delta E = E_0 - 12\gamma \propto \vec{\delta}^2 \qquad (5.52)$$

The density of states is thus also parabolic at the upper limit of the band. Its general shape in the entire Brillouin zone is represented on Fig. 5.16.

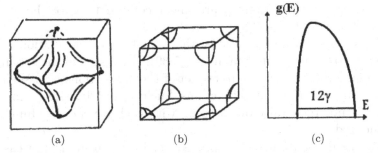

Fig. 5.15.   Equal energy surfaces corresponding to: (a) $E = E_0 - 6\alpha - 2\gamma$ and (b) to E close to the upper limit of the energy band. (c) Density of states relative to a band in the strong coupling scheme.

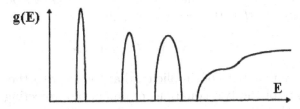

Fig. 5.16.   Electronic density of states in a crystal. The bands originating from the atomic states with the highest energy levels can overlap and will determine a quasi-continuous last band.

Consideration of several wavefunctions

In most crystals, the situation is more complicated than that considered above.

In the first place one must take into account the fact that each energy level of the atoms is degenerate and thus associated to several distinct wavefunctions $\varphi, \varphi', \varphi'', \ldots$ The probability amplitudes of these different functions have different spatial distributions. For instance, p_x, p_y, and p_z orbitals have cylindrical symmetry around respectively the x-, y-, and z-axes. In a given crystal structure, the overlap between neighbouring functions can therefore display a variety of situations. As a consequence, the energy bands associated to $\varphi, \varphi', \varphi'', \ldots$ will be differently shifted and widened although they have a common origin in the atomic level E_0.

On the other hand, if the crystal contains several atoms in its primitive unit cell, one has to consider a "molecular unit" in each unit cell, whose wavefunctions φ are linear combinations of the atomic orbitals.

In the two former situations, the tight binding calculation is not reduced to the computation in Eq. (5.39). It consists in the solving of a secular equation relative to the set of Bloch functions ψ_k associated to the different relevant φ functions. Such a procedure determines as many bands (generally overlapping with each other) as φ functions considered.

As an illustration of the procedure, let us examine the case of two atomic orbitals φ_1 and φ_2 which are either different orbitals of the same atom, or orbitals of two distinct atoms of the primitive unit cell. Let $\psi_1(\overrightarrow{k}, \overrightarrow{r})$, and $\psi_2(\overrightarrow{k}, \overrightarrow{r})$ be the Bloch functions (5.36) associated to φ_1 and φ_2. The state of the electron can be expressed as $(\psi_1 + \lambda\psi_2)$. If $\widehat{H}$ is the electron-Hamiltonian, we can write

$$\widehat{H}(\psi_1 + \lambda\psi_2) = E(\psi_1 + \lambda\psi_2) \tag{5.53}$$

Assume, for the sake of simplicity, that the ψ_i are orthogonal. The compatibility of the two equations obtained by projecting (5.53) on the ψ_i provides the secular equation:

$$\begin{bmatrix} H_{11} - E & H_{12} \\ H_{21} & H_{22} - E \end{bmatrix} \tag{5.54}$$

with $H_{ij} = \langle \psi_i | \widehat{H} | \psi_j \rangle$. For each $\overrightarrow{k}$-value, this secular equation determines two energy values. When $\overrightarrow{k}$ varies, two bands are thus generated, whose relative positions depend on the Hamiltonian (i.e. of the potential of the neighbouring atoms) and of the nature of the two φ_j functions.

5.4 Band Structure of Real Crystals

5.4.1 Bandwidth and electron-localization

The two methods described in Sections 5.2.1 and 5.3 in the view of determining the energy spectrum of the electron in a crystal have shown that in a given direction of the reciprocal space, this spectrum is formed by energy bands separated by gaps which contain no energy levels.

In the approximation of a *quasi-free electron*, the reference situation is that of an electron with uniform spatial distribution whose energy spectrum is continuous and involves no gaps. A limited spatial-confinement of the electron is induced by the small periodic potential, through reflections of the electron, which transform the probability plane wave of the reference situation into a stationary wave displaying maxima and minima of amplitude. *Narrow gaps result from this relative confinement of the electron to certain regions of space* (the maxima of amplitude).

In the tight-binding approximation, the reference situation is that of a total confinement in the atoms constituting the crystal. The energy spectrum consists of discrete energy levels separated by wide gaps. The slight overlap of the potentials (and of the orbitals) of neighbouring atoms determine a *spatial delocalization of the electron*. One obtains a limited widening of the discrete energy levels into *narrow bands*.

Hence, the occurrence of wide or narrow *gaps* is correlated, respectively, to the more pronounced, or less pronounced *spatial-confinement* of the electron. Likewise, the occurrence of wide or narrow energy bands is correlated to the spatial *delocalization of the electron*.

Since the two examined models refer to opposite approximations of the actual situation of the electron, we can conclude that the existence of energy-bands and of gaps is a general property of the electronic spectrum of real crystals.

Range of validity of the two approximations

From a qualitative standpoint, the *electrostatic* crystal potential $V(\vec{r})$ is *attractive*, in accordance with the fact that electrons have a *binding energy* which confines them within the volume of the solid. It has a dominant part resulting from the attraction of the positive atomic nuclei forming the structure of the crystal, and a repulsive part due to the other electrons. Figure 5.1 shows the expected schematic shape of $V(x)$ in the case of a one-dimensional crystal, with potential "wells" at the sites of the positive nuclei.

From a quantitative standpoint, the repulsive part of $\widehat{V}(\vec{r})$ is determined by the spatial distribution of the electrons which, in turn, derives from the expression of the functions $u_{\vec{k},n}(\vec{r})$. Conversely, the $u_{\vec{k},n}(\vec{r})$ are eigenfunctions of (3.17) and thus depend on the form of $V(\vec{r})$. Hence, in a specific crystal, the knowledge of $V(\vec{r})$ and of the $u_{\vec{k},n}(\vec{r})$ result from a common iterative "self-consistent" procedure. In certain solids, such a complex procedure shows (cf. Chapter 10) that a fraction of the total number of electrons occupy quantum states (e.g. with the energy E_2 on Fig. 5.1) for which the probability of presence of the electron is almost evenly distributed in the volume of the crystal. These "delocalized" states resemble the states of a "free-electron" in a uniform potential. This potential is the spatially averaged potential created by the constituents of the crystal. This is for instance the case of the alcaline or alcaline-earth metals (e.g. sodium and calcium) in which the delocalized electron states "see" attenuated variations of the potential exerted by the nuclei. This attenuation is due to the screening of this potential by the electron density of other more bounded electron states.

In all solids, there are such more bounded electron states (e.g. with energy E_1) whose probability of presence is spatially localized in the neighbourhood of the nuclei ("core electrons"), thus forming, with the nuclei, *positive ions* within the solid. The quantum states of these strongly bounded electrons resemble those of an electron in the isolated ion. The two former situations constitute respectively the natural frameworks of application of the almost-free electron approximation and of the tight-binding one.

Note that in the former approximation, the wavefunction of the electron is represented as a sum of plane waves $\exp i(\vec{k} + \vec{G}).\vec{r}$. This method will be useful if this sum, which is a Fourier expansion, contains a small number of relevant terms. Whenever a wavefunction has significant variations on a scale much smaller than the dimensions of the unit cell, its Fourier transform will contain components defined by $\vec{G}$-vectors much larger than the first-Brillouin zone. Hence, the above sum of plane waves will contain a large number of terms. It therefore appears that the almost-free electron approximation applies to the converse situation in which the *spatial variations of the wavefunction occur on a scale larger than the dimensions of the primitive unit cell.*

On the other hand, examining more carefully the validity of the tight-binding method, we can observe that the linear combinations of atomic orbitals in Eq. (5.36) are not strict eigenfunctions of the

Hamiltonian (5.38). This would only be the case if the potentials w_l ($l \neq j$) were vanishingly small in the spatial range of extension of φ_j. The tight-binding Bloch functions ψ_k are therefore acceptable as approximate wavefunctions of the electron, if the potential exerted on the electron in each atom is only slightly modified by the presence of the neighbouring atoms. Also note that it is necessary that the lifting of degeneracy of the atomic levels determined by the model must be small in comparison of the distance between atomic levels. Otherwise, the overlap of the bands would determine a situation of quasi-degeneracy (cf. Section 5.2.4) thus requiring a more complex calculation.

The two preceding conditions imply a small overlap of the neighbouring orbitals and of the neighbouring potentials. The tight-binding method is therefore well adapted to the description of electronic states related to atomic orbitals whose spatial extension is small as compared to the size of the unit cell. These are the orbitals (1s,1p,2p, ...) of lowest energies of the constituents.

The densities of state corresponding to the bands arising from the most localized and the delocalized states are represented respectively at the left and at the right of Fig. 5.16. The band width $\Delta E \propto \gamma$ increases with the energy of the atomic orbital considered.

For the electronic states whose spatial extension is intermediate between the two former types of bands the two approximation methods considered here are not adequate. The study of more elaborate methods permitting a satisfactory description of these states is out of the scope of this book. Let us however specify the two main aspects by which they differ from the considered approximations.

(a) In the first place these methods rely on the use of a more adapted base of wavefunctions, than plane waves or atomic orbitals. For instance, such a base can be constituted, on the one hand of atomic orbitals, as in the tight-binding approximation, and, on the other hand, by combinations of plane waves orthogonal to the former atomic orbitals.

(b) Potentials are constructed, through the iterative procedures mentioned above, with the view of representing in a more "realistic" manner the presence of the nuclei and of the electrons.

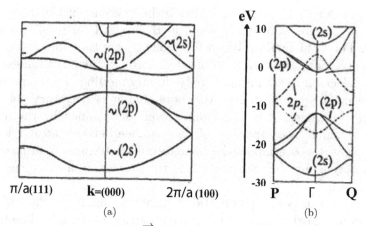

Fig. 5.17.   Dispersion curves $E(\vec{k})$: (a) for diamond in the directions of the sides and the diagonals of the cubic reciprocal cell; (b) for graphite in the directions of the hexagonal planes.

5.4.2   Examples of the band structure of real crystals

Figures 5.17(a) and 5.17(b) and 5.18 show the results of calculations of the band structure, on the one hand of "simple" crystals (diamond and graphite) containing a single element (carbon), and on the other hand, of a more complex crystal (a substance displaying the property of superconductivity) containing several types of atoms.

The bands of lowest energy in diamond and graphite are both associated to the lowest-energy atomic orbitals of the carbon atom, namely the orbitals (1s), (2s) and (2p). However, owing to the different structures of these two substances, the bands arising from these orbitals in the tight binding approximation are different.

Diamond has a FCC cubic Bravais lattice with *two atoms* in the primitive unit cell (cf. Chapters 1 and 2). It is therefore relevant to apply the tight binding method to φ-orbitals consisting of "molecular" orbitals obtained by linear combinations of the orbitals of a pair of carbon atoms. For reasons which will be clarified in Chapter 6, it is relevant to consider the five atomic orbitals of the carbon atom which have the lowest energy. These are (1s), (2s), $(2p_x)$, $(2p_y)$ and $(2p_z)$. The non-degenerate (1s) orbitals have a small spatial extension and therefore a small overlap with the corresponding orbital of neighbouring atoms. The (1s)-orbitals of a pair of carbon atoms are used to generate, in agreement with the procedure in Eq. (5.54) two narrow bands of lowest energy. The p_i-orbitals are degenerate and close in energy to the (2s)-orbital. Hence, one must consider them jointly in the more complex tight-binding framework of Section 5.3.2. Since each carbon atom is at the center of a regular tetrahedron formed by four other carbon atoms, it is appropriate to base the tight-binding method on the use of "molecular" orbitals of the tetrahedral unit, formed by linear combinations of (2s, $2p_x$, $2p_y$, $2p_z$), labelled sp³-orbitals, and pointing in the direction of the four neighbouring atoms. Similarly to the qualitative analysis in Section 5.3.1, one is lead to distinguish a band associated to "in-phase" sp³-orbitals (bonding-wavefunction) on neighbouring atoms, and "out-of phase" ones (antibonding-wavefunction) (Fig. 5.12). As in Fig. 5.17(a), the bonding bands have lower energy than the antibonding ones. One thus generates two bands separated by an energy-gap which is observed to be of the order of six electron-volts.

Graphite can be considered (cf. Fig. 1.5) as a stacking of parallel planes along the z-axis, each plane containing an hexagonal structure with two atoms per unit cell. Unlike the case of diamond, the neighbours of a carbon atom within a plane are much closer than the neighbours lying in another plane. This circumstance does not change the method used for the (1s) orbital, but it leads to dissociate the case of the ($2p_z$) orbital, which points along z from the (2s, $2p_x$, and $2p_y$) which point within a hexagonal plane. The ($2p_z$) orbital gives rise to a narrow band, owing to the weak overlap between atomic orbitals in the z-direction. Similarly to the case of diamond, it is relevant to construct molecular orbitals, combinations of (s, p_x, and p_y), pointing towards the neighbouring atoms within a plane. Pairs of these sp²-orbitals, associated to pairs of carbon atoms give rise, again, to binding and antibinding bands. However, unlike the case of diamond, it appears that, partly for symmetry reasons, the two preceding bands overlap, and that no gap exists between them Fig. 5.17(b) (cf. also Appendix A.1).

In the case of the superconducting substance $YBa_2Cu_3O_7$, which has a complex structure, a relatively large number of atoms in the unit cell (13), and henceforth of orbitals, which must be taken into account in a tight-binding calculation, one expects as shown on Fig. 5.18, that an irregular-shaped density of states will result from the overlap of several bands.

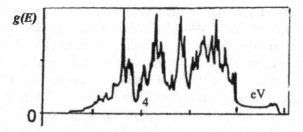

Fig. 5.18. Density of states resulting from the band structure of the superconducting material $YBa_2Cu_3O_7$.

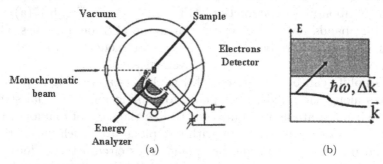

Fig. 5.19. (a) Schematic design of a photoemission experiment. (b) Nature of the energy transfers in such an experiment.

5.4.3　Experimental studies of the band structure of a solid

In the case of isolated atoms, the energy-level scheme is often deduced from spectroscopic measurements. One examines the characteristics (energy, momentum, ...) of photons or other particles which have been absorbed or emitted by the considered atom. From these energy exchanges, one deduces the energy intervals in the spectrum of the atom. The spectrum of a solid is investigated according to the same principles. Hence, the energy-differences between the quantum levels of a solid are deduced from measurement of the energies, carried by various types of particles, which are likely to be absorbed or emitted by the solid. However, as seen in the former sections, a peculiarity of solids is the existence of a continuum of energy-levels indexed by $\vec{k}$ and by the band index n.

A prominent experimental method used to study the band structure of solids consists in studying the frequency of photons (belonging to the range of X-rays or to the ultraviolet range) (Fig. 5.19).

Bibliography

Photoelectron Spectroscopy, *S. Hüfner*, Springer (2003).

Chapter 6

Equilibrium Electronic
Properties of Solids

Main ideas: Fermi level and band occupation (Section 6.2.2). Relationship between the occupation of energy bands and the insulating or conducting character of a solid (Section 6.4). Irrelevance of the fully occupied energy levels of the constituting atoms. Elementary and composite bands. Relationship between the characteristics of the crystal potential and the overlap of elementary bands (Section 6.4.3). Static effective mass (Section 6.5.2).

6.1 Introduction

In this chapter, we use the results of Chapter 5 relative to the possible quantum states of an electron, in order to describe the "electronic" properties of crystals in situations of thermodynamic equilibrium. Such a description has two purposes. The first one is of a "general" nature. It consists in showing that the physical electron-model of solids based on independent electrons having a quantum spectrum composed of *bands*, accounts satisfactorily for the large variations, mentioned in Chapter 1, between the properties of different solids. It is of importance, in particular, to account for the existence of insulating solids and conducting solids. The second objective is to understand the reasons which determine the specific electronic properties of a given solid defined by the nature of its constituents and by its crystal structure. For instance, one wishes to understand why diamond is insulating while graphite is conducting, or why the electronic specific-heat capacity of a bismuth crystal is 20 times smaller than that of a sodium crystal and 200 times smaller than that of a manganese crystal.

In Chapter 5, we had determined the quantum spectrum of a single electron in a solid. Since the solid contains a large number of electrons, two types of considerations, aside from the energy spectrum, will underly the determination of the equilibrium electronic properties of the solid. On the one hand, electrons being undistinguishable quantum objects, their global quantum states must comply with the constraints imposed by the *Pauli principle*. From this standpoint, electrons are *Fermions* (particles with half-integer spin value). Their thermodynamic equilibrium is thus determined by the *Fermi–Dirac* statistics, whose main features will be recalled in the next section. On the other hand, as implicitly admitted in the previous chapters, we will consider that the electrons *do not interact* with each other. Hence, the quantum spectrum of a single electron determines entirely the spectrum of the set of electrons of the crystal.

In Section 6.3, we will first discuss the *Sommerfeld model* of electrons in solids, which relies on the quantum spectrum of the electron in a *uniform potential* (cf. Section 5.2.1). We will observe that this model applies satisfactorily to *certain classes of metals*, but that it is not able to explain the diversity of solids, and in particular the existence of insulators. In this aspect, we will then show that a more general and successful model can be based on the energy-band spectrum discussed in the previous chapters.

6.2   Thermodynamic Equilibrium: Fermi Energy

6.2.1   Electron states and Pauli principle

We assume that the solid contains N *non-interacting* electrons. The electrons are all submitted to the same crystal potential, and therefore their individual quantum spectra are identical. We know that the orbital part of the quantum states of a single electron is a *Bloch function* $\phi_{\vec{k},n} = \exp(i\,\vec{k}\cdot\vec{r}).u_{\vec{k},n}(\vec{r})$, indexed by a reciprocal-space vector $\vec{k}$ and a band index n. The single-electron energy-spectrum $E(\vec{k},n)$ forms quasi-continuous *energy bands*. Taking into account the state of the spin of the electron, a complete wavefunction can be denoted $\psi_{\vec{k},n} = \phi_{\vec{k},n}.\alpha$, where α specifies the spin-state. Note that since the Hamiltonian equation (5.1) of the electron does not depend of the spin, the energy of $\psi_{\vec{k},n}$ is $E(\vec{k},n)$.

The possible wavefunctions $\psi(\vec{r_1},\vec{r_2},\ldots,\vec{r_n})$ of N electrons are linear combinations of the form:

$$\psi = \sum_{v_1,\ldots,v_N 0} a_{1,2,\ldots,N}\,\psi_{v_1}\cdot\psi_{v_2}\cdots\cdots\psi_{v_N} \tag{6.1}$$

where $\psi_i = \psi_{\vec{k_i},\nu_i}(\vec{r_i}).\alpha_i$ in which $\psi_{\vec{k_i},\nu_i}(\vec{r_i})$ is a single-electron Bloch function describing the orbital state of the i_{th} electron $(i = 1,\ldots,N)$. *The Pauli principle* restricts the possible linear combinations by imposing to the functions (6.1) to be *antisymmetric* with respect to the permutation of the particles (i.e. of their coordinates and, simultaneously, of their spin states). Each possible state ψ can be expressed as an $N \times N$ determinant,[1] necessarily associated to a set of N *distinct* quantum states ψ_i of the single electron:

$$\psi = \frac{1}{\sqrt{N!}}\sum(-1)^{P(v_1,v_2,\ldots,v_N)}\psi_{v_1}\cdot\psi_{v_2}\cdots\cdots\psi_{v_N} \tag{6.2}$$

where $(v_1,v_2,\ldots,v_N)$ is a permutation of the set of numbers $(1,2,\ldots,N)$ and where $P(v_1,v_2,\ldots,v_N)$ is the signature of this permutation (i.e. P is equal to 0 for even permutations and 1 for odd permutations). The sum in (6.2) is over all the $(N!)$ permutations of the set $(1,2,\ldots,N)$. For a set of N *non-interacting electrons*, it is straightforward to see that the total energy is the sum of the energies of each electron. Hence

$$E = E(\vec{k_1},n_1) + E(\vec{k_2},n_2) + \cdots + E(\vec{k_N},n_N) \tag{6.3}$$

where the $E(\vec{k_i},n_i)$ are the energies of the N distinct quantum states ψ_i involved in the determinant (6.2). The fact that all the functions ψ_i in (6.2) are distinct can be expressed by stating that *each quantum state of the single electron is either empty* (if this state does not

---

[1]As in atoms and in molecules, this determinantal form of the N-electron wavefunction is termed a Slater determinant. Remind that a determinant vanishes if two lines or columns are identical. Hence, the linear combination of functions cannot contain twice the same state.

appear in (6.2)) or is *occupied by one electron* (if this state appears in (6.2)).

6.2.2   Fermi factor: Fermi level

Thermodynamic equilibrium

In statistical thermodynamics, each possible state ψ_j of a system is assigned a probability of occurrence p_j. One then determines the equilibrium properties of the system,[2] within the framework of a *statistical model* called *ensemble*. An ensemble is defined by "constraints" on the global state of the set of particles composing the system.

For instance, the *canonical ensemble* is defined by the fact the average total energy of the system has an imposed value $\langle E \rangle$. The total energy itself, E, which is a function of the energy of each state, and of the probability assigned to this state, is allowed to fluctuate by exchanging energy with an energy reservoir (a thermostat) defined by its temperature T. Given the imposed value of the average energy $\langle E \rangle$, the equilibrium of the system corresponds to the maximum of the statistical entropy $S = -k \sum p_i Ln(p_i)$ which is a function of the probability distribution. This maximum corresponds to the *maximum disorder compatible with the imposed condition*. It can be shown that one then obtains for the "canonical" equilibrium the following expression for the probability of *each state* having the energy E_j:

$$p_j^{\text{eq}} = \frac{\exp(-iE_j/k_B T)}{Z_C} \tag{6.4}$$

With Z_C, the canonical partition function:

$$Z_C = \sum_{\text{states}} \exp(-iE_j/k_B T) \tag{6.5}$$

---

[2]In macroscopic physics, the concept of equilibrium is not very well defined. To quote approximately a famous definition by the physicist R. Feynman, thermodynamic equilibrium is installed when "rapid events have ended, and slow ones are still in progress".

in which k_B is the Boltzmann constant. The temperature T has the property that if two systems can exchange energy, their temperatures are equal at equilibrium.

For systems of N Fermions, e.g. electrons, the usual statistical ensemble is the *grand canonical ensemble* in which *two* average quantities have imposed values: the statistical average of the total energy $\langle E \rangle$ (as in the canonical ensemble) *and* the value of the statistical average $\langle N \rangle$ of the number of particles composing the system. In this statistical model, the number N can fluctuate by exchanging particles with a particle reservoir, characterized by the quantity $(\mu/k_B T)$ in which μ is the *chemical potential*. Each state of the system is associated to a given energy value E_j and a given number of particles N_j. The thermodynamical equilibrium of the system, again, corresponds to the probability distribution of the states of the system which makes maximum its statistical entropy. One then obtains for the "grand canonical" equilibrium the probability for a state to possess the energy E_j and the number of particles N_j:

$$p_j^{\text{eq}} = \frac{\exp[i(\mu N_j/k_B T - E_j/k_B T)]}{Z_G} \tag{6.6}$$

with the Z_G the grand-canonical partition function:

$$Z_G = \sum_{\text{states}} \exp[i(\mu N_j/k_B T - E_j/k_B T)] \tag{6.7}$$

Similarly to the case of the temperature, if two systems (having the same temperature) can exchange particles, their chemical potentials μ are equal at equilibrium.[3]

Fermi energy

In this framework, using Eqs. (6.2) and (6.3), it can be shown that the equilibrium properties of the N electrons are entirely determined *at given absolute temperature T*, and *given chemical potential μ*. A given single-electron quantum state $\psi_{\vec{k},n}$ will be assigned an *occupation factor* $f(E_{\vec{k},n} \,|\, \mu, T)$. The occupation factor f (which is also termed the Fermi factor) has its value necessarily in the interval $[0, 1]$. It is

---

[3]Note that the choice of sign for T is opposite to that of the chemical potential μ.

equal to zero if the state is empty, to one if it is occupied and takes values between zero and one depending of the probability of occupation of the considered state. In the context of electrons in solids, the chemical potential μ, which has the dimension of an energy, is termed *the Fermi energy* and is denoted E_F. The occupation factor f is

$$f(E_{\vec{k},n} \mid E_F, T) = \frac{1}{1 + \exp\{(E_{\vec{k},n} - E_F)/k_B T\}} \qquad (6.8)$$

Hence, this factor has the same value for quantum states possessing the same energy. Since a solid contains a large number of electrons (of the order of the Avogadro number), the relative fluctuations of N is negligible, and the statistical average $\langle N \rangle$ is interpreted as the actual number N of electrons in the solid. The occupation factor then complies with the relation:

$$N = \int_{E_{\vec{k},n}} g(E).f(E \mid E_F, T).dE \qquad (6.9)$$

which expresses the fact that the total number of particles is the sum of the number of states $g(E)$ in an energy interval dE multiplied by the occupation factor of these states, integrated over the energy spectrum. Equation (6.9) determines the Fermi energy $E_F(T, N)$ as a function of the temperature T and of the number of electrons N. Note that the Fermi energy, being a chemical potential, is a quantity which has the same value, at equilibrium, for systems which can exchange electrons.

The variations of the occupation factor $f(E \mid E_F, T)$ as a function of the energy E are recalled on Fig. 6.1, at zero absolute temperature $T = 0$, and at finite temperature $T \neq 0$:

(a) When $T = 0$, the Fermi factor f has the form of a step which can be expressed by use of the Heaviside function[4] $\mathcal{H}$ as $\mathcal{H}(E_F - E)$. At this temperature, the Fermi energy E_F separates the quantum states which are occupied by one electron $(E < E_F)$ from the empty states $(E > E_F)$. The Fermi energy is then *the energy of*

---

[4]The Heaviside function $H(x)$ is defined by $H(x) = 0$ for $x < 0$, and $H(x) = 1$, for $x > 0$.

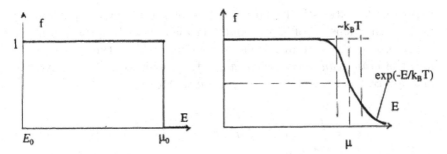

Fig. 6.1.   Variations of the Fermi factor for $T = 0\,$K and for $T \ll \Theta_F$ where $k\Theta_F = E_F$ (Fermi energy).

the occupied state having the highest energy. Given the variations of $f(E \mid E_F, T = 0)$, Eq. (6.9) takes the form:

$$N = \int_{E_0}^{\infty} g(E).\mathcal{H}(E_F - E).dE = \int_{E_0}^{E_F} g(E).dE \qquad (6.10)$$

where E_0 is the energy of the ground state (Lowest energy level) of the single-electron spectrum. For given value of N, the zero-temperature value of the Fermi energy $E_F(T = 0, N)$ defines a temperature called the *Fermi temperature* and expressed by the relation $k\Theta_F = E_F(T = 0, N)$.

(b)   When $T \neq 0$, different situations must be distinguished depending if the temperature is much smaller than Θ_F ($T \ll \Theta_F$) or if $T \gg \Theta_F$. In the former case, the Fermi energy is close to its zero temperature value $k\Theta_F$, and the occupation factor is equal to $\mathcal{H}(E_F - E)$ except in the vicinity of $E_F(T, N)$ (Fig. 6.1(b)). In the latter case, the Fermi energy is lower than the energy of the ground state, and the occupation factor $f \ll 1$ (the exponential in the denominator of Eq. (6.8) is much larger than 1) is a decreasing function of the energy whose expression (6.8) reduces to that of the Boltzmann thermal factor which governs the classical statistical equilibrium:

$$f = \exp\left(-\frac{(E - E_F)}{k_B T}\right) \qquad (6.11)$$

6.2.3   Calculation of the equilibrium properties

At a given temperature, the physical properties of the N electrons can be calculated as a function of the occupation factor f (which expresses the statistical characteristics of the thermal equilibrium) and of the density of states $g(E)$ (which describes the quantum spectrum of the system). Hence, the energy U of the N electrons and the *grand-canonical potential* A, from which one can derive the entropy S, the pressure P, etc. can be written as:

$$U = \int_{E_0}^{\infty} E.g(E).f(E \mid E_F, T).dE \qquad (6.12)$$

$$A = k_B T \int_{E_0}^{\infty} g(E).Ln\{1 - f(E \mid E_F, T)\}.dE \qquad (6.13)$$

For given number N of electrons, the Fermi energy E_F is determined by Eq. (6.9). The former expressions are then functions of the external control parameters of the system $(T, \Omega, \ldots)$. When there are several conducting solids in contact with each other, the value of N is not specified separately for each solid, owing to the possibility for the solids to exchange their electrons. In that situation, in order to determine the Fermi energy E_F^i of each solid one has to replace Eq. (6.9) by a condition of equality $(E_F^1 = E_F^2 = \cdots)$ of the Fermi energies of the solids in contact.

6.3   Sommerfeld Model for Metals

6.3.1   Degenerate free-electrons quantum gas

The Sommerfeld model of electrons in a solid assumes that each of the N electrons which ensure the conduction of electricity (cf. Chapter 1) is a *free electron* i.e. it corresponds to the situation analyzed in Section 5.2 in which the electron is submitted to a *spatially uniform* binding potential (the same for all the electrons of the solid). In this *electron gas*, the energy spectrum of an electron, considered in Section 5.2, determines a parabolic density of states $g(E) \propto \sqrt{E}$. Each state being statistically occupied according to (6.4), the energy distribution of electrons is determined by the function $\rho(E) = g(E).f(E|E_F, T)$ and it varies as a function of the energy E as represented in Fig. 6.2.

Inserting (Eq. (5.7)), $g(E) \propto 4\pi\Omega(2m)^{3/2}\sqrt{E}/h^3$ in Eq. (6.9), determines an explicit form of the Fermi energy E_F at zero

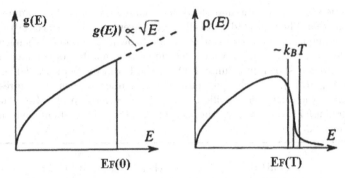

Fig. 6.2. Statistical distribution of the electronic energy in a free-electron gas $\rho(E) = g(E).f(E)$ at zero temperature and at a temperature satisfying the condition $T \ll \Theta_F$.

temperature:

$$\mu_0 = E_F(0) = \frac{\hbar^2}{2m} \left[3\pi^2 \frac{N}{\Omega} \right]^{2/3} \tag{6.14}$$

The Fermi temperature Θ_F characterizing the electron gas has the value:

$$k_B \Theta_F = E_F(0) \tag{6.15}$$

We know (Section 5.2.1 and Fig. 5.2) that the equal-energy surfaces $U = E$ are spheres of radius $k(E)$ in the reciprocal space. The surface corresponding to $U = E_F(0)$ is the *Fermi sphere* associated to the electron gas. In agreement with the variations of the occupation factor (Fig. 6.1), the surface limiting this sphere contains the occupied electron quantum states possessing the highest energy. Its radius is

$$k_F = \sqrt{\frac{2mE_F(0)}{\hbar^2}} = \left[3\pi^2 \frac{N}{\Omega} \right]^{1/3} \tag{6.16}$$

In the Sommerfeld model of a metal (e.g. copper) the relevant number of electrons to consider in the preceding formula is equal to the number of *valence electrons*, i.e. $N = N_A.z_{\text{eff}}$ where N_A is the number of atoms and z_{eff} is the valence of the atom. For all monoatomic metals, the value of the electronic density (N/Ω) determines a Fermi temperature Θ_F of the order of several 10^4 degrees (cf. Table 6.1). Hence, the accessible experimental range of temperatures

Physics of Electrons in Solids

Table 6.1.   Structural and electronic characteristics of three families of metals. The indicated unit cell is the conventional cell, whose lattice parameter and number of atoms Z are shown. The atomic chemical valence and the valence which determines the number of conduction electrons is shown in column 4. Θ_F is the Fermi temperature defined by Eq. (6.15). γ is the prefactor of the temperature in the expression of the electronic specific heat. γ_S is the value expected from the Sommerfeld model. γ_M is the measured value. The units of γ are in J/mole/degree.

Metal	Lattice	Cell Å/Z	Valence chem/met	Θ_F $\times 10^4$ K	γ_S	γ_M
Sodium	BCC	4.2/2	1/1	3.8	11	15
Potassium	BCC	5.2/2	1/1	2.5	17	20
Rubidium	BCC	5.6/2	1/1	2.1	19	24
Aluminum	FCC	4.1/4	3/3	13	9.2	12.5
Copper	FCC	3.6/4	2/1	8.2	5	6.7
Silver	FCC	4.1/4	1/1	6.4	6.3	6.7
Gold	FCC	4.1/4	3/1	6.4	6.3	6.7
Calcium	FCC	5.5/4	2/2	5.4	15	27
Strontium	FCC	6.1/4	2/2	4.6	18	36
Barium	BCC	5.0/2	2/2	4.2	20	27
Zinc	H	(2.7–5.0)/2	2/2	11	7.5	5.8
Iron	BCC	2.9/2	2/2	13	6.7	50
Niobium	BCC	3.3/2	3-5/1	6.2	6.7	84
Manganese	BCC	2.9/2	2/2	13	6.3	170
Bismuth	Rhomb	6.676/2	5/5	12	18	0.8
Antimony	Rhomb	3.1/2	5/5	13	16	6.3

satisfies the condition $T \ll \Theta_F$. In this case, the energy dependence of the Fermi factor differs little from its zero temperature behaviour (see Fig. 6.1), and it can be shown that this dependence can be approximated by the function:

$$f(E|E_F, T) \approx H(E_F - E) - \frac{\pi^2 (k_B T)^2}{6} \delta'(E - E_F) \qquad (6.17)$$

where $\mathcal{H}(x)$ is the Heaviside function already considered in Eq. (6.10) and $\delta'(x)$ is the first derivative of the Dirac function $\delta(x)$. Using this expression, it can be shown that the value of physical quantities (e.g. the Fermi energy E_F, or the statistical average of the electron-energy), only differ from their zero temperature values by corrections of second order in (T/Θ_F). *Hence, if the temperature of the solid is not of the same order of magnitude as the Fermi temperature* Θ_F, *the* behaviour of the electron gas is close to that of its zero temperature behaviour. As will be shown in the next paragraph, the properties of such a gas are essentially attached to the electronic energy levels which are close to the Fermi energy. Hence, the "active electrons" all have approximately *the same energy*, and accordingly the electron gas is termed *degenerate*.

6.3.2 Properties of the degenerate free-electron gas

Let us stress the main specific properties of a free-electron gas at temperatures T such as $T \ll \Theta_F$.

Electron kinetic energy

The fact that the Pauli principle restricts the occupation of every single quantum state to one electron at most has the consequence that an important fraction of the N electrons occupies energy levels which lye much above the ground state. Hence, the statistical average of the energy is of the same order of magnitude (electron-volts) as the Fermi energy $k_B \Theta_F$, and is therefore much larger than the thermal energy $\frac{3}{2} k_B T$ of a particle in an ordinary gas of classical particles ($\sim 25\,\text{meV}$ at room temperature). Since this energy is of kinetic nature ($E = \hbar^2 k^2 / 2m = m\mathbf{v}^2 / 2$), the average velocity $\langle \mathbf{v} \rangle$ of the electrons ($\sim 10^5$–10^6 m/s) is also an order of magnitude larger than if electrons formed a classical gas.

Active electron states

In studying the physical equilibrium properties of the electrons of a solid, one is usually interested in the evolution of the electrons resulting from *small* external perturbations of the system (e.g. heat exchange, application of an external electric or magnetic field) involving an amount of energy ϵ per electron, which is small as compared to $k_B \Theta_F$. It then appears that, for a degenerate electron gas, only *a small number of electrons* (of the same order as the largest of the two numbers $[N\epsilon / k_B \Theta_F]$ and $[NT / \Theta_F]$) contributes to the physical properties.

Indeed, the values of the macroscopic physical quantities characterizing the electron gas are equal, in the microscopic and statistical description of this system, to the statistical average of the

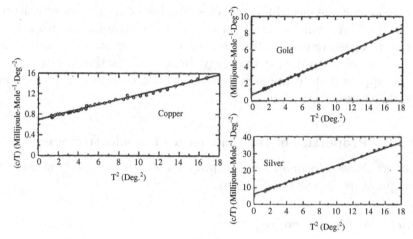

Fig. 6.3.   Temperature dependence of the electronic specific heat of three metals.

corresponding microscopic quantities. Modifying these macroscopic values implies *changing the occupation factors* of the different quantum states. Such changes can *neither concern the fully occupied states distant from the Fermi level* (this would require the forbidden transfer of electrons to neighbouring states which are already occupied), *nor empty states distant from the Fermi level* (since there are no electrons to be transfered in the neighbouring states which are also empty). These changes will therefore only affect the *fringe of partly occupied states* situated in an energy interval $\Delta E \approx \epsilon$ or $\Delta E \approx k_B T$ on either sides of the Fermi level (Fig. 6.2). The number of levels in this interval determines the number of *active electrons*. This number is of the order of $\Delta E.g(E_F) \approx N.(T/\Theta_F)$. For $T \sim 300\,\mathrm{K}$, this number is of the order of $10^{-2}N$.

For instance, if we assume that each of the active electrons behaves as the independent particles of a *classical gas, and* contributes to the specific heat by the amount $3k_B/2$, we can deduce the *electronic specific heat of the solid*:

$$C_\epsilon = \frac{3}{2}k_B^2\, T.\, g(E_F) = \gamma T = \frac{\pi^2}{2} N k_B \left(\frac{T}{\Theta_F}\right) \qquad (6.18)$$

which grows linearly as a function of temperature in agreement with experiments (Fig. 6.3).

A similar qualitative argument can be developed to evaluate the magnetic susceptibility of the electron gas. The spins of the

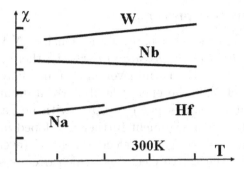

Fig. 6.4. Schematic temperature dependance of the magnetic susceptibility of a few selected metals.

active electrons (each of which is associated to a magnetic moment $(\overrightarrow{\mu} = -\mu_B . \overrightarrow{s})$) can be assumed to orient *independently from each other* in a magnetic field. Their equilibrium is therefore determined by the Boltzmann statistics: the probability for these magnetic moments oriented parallel to the field or opposite to it are respectively $\propto \exp(\mp \overrightarrow{\mu} . \overrightarrow{B}/k_B T)$, thus determining the statistical average per active electron

$$\langle \overrightarrow{\mu} \rangle = 2\mu_B \tanh(B\mu_B/k_B T) \qquad (6.19)$$

The susceptibility $\chi = [\partial \langle \overrightarrow{\mu} < /\partial \overrightarrow{B}]_{B=0}$ is governed by the *Curie law* $\chi = (2\mu_B^2/k_B T) = A/T$. Since the number of active electrons is NT/Θ_F, we obtain an almost *temperature independent susceptibility:*

$$\chi = \frac{2N\mu_B^2}{k_B \Theta_F} \qquad (6.20)$$

At room temperature, the magnitude of this susceptibility is in the ratio $(T/\Theta_F) \approx 10^{-2}$ with the paramagnetic susceptibility of N independent spins. This magnetic behaviour characterizes the *Pauli susceptibility* of metals, in agreement with experimental observations in certain metals (Fig. 6.4).

Note that the preceding qualitative arguments rely on the assumption that the active electrons behave as statistically independent electrons not correlated by the Pauli principle. The legitimacy of this assumption derives from the fact that the corresponding quantum states are partly occupied. Hence, a small variation of the occupation of each such state is always possible thus weakening the constraint set by the Pauli principle.

Electric conducting property

In the Sommerfeld model, whatever the density (N/Ω) of the electrons, that is, *whatever the nature of the solid,* there is a continuum of empty states above the Fermi level, in its immediate vicinity. This is due to the fact that the energy levels form a unique band (with parabolic density of states $g(E) \propto \sqrt{E}$), with no gaps. Anticipating qualitatively on an argument further developed in Chapter 8, if we apply to the electron gas an electric field $\vec{e}$ (even of infinitesimal magnitude), the active electrons, which occupy states close to the Fermi level, can absorb the excess energy $\Delta E = |e^2 \vec{e} . \vec{r}|$, and occupy, as measure as the former energy grows, consecutive empty states above the Fermi level. Accordingly, an overall motion is communicated by the field to the electron gas, thus producing an electric current: the considered solid is *a conductor.*

Non-universality of the Sommerfeld model

The shortcomings of the Sommerfeld model have been partly pointed out in Chapter 1.

(a) Qualitative shortcomings. In the first place, as shown above, this model is only able to account for the electronic properties of conducting solids. Not all solids being conductors, it is of interest to understand why the electrons of insulating solids do not constitute a free-electron gas.

On the other hand, even if we restrict our attention to metallic substances, the number N of electrons used in the model, and assigned to the valence electrons of the solid does not always have a clear justification. In alcaline metals (sodium, potassium,...), the Sommerfeld electron gas is consistently associated to one electron per atom, i.e. equal to the chemical valence of these atoms. By contrast, in copper and gold the adequate number of electrons in the Sommerfld model is also one electron per atom while these elements display several chemical valences (1 and 2 for copper, 1 and 3 for gold). Likewise, niobium is chemically trivalent or pentavalent whereas its "metallic" valence is one.

Still another shortcoming, already pointed out in Chapter 1, resides in the fact that in certain metals (e.g. zinc), the carriers forming the electric current appear to possess a positive charge, as evidenced by Hall effect measurements (cf. Chapter 1).

(b) Quantitative shortcomings. As shown by Eq. (6.18), the Sommerfeld model predicts a linear temperature dependence $C_\epsilon = \gamma T$ of the electronic specific heat. Experimentally, such a linear dependence has indeed been deduced for many metals from low temperature measurements of the solid's specific heat. The corresponding γ values are reproduced on Table 6.1 together with the theoretical value $\pi^2 N k_B / 2\Theta_F$ derived from the Sommerfeld model. It appears that three classes of metals can be distinguished. For the first class, which comprises essentially the alcaline metals (Na, K, ...) as well as the noble metals (Au, Ag, ...) aluminum and copper, there is a qualitative agreement (experimental values being slightly larger by 10–50%). In the second class, which comprises alcaline-earth metals (Be, Ca, Sr, ...) and certain transition metals (Zn, Cd, ...), the experimental and theoretical values differ by a factor of 2. Finally, in certain elements as iron, niobium, manganese or bismuth, the difference is by more than an order of magnitude.

6.4 Energy Bands: Conductors and Insulators

In this section, we consider the more elaborate description of the quantum spectrum of an electron, developed in Chapter 5, and consisting in a succession of energy bands arising from the crystal's spatially periodic potential.

6.4.1 Distinction between conductors and insulators

The conducting or insulating character of a given solid can be derived from its electronic state at zero temperature $(T = 0\,\mathrm{K})$. This character results, on the one hand, from the shape of the density of states associated to the quantum spectrum of the single electron, and, on the other hand, from the mode of occupation of the quantum states.

Location of the occupied state of highest energy

As shown in Chapters 3 (Section 3.4) and 5 (Section 5.2), the electronic density of states $g(E)$ in a solid comprises *energy bands* formed

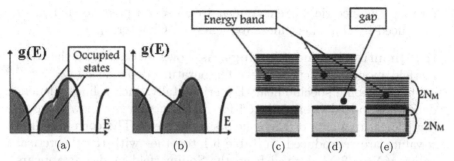

Fig. 6.5. Band occupation in the density of states of (a) a conductor and (b) of an insulator. Relative positions and occupation of adjacent bands containing $2N_M$ states each. Images (c) and (d) have disjoint bands one of which is (c) partly occupied (conductor), or (d) completely occupied (insulator). In (e), though the lower band is totally occupied, the slight overlap of bands determines a conductor.

by a quasi-continuum of quantum states, and *forbidden bands* (or gaps) deprived from quantum states and separating the energy bands (Fig. 6.5).

As recalled in Section 6.2.2, at $T = 0$ K, the occupation factor $f(E|E_F, T) \approx 1$ for $E < E_F$. In other terms, the N electrons of the solid occupy the N quantum states having the lowest energy. Two different situations can then arise.

If the occupied state with the highest energy does not coincide with the upper limit of one of the energy bands (Fig. 6.5(a)), there will be a continuum of unoccupied quantum states above the highest occupied state. The same argument used in Section 6.3.2 can then be developed here to show that any electric field, even of infinitesimal magnitude will induce the flow of current in the solid. *The considered solid is therefore a conductor.*

The partly occupied band which contains the empty quantum states at the origin of the conduction is termed the *conduction band of the solid*.

If instead the occupied state with the highest energy *coincides* with the upper limit of a band (Fig. 6.5(b)) there are no empty states surmounting the band and lying in its immediate vicinity. The electrons cannot absorb energy from an infinitesimal electric field since their transfer to a neighbouring unoccupied quantum state is not possible. No current flow is thus induced: the solid is an electric insulator.[5]

In an insulator, the highest occupied band is termed the *valence band*, the next band, unoccupied for $T = 0\,\text{K}$, being the conduction band.

Location of the Fermi energy at $T = 0\,\text{K}$

The Fermi energy $E_F(0)$ separates the fully occupied energy levels from the empty ones. In the case of a conductor $E_F(0)$ clearly coincides with the highest occupied level of the conduction band. In contrast, in an insulator, the highest occupied level E_V and the lowest unoccupied one E_C being separated by a gap, the possible locations of $E_F(0)$ can be at any energy value within the gap. Actually, as will be shown in Chapter 9 (Section 9.2.1), its position is *in the middle of the gap* at $T = 0\,\text{K}$ and within a few $k_B T$ from this position at other temperatures.

6.4.2 Factors determining the band occupation

The above distinction between conductors and insulators does not entirely clarify the microscopic origin of this important qualitative difference. The complete or partial occupation of a band is not a random circumstance. It is the result of two types of definite characteristics of a solid.

One is qualitative and relatively simple. It consists in the fact that both the number pN_M of quantum states in an energy band (cf. Chapter 3, Section 3.4), and the number zN_M of electrons which can occupy these states *are multiples of the number N_M* of elementary unit cells of the crystal. *The multiplying factors p and z are generally different*, and in particular, the number of electrons depends of the total number of valence electrons in a unit cell, which can be even or

---

[5]If the field is large enough, an electron can transfer to the empty states of the next band. The distance between bands being of the order of several electron Volts, this requires a field of magnitude $10^5\,\text{V/cm}$. The corresponding flow of current is not associated to the conduction (which occurs for small values of the field) but to the electric breakdown of the insulator.

odd, while, owing to the spin degeneracy, p is necessarily even. Hence, various qualitative situations can occur, depending of the crystal structure, leading either to exactly filled bands at $T = 0\,\mathrm{K}$, or to one partly filled band.

The second characteristics is quantitative, more elaborate, and tributary of the precise shape and magnitude of the crystal potential $V(\vec{r})$ acting on an electron. This potential can induce complicated overlaps between bands and thus influence the multiplying factor p and accordingly the conducting or insulating character of the solid.

Occupation of the bands having the lowest energy

As stressed in Section 5.3.1, the lowest lying electron levels of the solid are well described by the tight-binding approximation. In this method, the narrow bands, which contain the most strongly bound states, are composed of the lowest energy levels of the atomic constituents. In the atoms, these levels are fully occupied. As made clear by Fig. 6.6, The corresponding bands, *which contain the same number of states as the atomic levels*, will also be fully occupied. Hence, these bands have no influence on the insulating or conducting character of a solid. This character relies entirely on the occupation of the bands lying at higher energies and related to the atomic levels which are fully or partly occupied by the *valence electrons of the atoms*.

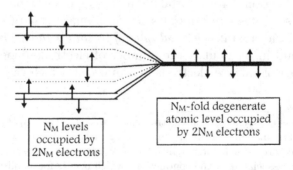

Fig. 6.6.   Correspondance between the energy and the occupation of bands, and on the other hand of the fully occupied atomic energy levels from which these bands are originated.

Number of states in a band: Elementary and composite bands

In the tight-binding model described in Section 5.3.1, and relative to a monoatomic solid involving a single, non-degenerate, atomic level, the band arising from this level contains N_M orbital states, N_M being the number of primitive cells of the crystal. Taking into account the spin degeneracy results in a band containing $2N_M$ states.

This property is general. Indeed as shown in Chapter 3 (Section 3.4), any band $E(\overrightarrow{k}, n)$ associated to Bloch states of given band index n, contains N_M orbital states ($2N_M$, if we take the spin into account).

In principle, the energy spectrum of the electron is therefore a succession of bands having $2N_M$ states each. If the number of valence electrons of the atoms in a primitive cell is z, the total number of valence electrons occupying the highest energy bands of a solid is zN_M. We could therefore conclude that if z *is even*, bands will either be fully occupied or empty, thus determining the situation of an insulator. If z is odd, one band with highest energy will be half-occupied, thus determining conducting properties.

This conclusion is partly incorrect. Indeed, several reasons can determine the constitution of bands having a number of states equal to a *multiple of* $2N_M$, namely a degeneracy of the atomic levels which remains unlifted by the crystal potential, or an overlap between bands of different indices n. Clearly, such a circumstance will not modify the conclusion of principle which arises from an odd number of valence electrons, but in the case of an even number of electrons, a solid can nevertheless be a conductor.

For the sake of clarity we will term *elementary band* a non-degenerate band relative to a given band index n and containing $2N_M$ states. Actual bands resulting from any of the mentioned complications will be termed a *composite band*. In all cases, a band will contain *a minimum* of $2N_M$ states.

6.4.3 Formation of composite bands: Degeneracy and overlap

The preservation of atomic degeneracies

Let us place ourselves in the framework of the tight-binding approximation, and consider the formation of a band arising from an *orbitally degenerate* atomic level. An example is provided by a $2p$ atomic level which has a threefold orbital degeneracy (principal quantum number $n = 2$, azimuthal quantum number $l = 1$). A basis of the corresponding eigenspace is constituted by the three functions $\phi = (2p)_x$, $\phi' = (2p)_y$, $\phi'' = (2p)_z$. Each of these functions gives rise, in conformity with the scheme described in Fig. 6.6, to an elementary band containing N_M orbital states. Depending on the *structure of the crystal*, and on the *width and position of the elementary bands* (which in turn depend on the overlap of the ϕ, ϕ', ϕ'' orbitals and on the nature of the crystal's potential), we can obtain three bands which either coincide, or partly overlap, or else, are well separated.

For instance, if the structure of the crystal is identical in the x, y, and z directions (e.g. in a cubic crystal), the three bands (ϕ, ϕ', ϕ'') will have the same width and the same energy shift since the spatial overlaps between neighbouring atomic orbitals are identical. In this case, the three bands coincide and we obtain a single *composite band* in which each state has three-fold degeneracy, and which therefore contains $6N_M$ states (including the spin degeneracy).

In a different structure, the three atomic orbitals (or their linear combinations) can give rise either to three separate bands with $2N_M$ states each (e.g. for a low symmetry structure, and a crystal potential determining differently shifted narrow bands), or to one band containing $4N_M$ states and one band containing $2N_M$ states (e.g. if the structure is identical in the x and y directions).

In addition to the various complex situations which can thus arise, the orbital degeneracy between bands can be lifted, or their overlap modified. One cause of such a degeneracy lifting is the so-called *spin–orbit interaction*. This is the interaction between the spin of the electron and its orbital momentum.[6]

The overlap of bands

We have pointed out in Chapter 5 (Section 5.3) that, in the tight-binding approximation, an overlap between bands originating from different atomic states is more likely for the states having the largest

---

[6]This interaction of relativistic nature will not be discussed in this book. Its contribution to the Hamiltonian is of the form $\propto \vec{L}.\vec{S}$, scalar product of the orbital momentum and the spin.

spatial extension, i.e. the states occupied by the valence electrons (and perhaps also the states just below). The set of such states gives rise, in the tight-binding approximation, to a single composite wide band. Let us denote this band **B**. The bands situated at lower energies being always fully occupied, it is the occupation of the band **B** which is the basis of the insulating or conducting character of the solid. If the atomic states giving rise to **B** are occupied in the atomic constituents of the primitive cell by z electrons, the number of electrons which occupy **B** is $N = N_M.z$.

Note that **B** is not likely to be described adequately by the tight-binding approximation since it concerns the states which are most perturbed by the interaction between atoms. Hence, certain gaps which are not obtained in this approximation may nevertheless exist. In order to examine on more safe grounds the possible insulating or conducting character of a solid, it is therefore preferable to consider the band **B** in the framework of the approximation of quasi-free electrons.

In the framework of this approximation, as emphasized in Section 5.2, the discontinuities in $E(\overrightarrow{k}, n)$ which may induce the existence of gaps in the density of states, are located, in the reciprocal space, on the planes separating the different Brillouin zones. Consequently, within **B**, the bands are formed by one or several overlapping elementary bands belonging to adjacent Brillouin zones. As already stressed, each elementary band comprises $2N_M$ states. The gaps are formed by energy intervals common to the discontinuities in $E(\overrightarrow{k}, n)$ occurring in the various directions in the reciprocal space.

As an illustration let us examine the case of a solid possessing a simple cubic Bravais lattice. Figure 6.7(a) represents the first Brillouin zone and the plane $\Gamma X R$ of the reciprocal space. This plane contains both the shortest distance ΓX and the longest one ΓR between the origin and the surface of the first Brillouin zone. Figure 6.7(b) depicts schematically a situation of "weak" discontinuities on the surface of the Brillouin zone ($|V_G| \ll (\hbar^2/2m)|\Gamma X|^2$) while Fig. 6.7(c) shows a situation of "strong discontinuities" ($|V_G| \lesssim (\hbar^2/2m)|\Gamma R|^2$).

It appears that, in the first situation, whatever the direction $\Gamma \mu$ in the plane (and therefore, for symmetry reasons, in the entire Brillouin zone) quantum states are available for any value U of the energy. If $U = E_1$ lies within the energy gap in the ΓX direction,

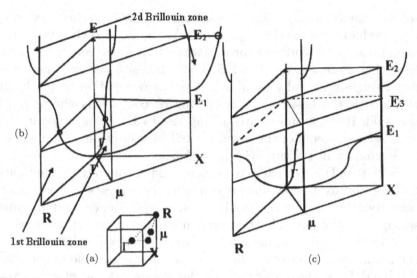

Fig. 6.7. Example of overlap between elementary bands for a simple cubic crystal in the almost free-electron model. Figures (a)–(c) correspond respectively to small and large discontinuities on the Brillouin planes. The first case is that of a conductor and the second of an insulator.

the available states belong to the first Brillouin zone in directions comprised between $\Gamma\mu$ and ΓR. If $U = E_2$ lies in the energy gap in the ΓR direction, the available quantum states belong to the second Brillouin zone in directions such as ΓX. Hence, the two elementary bands corresponding to the two first Brillouin zones have some overlap. They form a **B**-band possessing at least $4N_M$ states (notwithstanding an analysis of the subsequent Brillouin zones).

In the situation of "large" discontinuities, there are values of the energy, such as $U = E_3$, associated to no quantum state. The elementary band relative to the first Brillouin zone (with $2N_M$ states) is separated from that relative to the second Brillouin zone (also with $2N_M$ states) by a gap, the density of states being qualitatively of the type in Fig. 6.5(b).

6.4.4 Simple examples of band occupation

Sodium

The structure of crystalline sodium is of the body centered cubic type (BCC) with one atom per primitive unit cell and a conventional cubic cell of edges $a \sim 4.2$ Å. The electronic configuration of the sodium atom, which possesses 11 electrons, is represented by the sequence of 11 occupied atomic orbitals $(1s)^2(2s)^2$, $(2p)^6(3s)^1$. In accordance with the argument developed in Section 6.4.3, the cubic symmetry preserves the three-fold degeneracy of the $(2p_x, 2p_y, 2p_z)$ orbitals. These states give rise to bands which are also degenerate. In the framework of the tight-binding approximation, the sequence of atomic states (1s),(2s),(2p),(3s) give rise to four bands which do not overlap. The two lowest bands (1s) and (2s) contain $2N_M$ states each, and the degenerate band (2p) contains $6N_M$ states. Hence $10N_M$ of the $11N_M$ electrons of the solid fill totally the three former bands of strongly bound states.

The (3s) band, which contains $2N_M$ states, is the first of a series of less bound bands (3p, 3d, ...) which overlap partly, owing to their larger spatial extension (the (3s) orbital contains the valence electron of the atom). This series of states form the **B** band which contains the N_M valence electrons. Whatever the relative position of the elementary bands composing **B** and corresponding to the successive Brillouin zones, the valence electrons will occupy at most half of the lowest elementary band. Sodium is therefore a conductor.

Let us examine in greater detail the occupation of band **B** in the framework of the quasi-free electron approximation. Referring first to the Sommerfeld model, let us recall that the N_M valence electrons occupy, in reciprocal space, all the states located inside the Fermi sphere whose radius k_F is defined by Eq. (6.16). Replacing in this equation (N/Ω) by $N_M/\Omega) = (2/a^3)$, we obtain:

$$k_F = 0.62 \left(\frac{2\pi}{a} \right) \tag{6.21}$$

On the other hand, the reciprocal lattice being of the FCC type (see Section 3.3.3), the shortest distance between the origin and other nodes of this lattice is $|\vec{G}| = \sqrt{2}(2\pi/a)$ (the edge of the conventional reciprocal cell being $(4\pi/a)$). The distance between the origin and the closest Brillouin planes is:

$$\frac{|\vec{G}|}{2} = 0.71 \left(\frac{2\pi}{a} \right) \tag{6.22}$$

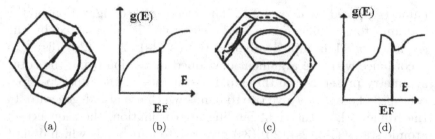

Fig. 6.8.   Schematic representations for sodium of the Fermi surface and the first Brillouin zone (a), and (b) of the density of states in the conduction band and the location of the Fermi energy. Images (c) and (d) represent these features for calcium.

Comparing Eqs. (6.21) and (6.22) we conclude that the Fermi sphere is entirely contained inside the first Brillouin zone (Fig. 6.8(a)). More precisely, the distance between the surface of the sphere and the Brillouin plane $\hbar^2(\Delta k)^2/2m$ is of the order of $1\,\mathrm{eV}$. In sodium, the gaps induced by the crystal potential being much smaller than $1\,\mathrm{eV}$, the situation in this material is the one depicted on Fig. 6.8(b). The quantum spectrum $E(\overrightarrow{k}, n)$ below the Fermi energy is close to that of the free electron. The electronic properties of sodium are therefore expected to be correctly predicted by the Sommerfeld model (Section 6.3.2).

Calcium

The structure of crystalline calcium is of the face-centered cubic type (FCC), with a conventional cell, of edge $a = 5.6\,\mathrm{\mathring{A}}$, containing four atoms. Each calcium atom has 20 electrons, and its electronic configuration is $(1s)^2(2s)^2(2p)^6$, $(3s)^2(3p)^6(4s)^2$. At $T = 0\,\mathrm{K}$, 18 N_M electrons of the solid fill the available states of bands originating from the strongly bound atomic levels (1s), (2s), (2p), (3s), (3p).

The band associated to the (4s) atomic level, is the first of a series of bands which form the composite band **B**, occupied by the remaining $2N_M$ electrons. In contrast to the case of sodium, it is crucial to determine if the elementary bands related to the two first Brillouin zone overlap. Indeed, without such an overlap, the $2N_M$ valence electrons will exactly fill the lowest elementary band whose upper limit would be separated from the next band by a gap: calcium would then be an insulator.

The experimentally observed conducting character of calcium (and of the other alcaline-earth metals) implies that there is an overlap between the lowest lying subbands of **B**. Hence, the states of **B** situated in the first Brillouin zone are only partly occupied, a fraction of occupied states of this band being in the second Brillouin zone.

As in the case of sodium, a more detailed analysis of the electronic state of calcium can proceed through a comparison of the radius of the Fermi sphere associated to the $2N_M$ valence electrons and of the distances to the various Brillouin planes (Fig. 6.8(c) and 6.8(d)).

The reciprocal lattice of calcium is of the BCC-type with a conventional cell of edge $(4\pi/a)$. The radius of the Fermi sphere determined by $(N/\Omega) = (8/a^3)$ is $k_F = 0.49(4\pi/a)$. On the other hand, the two families of Brillouin planes closest to the origin are respectively at distances $(\sqrt{3}/4)(4\pi/a) \sim 0.44(4\pi/a)$ and $0.5(4\pi/a)$. The Fermi sphere cuts the planes limiting the first Brillouin zone. Hence, within the approximation of the Sommerfeld model, part of the occupied electron-states are situated outside the first Brillouin zone, in agreement with the expectation of an overlap between elementary bands.

Beyond the Sommerfeld description, the Fermi surface keeps an almost spherical shape (Fig. 6.8(c)). The occupied states of the second Brillouin zone are in the directions of the cube diagonal while the empty states of the first Brillouin zone are located in the directions of the cube axes.

Diamond and graphite

Each carbon atom possesses six electrons in the quantum states $(1s)^2(2s)^2(2p)^2$. The structures of diamond and graphite both have two atoms per elementary unit cell. Among the $12N_M$ electrons of the solid, $4N_M$ fill the lowest energy bands formed by the $(1s)$ states.

As shown in Chapter 5 (Section 5.3) the next bands, which form the lowest end of the **B** band, are two sp^3 bands separated by a gap of magnitude $6\,\text{eV}$ (Figs. 6.9 and 6.10). Each of these bands contains $8N_M$ states. At $T = 0$ K, the remaining $8N_M$ valence electrons of the solid will exactly fill the lowest sp^3 band, thus explaining the insulating character of diamond, mentioned in Chapter 1. Silicon which possesses the same crystal structure as diamond, a similar band structure, and the same number of valence electrons is also an insulator. Its case will be discussed in greater details in Chapter 9.

Fig. 6.9. sp^2 and sp^3 orbitals of carbon.

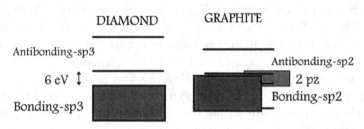

Fig. 6.10. Schematic valence and conduction bands structures in diamond and graphite.

As also indicated in Chapter 5 (Section 5.3), the composite **B** band in graphite comprises 3 subbands. Two sp^2 bands containing $6N_M$ states each are almost adjacent (cf. Appendix A.1). A third one $(2p_z)$ contains $4N_M$ states, and overlaps with the two others. There is a continuum of $16N_M$ states which will only be partly occupied by the $8N_M$ valence electrons. In agreement with the result anticipated in Chapter 1, graphite is therefore a conductor.

6.5   Diversity of the Equilibrium Properties in Solids

In this paragraph, we examine from a qualitative standpoint two questions raised in Section 6.3.2, namely, the determination of the effective number of valence electrons, and the quantitative inadequacy of the Sommerfeld model in accounting for the value of the electronic specific heat of certain metals.

6.5.1   Effective valences in metals

Alcaline and alcaline earth metals

The electronic states of sodium and calcium have been studied in Section 6.4.4. The two former solids are representative, respectively of the properties of the alcaline and of the alcaline-earth metals.

Alcaline metals (Na, K, Rb, Cs) have N_M electrons in the conduction band. The effective valence related to the number of conduction electrons has a clear meaning since it is equal to the chemical valence of the atoms (1 electron per atom). The occupied states of the conduction band have their $\overrightarrow{k}$-vectors *inside* the first Brillouin zone, distant from the planes limiting this zone. Accordingly (cf. Section 5.2.3) the N_M conduction electrons have properties very similar to a free-electron gas: isotropy of $E(\overrightarrow{k}, n)$, parabolic density of states $(g(E) \propto \sqrt{E})$, spherical Fermi surface. This explains the good agreement between the value of their electronic specific heat (cf. Table 6.1) and the value expected from the Sommerfeld model (the small difference observed is discussed in Section 6.5.2).

In the case of alcaline-earth metals (Ca, Sr, Ba), the conduction band contains $2N_M$ electrons, in agreement with the chemical valence of the atoms. The conduction properties have their origin in the overlap between elementary bands associated to the two first Brillouin zones. In turn (see Fig. 6.8(c) and 6.8(d)) such an overlap derives from the smallness of the discontinuities in $E(\overrightarrow{k}, n)$ on the Brillouin planes closest to the origin of the reciprocal lattice. These discontinuities affect the properties of the electronic states near the Fermi surface, and therefore induce significant deviations from the properties of free electrons.

Transition metals

Transition metals are solids constituting a large fraction of elements of the Mendeleiev table (cf. Fig. 2.20). They comprise the elements of three rows of this table, with atomic numbers between 21 and 30, 39 and 48, 72 and 80. They have in common the characteristics of possessing certain occupied states originating from **d**-atomic states (i.e. atomic states corresponding to an angular momentum $l = 2$). In a broad sense, one can also include in this set *rare earth* metals

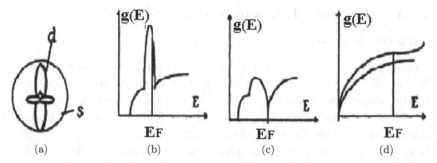

Fig. 6.11.   (a) Compared spatial extensions of (nd) and $(n + 1, s)$ states. (b) Contribution to the density of states resulting from the high degeneracy and the narrow width of a d-band. (c) Small overlap between two bands determining a small value of the density of states at E_F. (d) Higher density of states in alcaline earth metals than in the free electron model.

(atomic numbers 57 to 71) as well as transuranides (atomic numbers above 90). The two latter families possess occupied states originating from **f**-atomic orbitals $(l = 3)$.

In atoms, the energy and the spatial extension of the electronic states generally increase with the values of the quantum numbers n and l. A (3s) state or a (2p) state have higher energies than a (2s) state and their spatial extension is larger. From this standpoint, the (d) and (f) states depart somewhat from the rule (Fig. 6.11(a)). Their spatial extension is abnormally small and their energy abnormally high. Accordingly, the energy of a (nd) state (e.g. 3d) is close to that of a $(n+1, s)$ state (e.g. 4s), and its spatial extension is smaller. This trend is even more pronounced in the case of (nf) levels.

These characteristics deeply affect the shape of the density of states in the conduction band of the noble and transition metals. In these atoms, the (d) or (f) states have energy in same range as the energy of the valence electrons. However, their spatial extension being small, their overlap with the states of neighbouring atoms is reduced. Consequently, they give rise to relatively narrow subbands lying within the range of the conduction band or in its neighbourhood. Besides, the degeneracies of a (d)-atomic level (10) or that of an (f)-level (14) being large their contribution to the density of states will be large in the range of this subband.

The great variations in the electronic properties of these metals are determined by the presence of such an abnormal density of states.

A small change in the position of the Fermi energy with respect to this subband will induce a large modification in the value of the density of states $g(E_F)$ at the Fermi level, a value which governs a number of important physical properties (e.g. the electronic specific heat (cf. Eq. (6.18) and next paragraph).

This specificity of noble and transition metals also constitutes an explanation of the correspondance, already pointed out in the comments on the quantitative shortcomings of the Sommerfeld model (Section 6.3.2), between the chemical valence of an atom and the number of valence electrons in the electron gas determining the conduction properties. Hence in copper, gold, and niobium, the (d) subband is situated slightly below the conduction band which contains a single electron (metallic valence equal to one). However, in the atom, the (d) states being close in energy will participate to the formation of chemical bonds. The number of valence electrons will comprise one (s) electron and one or several (d) electrons (one in copper, two in gold, and either two or four in niobium).

6.5.2 Electronic specific heat: Static effective mass

The expression of the specific heat in the second member of Eq. (6.18), has a general validity since it only relies on the value of the density of states at the Fermi energy. The γ coefficient appearing in Table 6.1 is proportional to $g(E_F)$. On the basis of the discussions in Section 6.5.1, it is now possible to understand the deviations between the experimental values of γ and the values deduced from the Sommerfeld model.

The very large γ values in iron, niobium and manganese, which are transition metals, are related to the fact that E_F is located within a d-band associated to a peak in the density of states (Fig. 6.11(b)).

In contrast, the small $\gamma \propto g(E_F)$ values in bismuth and antimony reflect the situation sketched in Fig. 6.11(c), and similar to that encountered in graphite (Fig. 6.10). These two materials possess $10N_M$ valence electrons which can fill five elementary bands. The Fermi energy is situated near the borderline between two elementary bands whose *overlap is small* (also see Appendix A.3). Consequently, there is a small density of states in this range. These elements are termed *semimetals*.

Figure 6.8 explains the quantitative departure from the Sommerfeld model for alcaline-earth metals: it is due to the increase of the density of states induced by the discontinuities of $E(\overrightarrow{k}, n)$ near a Brillouin plane.

Finally, let us consider the moderate but nevertheless significant excess of γ values observed in alcaline metals and noble metals. In both cases, the Fermi energy is situated in the lower part of a conduction band of s-type, displaying a parabolic density of states $g(E) = K\sqrt{E}$. The inverse curvature K of the parabola is not necessarily equal to its value in a free electron gas $[K_0 = (8\sqrt{2}\pi\Omega m^{3/2}/h^3)]$. Indeed, within the tight binding approximation, K is proportional to the inverse of the band width. In the alcaline metals, or the noble metals, the conduction electrons, though behaving almost as free electrons are *more localized* than such electrons by the effect of the crystal potential. Bands are narrower resulting in a higher value of K. Consistently the density of states (Fig. 6.10) and the values of γ for these good metals are larger than for free electrons.

Note that we can set $K = (m^*/m)^{3/2}.K_0$, where K_0 is the expression of K for a free-electron gas. m^* is the *static effective mass* of the electron in the conduction band of the considered metal. We will see in Chapter 7 that m^* also controls the dynamics of an electron of the considered solid submitted to an electric or a magnetic field, in the same way as m controls the dynamics of a free-electron.

Bibliography

Fundamental of Statistical and Thermal Physics, *F. Reif*, Mc Graw Hill (1965).

Chapter 7

The Dynamics of Electrons in a Crystal

Main ideas: Nature of the semi-classical approximation describing the dynamics of free electrons (Section 7.2) and of Bloch-electrons (Section 7.3). Influence of the band structure on the semi-classical electron dynamics (Section 7.3.2). Concepts of dynamic effective mass (Section 7.3.3) of electrons and of holes (Section 7.3.5).

7.1 Introduction

In this chapter, we examine the *collective dynamics* of electrons in a solid, i.e. the collective evolution of their states as a function of time, which is induced by the application of external fields (electric or magnetic). A clarification of this evolution is required before addressing the question, treated in the next chapter, of the determination of the electronic *transport properties* of solids, such as the electrical or thermal conductivity of metals.

The dynamics of electrons is studied in the framework of the so-called *semi-classical approximation* which consists in studying the motion of electrons by means of the equations of classical mechanics, although their states are described, as in Chapters 3 and 5, by quantum mechanics. This hybrid nature of the theory is at the origin of odd but useful concepts such as that of *effective mass* of the electron, or that of *hole*. In the first section, we focus our attention on the case of a gas of *free electrons* submitted to an electric or a magnetic field. We examine, in addition to the semi-classical treatment, the results obtained by a complete quantum derivation of the dynamics. The second section is devoted to the semi-classical collective dynamics of *Bloch electrons*.

7.2   Collective Dynamics of a Free-Electrons Gas

Let us consider the situation in which, within the volume Ω of a solid, the electrons form a quantum gas of free electrons. As shown in Chapter 5 (Section 5.2.1) the electrons are in quantum states consisting of plane waves $\exp(i\,\vec{k}.\vec{r})$. These are defined by a vectorial quantum number $\vec{k}$ taking discrete values. For such free electrons, $\vec{k} = (\vec{p}/\hbar)$ is proportional to the particle momentum $\vec{p}$.

We have seen in Chapter 6, Section 6.3.1, that, at zero temperature, and in the absence of external fields, the occupied quantum states of an electron gas are defined, in the reciprocal space, by the set of discrete $\vec{k}$-values lying inside the Fermi sphere of radius k_F (Fig. 7.1). In the situation realized in most metals, a similar collective state of the gas also prevails at finite temperatures $(T \ll \Theta_F)$. The only difference consists in the fact that the surface of the Fermi sphere has a "thickness", because some of the states internal or external to the sphere, and close to the surface are partly occupied. The distribution of occupied states is *isotropic* in the reciprocal space.

The application of an electric or a magnetic field modifies this distribution.The quantum study of this modification would require examining the form of the eigenfunctions of a Hamiltonian which takes into account the potential of the external field. This will be performed in Section 7.2.5 for the magnetic field, and in Section 7.3.4 for

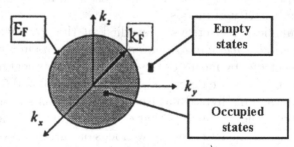

Fig. 7.1.   Isotropic distribution of the occupied $\vec{k}$ states in a metal for zero electric field.

the electric field. Let us first examine an approximate solution of this problem, valid for temperatures which are not too low ($T > 10$ K).

7.2.1 Classical approximation: Wavepacket and group velocity

We place ourselves in the framework of the classical approximation of quantum mechanics, in which one can speak of the *motion of an electron submitted to an external force*. The instantaneous position of the electron is then assimilated to the position of the centre of the wave-packet associated to the electron.

It can be shown that if the stationary state of a particle is the plane wave $\exp(i\vec{k}.\vec{r})$ at time $t = 0$, this state evolves as a function of time and becomes $\psi(\vec{r}, t) = \exp[i(\vec{k}.\vec{r} - E_k.t/\hbar)]$ at time t, with $E_k = (\hbar^2 \vec{k}^2/2m)$. Such a state determines, at any time, a *spatially uniform* probability of presence $\propto |\psi|^2$, which is clearly inadequate to represent a *spatially localized* particle. However, a localized particle can be satisfactorily represented by a weighted sum of planewaves having their energies in a narrow energy-interval, and forming together a *wavepacket*. For instance, a *normalized Gaussian wavepacket* has the form (Fig. 7.2):

$$\psi(\vec{r}, t) = \left[\frac{2^{1/3}}{(2\pi)^{3/4}\delta_k^{3/2}}\right] \int \exp\left(-\frac{(\vec{k} - \vec{k}_0)^2}{\delta_k^2}\right)$$

$$. \exp\left(i[\vec{k}.\vec{r} - \frac{E_k.t}{\hbar}]\right) . dk_x dk_y dk_z \qquad (7.1)$$

The $\vec{k}$-vectors of the waves which contribute with significant amplitude to ψ have their modulus in the interval $|\Delta \vec{k}| = 2\delta_k$ centered on $\vec{k}_0$ ($\delta_k \ll |\vec{k}_0|$). One can show that, in the direct space, the amplitude of the function $\psi(\vec{r}, t)$ is essentially concentrated around the point $\vec{r} = \vec{V}_g.t$ which, therefore, moves with the velocity:

$$\vec{V}_g = \left(\frac{1}{\hbar}\right) \vec{\nabla}_k E(\vec{k}_0) \qquad (7.2)$$

where $\vec{\nabla}_k$ is the gradient in reciprocal space. $\vec{V}_g$ is termed the *group-velocity of the wave-packet* and is equal, for a free-electron, to $\hbar\vec{k}/m$.

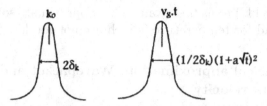

Fig. 7.2. Gaussian wavepacket in reciprocal space (left) and in direct space (right). The width of the latter grows as a function of time t.

At time $t = 0$, the spatial width of the wave-packet is:

$$|\Delta \vec{r}| \cong \left(\frac{1}{2\delta_k} \right) \qquad (7.3)$$

Let us derive Eqs. (7.2) and (7.3) in the more general situation of Bloch electrons in which $E_k = E(\vec{k})$ is not necessarily proportional to $\vec{k}^2$. If $E(\vec{k})$ has no singularity for $\vec{k} = \vec{k}_0$, and if $\delta_k \ll |\vec{k}_0|$, we can write, with $E_0 = E(\vec{k}_0)$ and:

$$E_k = E_0 + (\vec{k} - \vec{k}_0).\vec{\nabla}_k E(\vec{k})|_{k_0} \qquad (7.4)$$

Using expression (7.2), the integral (7.1) then takes the approximate form:

$$\psi(\vec{r}, t) \propto \exp\left(\frac{\delta_k^2}{4} [\vec{r} - \vec{\mathbf{v}}_g.t] \right) . \exp\left(i[\vec{k}_0.\vec{r} - \frac{E_0.t}{\hbar}] \right) \qquad (7.5)$$

The probability of presence $|\psi|^2$ is concentrated in an interval of width $\approx (2/\delta_k)$ centered on the position $\vec{r} = \vec{\mathbf{v}}_g.t$.

7.2.2 Classical approximation: Dynamical equation

In quantum mechanics, the time dependence of the quantum average $\langle A \rangle$ of a physical quantity A is governed by the Ehrenfest equation:

$$i\hbar \frac{d\langle \widehat{A} \rangle}{dt} = \langle [\widehat{A}, \widehat{H}] \rangle \qquad (7.6)$$

where $\widehat{H}$ is the Hamiltonian of the system. For instance, if a particle is submitted to a potential $V(\vec{r})$, its Hamiltonian is $\widehat{H} = (\widehat{\vec{p}}^2/2m) + \widehat{V(\vec{r})}$. Applying, with this form of $\widehat{H}$, the Ehrenfest equation to each component $\widehat{x}_i$ of $\widehat{\vec{r}}$ we obtain, using the commutation relation $[\widehat{x}_i, \widehat{p}_i] = i\hbar$, and $[\widehat{x}_i, \widehat{p}_j] = 0$:

$$\frac{d\langle \vec{r} \rangle}{dt} = \left\langle \frac{\vec{p}}{m} \right\rangle \tag{7.7}$$

Applying, likewise, the Ehrenfest equation to $\widehat{\vec{p}}$, we obtain

$$\langle [\widehat{p}_i, \widehat{H}] \rangle = \langle [\widehat{p}_i, V(\vec{r})] \rangle = \frac{\hbar}{i} \int \psi^* \left\{ \frac{\partial(V\psi)}{\partial x_i} - V \frac{\partial \psi}{\partial x_i} \right\} dx_i$$

$$= \frac{\hbar}{i} \left\langle \frac{\partial V}{\partial x_i} \right\rangle \tag{7.8}$$

Hence:

$$\frac{d\langle \vec{p} \rangle}{dt} = \langle -\vec{\nabla}_r V \rangle = \langle \vec{F}(\vec{r}) \rangle \tag{7.9}$$

$\vec{F}(\vec{r})$ is a force deriving from the potential $V(\vec{r})$. These two equations have a similar form as the classical equations of motion. It can be shown on the basis of the former equations, that, as long as the width of the wavepacket associated to a particle remains small, the motion of the particle is well described by the *classical dynamical equation:*

$$\frac{d\langle \vec{p} \rangle}{dt} = \frac{d^2 \langle \vec{r} \rangle}{dt^2} = \vec{F}(\langle \vec{r} \rangle) \tag{7.10}$$

The approximate equation (7.10) involves a force exerted on the center $\langle \vec{r} \rangle$ of the wavepacket. We will accept the general validity of this equation for forces $\vec{F}$ which do not necessarily derive from a scalar potential $V(\vec{r})$ (i.e. magnetic forces). For a free-electron submitted to electric and magnetic fields, since $\langle \vec{p} \rangle = \hbar \vec{k}_0$, with $\vec{k}_0$ the center of the wavepacket, we can write:

$$\hbar \frac{d\vec{k}_0}{dt} = \vec{F} = -e\vec{\mathcal{E}} - e\vec{\mathbf{v}}_g \times \vec{\mathcal{B}} \tag{7.11}$$

Note that the latter equation can be applied to spatially non-uniform fields. It is therefore of interest to underline that it is only valid if *the spatial variations of the fields* occur on a *scale l much larger than the size of a crystal unit cell.* This criterion results from two considerations. First, we can observe that the concept of a force acting on a localized particle has a meaning only if the force, or the potential, is almost uniform on the spatial range ($\approx 1/\delta_k$) of the wavepacket associated to the particle. On the other hand, as stressed in Chapter 6 (Section 6.3.2), the quantum states which are involved in the evolution of a degenerate electron gas submitted to external forces, are states close to the Fermi-surface $E(\vec{k}) = E_F$. Hence, for a free-electron gas, $|\vec{k}_0| = k_F \approx (1/a)$, a being the size of the unit cell. The required smallness of the width of the wavepacket $\delta_k \ll |\vec{k}_0|$ implies that the characteristic length l of variation of the fields satisfies the condition

$$l \gg (1/\delta_k) \gg a \qquad (7.12)$$

Also note that Eq. (7.12) shows that the wavepacket associated to an electron-state close to the Fermi surface is spread over many unit cells.

Condition (7.12) imposes an upper limit to the value of the electric field. For instance, the uniformity of the field on the size of a unit cell can be expressed by stating that the corresponding variation of the potential $|e\Delta V|$ is much smaller than the thermal energy $k_B T$. Thus, $|ea\mathcal{E}| \ll k_B T$. At room temperature, and for unit cell $a \approx 5\text{Å}$, we obtain $\mathcal{E} \ll 10^6$Volts/cm. Such a condition is generally fulfilled in practice.

The width in position-space of a wavepacket does not remain constant as a function of time. It can be shown to increase progressively, and reach, for instance, twice the initial value after a time $t \approx (2m\Delta r^2/\hbar) \approx 10^{-12} - 10^{-13}s$. Equation (7.11) will therefore remain valid during a limited time-interval Δt. However, this restriction is irrelevant in the determination of the transport properties of solids. Indeed, as will be explained in the next chapter, the validity of the dynamical equation is only required during the time-interval τ separating two consecutive collisions involving the electron. The characteristic time τ being in the range 10^{-14}–10^{-12}s, the wavepacket will remain narrow between two collisions.

7.2.3 Dynamics induced by an electric field

Let us assume that the electron gas is submitted to a time independent and spatially uniform field $\overrightarrow{\mathcal{E}}$. For such a field, Eq. (7.11) determines a uniform motion of the center $\overrightarrow{k}_0$ of each wavepacket, parallel to the field, and identical for all wavepackets:

$$\overrightarrow{k}(t) - \overrightarrow{k}_0 = \frac{-e\overrightarrow{\mathcal{E}}}{\hbar}t = \Delta\overrightarrow{k}(t) \qquad (7.13)$$

The initial collective state of the electron gas consists of the individual electron states associated to the $\overrightarrow{k}$-values situated inside the Fermi sphere centered on the origin (Fig. 7.1). Hence, the collective state of the electron gas at a given time t_0 is represented by the set of quantized $\overrightarrow{k}$-values in a *shifted Fermi sphere* whose center is displaced by $\Delta\overrightarrow{k}(t_0)$ with respect to the origin of the reciprocal space (Fig. 7.3). The isotropy of the $\overrightarrow{k}$-distribution is destroyed by $\overrightarrow{\mathcal{E}}$, in agreement with the fact that $\overrightarrow{\mathcal{E}}$ defines an asymmetry of the system.

Let us show that, corresponding to this asymmetry, the collective motion of all the electrons of the solid determines the flow of an electric current parallel to the field through the system.

In the first place, we can note that if a solid contains, per unit volume, $\rho(\overrightarrow{k})d^3\overrightarrow{k}$ electrons having wavevectors within the infinitesimal reciprocal volume $(\overrightarrow{k}, d^3\overrightarrow{k})$, these electrons determine, per unit time, a flow δQ of electric charge across an elementary surface δS:

$$\delta Q = -e(\overrightarrow{v}.\overrightarrow{n})\rho(\overrightarrow{k})d^3\overrightarrow{k}.\delta S \qquad (7.14)$$

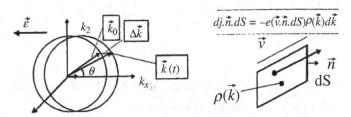

Fig. 7.3.   Left: Drift of the occupied states induced by an electric field. Right: Electric charge flow across a surface dS during a time unit.

$\overrightarrow{v} = \hbar \overrightarrow{k}/m$ is the average velocity of the electrons. The resulting infinitesimal current density $d\overrightarrow{j}$, defined by $\delta Q = (d\overrightarrow{j} \cdot \overrightarrow{n})\delta S$ is:

$$d\overrightarrow{j} = -e\overrightarrow{v}\rho(\overrightarrow{k})d^3\overrightarrow{k} \qquad (7.15)$$

The global current density $\overrightarrow{j}$ within the solid is obtained by integration of $d\overrightarrow{j}$ over the reciprocal space, and therefore reflects the characteristics of the electron distribution $\rho(\overrightarrow{k})$. Note that $\rho(\overrightarrow{k})$ is related to the number (N/Ω) of electrons per unit volume by

$$\int_{\overrightarrow{k}} \rho(\overrightarrow{k})d^3\overrightarrow{k} = \frac{N}{\Omega} \qquad (7.16)$$

As recalled above, in the absence of applied field, $\rho(\overrightarrow{k}) = \rho_0(\overrightarrow{k})$ is isotropic. Thus, since $\rho_0(\overrightarrow{k}) = \rho_0(-\overrightarrow{k})$, and $\mathbf{v}(-\overrightarrow{k}) = -\mathbf{v}(\overrightarrow{k})$, the states $\overrightarrow{k}$ and $(-\overrightarrow{k})$ determine opposite contributions to $\overrightarrow{j}$. The global current density is therefore $\overrightarrow{j} = 0$. Referring to Eq. (6.6) and to Eq. (3.29), we also infer, that, in the absence of applied field, the $\overrightarrow{k}$-distribution of free electrons, in the unit-volume, is related to the $\overrightarrow{k}$-density of states $g(\overrightarrow{k}) = (\Omega/4\pi^3)$ and to the Fermi-factor $f(\overrightarrow{k})$ by

$$\rho_0(\overrightarrow{k}) = \frac{f(\overrightarrow{k})}{4\pi^3} \qquad (7.17)$$

The vanishing of $\overrightarrow{j}$ for symmetry reasons can be expressed as:

$$\overrightarrow{j} = -\frac{e\hbar}{m} \int_{\overrightarrow{k}} \overrightarrow{k}\rho_0(\overrightarrow{k})d^3(\overrightarrow{k}) = -\frac{e\hbar}{4m\pi^3} \int_{\overrightarrow{k}} \overrightarrow{k}f(\overrightarrow{k})d^3\overrightarrow{k} = 0 \qquad (7.18)$$

For $T \ll \Theta_F$, the Fermi factor satisfies $f(\overrightarrow{k}) \sim 1$ inside the Fermi sphere and $f(\overrightarrow{k}) \sim 0$ outside this sphere.

Let us now examine the effect of the field $\overrightarrow{\mathcal{E}}$. The application of Eq. (7.11) yields:

$$\overrightarrow{\mathbf{v}}_g(t) = \overrightarrow{\mathbf{v}}_g(0) - \frac{et}{m}\overrightarrow{\mathcal{E}} \qquad (7.19)$$

Using this expression in Eq. (7.15) and integrating, we obtain:

$$\overrightarrow{j} = -e \int_{\overrightarrow{k}} \overrightarrow{\mathbf{v}}_g(t).\rho(\overrightarrow{k})d^3\overrightarrow{k} \qquad (7.20)$$

We know that the contribution of $\overrightarrow{\mathbf{v}}_g(0)$ vanishes due to the isotropy of the zero-field distribution $\rho_0(\overrightarrow{k})$. Thus, based on Eqs. (7.19) and (7.16):

$$\overrightarrow{j} = \frac{e^2 t \overrightarrow{\mathcal{E}}}{m} \int_{\overrightarrow{k}} \rho(\overrightarrow{k})d^3\overrightarrow{k} = \frac{Ne^2 t}{m\Omega}\overrightarrow{\mathcal{E}} \qquad (7.21)$$

The current density induced by the field increases linearly with time.

7.2.4 Irrelevance of occupied and empty states

It can be noticed on Fig. 7.3 that, in the presence of a field, the distribution $\rho(\overrightarrow{k})$ of the occupied states in $\overrightarrow{k}$-space is equal to:

$$\rho(\overrightarrow{k}) = \rho_0(\overrightarrow{k} - \Delta\overrightarrow{k}) = \frac{f(\overrightarrow{k} - \Delta\overrightarrow{k})}{4\pi^3} \qquad (7.22)$$

The current density $\overrightarrow{j}$ induced by the field can then be obtained through Eq. (7.18) by using Eq. (7.22). Taking into account the vanishing of the zero-field current density, we have

$$\overrightarrow{j} = -\frac{e\hbar}{m} \int_{\overrightarrow{k}} \overrightarrow{k}[\rho - \rho_0]d^3\overrightarrow{k} \qquad (7.23)$$

As mentioned above, the validity of the dynamical equation (7.11) is limited to a certain duration δt, due to the progressive widening of the wavepacket. This limitation is equivalent to the condition:

$$|\Delta \vec{k}| \ll k_F \qquad (7.24)$$

Consequently, the difference $[\rho - \rho_0]$ is non-zero only in a small range of $\vec{k}$-values close to the Fermi surface. It can be developed as

$$\rho - \rho_0 = \frac{1}{4\pi^3}[f(\vec{k} - \Delta\vec{k}) - f(\vec{k})]$$

$$= -\frac{1}{4\pi^3}\Delta\vec{k}.\vec{\nabla}f = -\frac{\hbar^2}{4m\pi^3}\frac{df(E)}{dE}\vec{k}.\Delta\vec{k} \qquad (7.25)$$

where the last expression is derived from $E = (\hbar^2 \vec{k}^2/2m)$. It appears that, since the derivative of $f(E)$ is non-zero only in the vicinity of the Fermi surface, *the current density is entirely determined by electron-states close to this surface*.

Note that expression (7.25) allows an alternate calculation of the current density. Anticipating the fact that the current is parallel to the field, we take the x-axis oriented along the field and opposite to it and project the current density (7.23) on this axis. We obtain, with $k = |\vec{k}|$, and the angle θ between $\vec{k}$ and $\vec{x}$:

$$j = -\frac{e^2\hbar^2 t}{2m^2\pi^2}\mathcal{E}\int_{\theta=0}^{\theta=\pi}\cos^2\theta\sin\theta d\theta \int_{k=0}^{\infty}\frac{df}{dE}k^4 dk \qquad (7.26)$$

For a degenerate quantum gas $(T \ll \Theta_F)$, the derivative of the Fermi factor can be approximated by a Dirac peak, $(df/dE)^{\tilde{}} - \delta(E - E_F)$. On the other hand, the integration over k can be replaced by an integration over the energy values using the relation

$$\frac{\Omega}{4\pi^3}d^3\vec{k} = \frac{\Omega}{\pi^2}k^2 dk = g(E)dE \qquad (7.27)$$

where $g(E)$ is the free-electron density of states. The density of current takes the form:

$$j = \frac{e^2\hbar^2 t}{3m^2\Omega}k_F^2 g(E_F).\mathcal{E} \qquad (7.28)$$

and, by replacing k_F and $g(E_F)$ by their values (6.12)–(6.14),

$$j = \frac{Ne^2t}{\Omega m}\mathcal{E} \tag{7.29}$$

identical to Eq. (7.21).

A simple qualitative argument also provides the same result. Indeed, the number of "active" electrons per unit volume with states close to the Fermi surface is (cf. Eq. 7.13):

$$\frac{\Delta k}{k_F}\cdot\frac{N}{\Omega} = -\frac{et\mathcal{E}}{\hbar k_F}\frac{N}{\Omega} \tag{7.30}$$

Each electron has the velocity $\mathbf{v} = (\hbar k_F/m)$. We deduce, again, on the basis of Eq. (7.15) that

$$j = -e\rho\mathbf{v} = -e[-\frac{et\mathcal{E}N}{\hbar k_F\Omega}][\frac{\hbar k_F}{m}] = \frac{Ne^2t}{\Omega m}\mathcal{E} \tag{7.31}$$

7.2.5 Individual dynamics induced by a magnetic field

Reciprocal trajectory

If a magnetic field is applied to a free-electron gas, the dynamical equation (7.11) becomes:

$$\frac{d\vec{k}}{dt} = -\frac{e}{m}\vec{k}\times\vec{B} \tag{7.32}$$

in which $\vec{v}_g$ has been replaced by $(\hbar\vec{k}/m)$. Multiplying the two members of Eq. (7.22) by $\vec{k}$ we obtain: $\vec{k}.d\vec{k}/dt = 0$. This implies that the endpoint of $\vec{k}$ is on the constant-energy spherical surface $\hbar^2\vec{k}^2/2m = E$. On the other hand, since $(d\vec{k}/dt).\vec{B} = 0$, the endpoint of $\vec{k}$ lies in a plane perpendicular to $\vec{B}$.

This endpoint follows a circular *reciprocal trajectory* perpendicular to $\vec{B}$, (Fig. 7.4) and determined by the initial electron state $\vec{k}_0$. Its uniform motion on the circle, counter-clockwise around the magnetic field, is periodic with the frequency $\nu_c = (eB/2\pi m)$, termed the *cyclotron frequency*. For a field of magnitude 1 Tesla, this frequency is equal to 28 GHz.

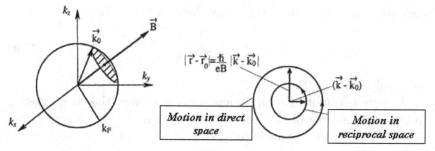

Fig. 7.4. Left: Reciprocal trajectory on the Fermi sphere. Right: Relationship between the reciprocal and direct trajectories.

When $\vec{k}_0$ is inside the Fermi sphere, the points of the trajectory determined by (7.32) correspond to occupied states. The evolution governed by Eq. (7.32) is then forbidden by the Pauli principle, which is not taken into account by the dynamical equation. Similarly to the case of an electric field, the only occupied states of the electron gas whose evolution is effectively described by (7.32) are the states lying close to the Fermi surface ($|\vec{k}| = k_F$).

Anticipating on Section 7.2.7, we can note that similarly to the case of a harmonic oscillator, the existence of a characteristic frequency ν_c in the classical dynamics denotes the occurrence of a characteristic energy-interval $\Delta\varepsilon = hf_c$ in the quantum spectrum of the system. On this basis, the validity of the classical approximation can be specified. It holds for temperatures T satisfying the condition:

$$T \gg \Theta_c = \frac{h\nu_c}{k_B} \qquad (7.33)$$

For $B = 1$ Tesla, $\Theta_c \approx 1$ K.

Direct trajectory

In direct space, the electron trajectory is a helix of axis $\vec{B}$. The electron moves along $\vec{B}$ with constant velocity $(m\vec{k}_0.\vec{B}/\hbar B)$. In a plane perpendicular to $\vec{B}$, its circular motion can be determined by integrating Eq. (7.32), with $\vec{v}_g = (\hbar\vec{k}/m)$:

$$\vec{k} - \vec{k}_0 = -\frac{e}{\hbar}(\vec{r} - \vec{r}_0) \times \vec{B} \qquad (7.34)$$

Denoting $\vec{r}_\perp$ the projection of $(\vec{r} - \vec{r}_0)$ on the plane perpendicular to $\vec{B}$, and observing that $(\vec{k} - \vec{k}_0) = \Delta \vec{k}$ is perpendicular to $\vec{B}$, we deduce from (7.34):

$$\vec{r}_\perp = -\frac{\hbar}{eB} \frac{\vec{B}}{|\vec{B}|} \times \Delta \vec{k} \qquad (7.35)$$

The circular projection of the direct trajectory is obtained (Fig. 7.4) by rotating the reciprocal trajectory around $\vec{B}$ by the angle $(+\pi/2)$, and applying a scaling factor $(\hbar/eB)$. Taking $|\Delta \vec{k}| \sim k_F$, and $B = 1$ Tesla, we find that the radius of the circular trajectory is $|\vec{r}_\perp| \approx (\sqrt{2mE_F}/eB) \approx 1$ micron. This value, being much larger than the size of the unit cell of the crystal, is therefore consistent with the classical approximation.

The circular component of the electron motion is associated to an electric current i and to a magnetic moment $\mu = i\pi r_\perp^2$ directed *opposite* to the magnetic field. This response of the electron gas is therefore of *diamagnetic* nature. This diamagnetism which derives from the classical motion of the electron expresses the *orbital magnetism* of the electron gas. As recalled in Chapter 6, Eq. (6.20), the electron gas also has a *paramagnetic* behaviour (i.e. generating a magnetic moment in the direction of the applied field) related to the electron spin, a quantity, which, being without a classical analog, is not taken into account in the classical approximation.

7.2.6 Collective dynamics of the electron gas

As emphasized by Eqs. (7.32) and (7.34) the solution of the dynamical equation in reciprocal or direct space implies the knowledge of initial states $\vec{k}_0$ or $\vec{r}_0$ for the electron. In a solid, these values are obviously unknown. The initial state is characterized by the equilibrium occupation factor $f(\vec{k})$, which specifies the statistical occupation of each $\vec{k}$-state, while in direct space the electron positions $\vec{r}_0$ are uniformly distributed. As shown by Eq. (7.21), the calculation of the current density as a function of the electric field does not make use of the individual initial states of the electrons and relies entirely on the knowledge of $\rho(\vec{k}) \propto f(\vec{k})$. Hence, Eq. (7.34) is unusefully

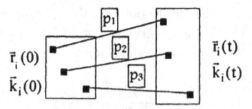

Fig. 7.5.   Conservation of the values of the probabilities during the statistical evolution of the electron gas.

accurate, and should be replaced by an equation governing the evolution of the statistical occupation factor (Fig. 7.5).

Let us derive such an equation in a more general framework in which the fields applied are not necessarily spatially uniform or time-independent. However, in order to preserve the validity of the classical approximation, the spatial variations of the field must occur on a spatial scale much larger than the size of the electron-wavepacket.

If an electron is localized, at time t, in the elementary volume $d^3\overrightarrow{r}$ in direct space (large as compared to the size of the wavepacket), one can associate to it a uniform density of states in $\overrightarrow{k}$-space equal to

$$g(\overrightarrow{k}) = \frac{d^3\overrightarrow{r}}{4\pi^3} \tag{7.36}$$

In the presence of external fields creating a situation differing from thermodynamic equilibrium, we can define an *instantaneous and space-dependent* occupation factor $\mathbf{f}(t, \overrightarrow{r}, \overrightarrow{k})$ characterizing the probability of occupation of the various $\overrightarrow{k}$-states. For instance, in the case considered in Section 7.2.4, of a constant and uniform applied electric field, we noted that the occupation of the $\overrightarrow{k}$-states is defined by $\mathbf{f}(t, \overrightarrow{r}, \overrightarrow{k}) = f(\overrightarrow{k} + \frac{e\overrightarrow{\mathcal{E}}}{\hbar}t)$, where f is the equilibrium Fermi-factor. The knowledge of $\mathbf{f}(t, \overrightarrow{r}, \overrightarrow{k})$ is sufficient to determine the physical quantities of interest. On the basis of Eq. (7.15) the current density can be expressed as

$$\overrightarrow{j}(t, \overrightarrow{r}) = -e \int_{\overrightarrow{k}} \overrightarrow{\mathbf{v}} . \mathbf{f}(t, \overrightarrow{r}, \overrightarrow{k}) \frac{d^3\overrightarrow{k}}{4\pi^3} \tag{7.37}$$

Let us derive the equation which determines $\mathbf{f}(t, \overrightarrow{r}, \overrightarrow{k})$. In this view, note that if an electron has the probability p to be in the initial state $[\vec{r}_0, \vec{k}_0]$, it will also have the probability p to be, at time t, in the state $[\vec{r}(t), \vec{k}(t)]$ determined by the dynamical equation (7.11) and by $[\vec{r}_0, \vec{k}_0]$. Consequently the value of the occupation factor $\mathbf{f}(t, \vec{r}(t), \vec{k}(t))$ is preserved during the motion of the electrons. By derivation of $\mathbf{f}(t, \vec{r}, \vec{k}) = f$ with respect to t, we obtain

$$\frac{\partial \mathbf{f}}{\partial t} + \frac{d\vec{r}}{dt} . \vec{\nabla}_r \mathbf{f} + \frac{d\vec{k}}{dt} . \vec{\nabla}_k \mathbf{f} = 0 \qquad (7.38)$$

where $\vec{\nabla}_r$ and $\vec{\nabla}_k$ are the gradients in direct and reciprocal space. Introducing the group velocity $\vec{v}_g = d\vec{r}/dt$ of the wavepacket and replacing $d\vec{k}/dt$ by its expression (7.11) as a function of the fields, we obtain the required evolution equation, which is the *Liouville equation*:

$$\frac{\partial \mathbf{f}}{\partial t} + \vec{v}_g . \vec{\nabla}_r \mathbf{f} + \frac{\vec{F}}{\hbar} . \vec{\nabla}_k \mathbf{f} = 0 \qquad (7.39)$$

As stressed above, its validity is the same as that of the classical approximation, i.e. it implies smooth variations of $\mathbf{f}$ and of the fields, at the scale of the size of the electron wavepacket.

Note that, in the situation of a constant and uniform electric field, $\vec{F} = -e\vec{\mathcal{E}}$ and $\vec{\nabla}_r \rho = 0$. One can check easily that Eq. (7.39) is satisfied by $\mathbf{f}(t, \vec{k}) = f(\vec{k} + et\vec{\mathcal{E}}/\hbar)$ in agreement with the result in Section 7.2.4.

7.2.7 Quantum levels of the electron gas in a magnetic field

In this paragraph, we consider the quantum properties of a free-electron gas. As anticipated in Eq. (7.33), such properties will only be observable at temperatures as low as a few degrees Kelvin.

Let us recall that (cf. Section 3.3.6), in the absence of applied magnetic field, the confinement of the electron gas in the volume of the solid determines quantized $\vec{k}$-states, whose density is uniform in reciprocal space. Their spacing along the i-direction of this space is $(2\pi/L_i)$, where L_i is the length of the solid in the i-direction.

Quantum states in a magnetic field

The electron gas in the absence of applied field being isotropic, let us take the axis $\overrightarrow{z}$ parallel to the magnetic field $\overrightarrow{B}$. It can be shown that the Hamiltonian of the electron submitted to a magnetic field depends indirectly of $\overrightarrow{B}$ through the vector potential $\overrightarrow{A}(\overrightarrow{r})$ by

$$\widehat{H} = \frac{(\widehat{\overrightarrow{p}} - e\overrightarrow{A})^2}{2m} \tag{7.40}$$

$\overrightarrow{B}$ is related to $\overrightarrow{A}$ through $\overrightarrow{B} = \overrightarrow{\nabla}_r \times \overrightarrow{A}$ (the notation in this paragraph being unambiguous, we will denote $\overrightarrow{\nabla}_r = \overrightarrow{\nabla}$). The preceding relationship defines $\overrightarrow{A}$ *up to the gradient of an arbitrary function of the space coordinates*. This arbitrariness, which has no consequence on the values of the measurable physical quantities, allows to make the choice $\overrightarrow{A} = (0, Bx, 0)$. The Schrödinger equation of the electron is then:

$$-\frac{\hbar^2}{2m}\left\{ \frac{\partial^2}{\partial x^2} + \frac{\partial^2}{\partial z^2} + \left[\frac{\partial}{\partial y} - i\frac{eB}{\hbar}x \right]^2 \right\}\psi = E\psi \tag{7.41}$$

In addition to Eqs. (7.41), the electron must satisfy, as for a free-electron, the Born–Von Karman conditions, which express its confinement within the volume of the solid.

A partial decoupling of variables in Eq. (7.41) is obtained by writing the solution in the form:

$$\psi = \exp(ik_z.z)\exp(ik_y y).u\left(x - \frac{\hbar k_y}{eB} \right) \tag{7.42}$$

The u function must then satisify the equation:

$$-\frac{\hbar^2}{2m}u'' + \frac{m\omega_c^2}{2}\left(x - \frac{\hbar k_y}{eB} \right)^2 u = \left(E - \frac{\hbar^2 k_z^2}{2m} \right)u \tag{7.43}$$

where $\omega_c = 2\pi\nu_c$ is associated to the cyclotron frequency ν_c defined in Section 7.2.5. One recognizes in Eq. (7.43) the Schrödinger equation of a linear harmonic oscillator (cf. Sections 1.2.2 and 10.3.2),

$$-\frac{\hbar^2}{2m}\phi''(X) + \frac{m\omega^2 X^2}{2}\phi = E'\phi \tag{7.44}$$

In which the origin of the displacements X is shifted by the quantity $(\hbar k_y/eB)$ and the origin of energies E is shifted by $(\hbar^2 k_z^2/2m)$. The quantized values of the energy of the linear atomic oscillator (Section 10.3.2) are $E'_n = (n + \frac{1}{2})\hbar\omega$. Hence, the quantized energies of the electron in a magnetic field must satisfy the relation:

$$E_n = \left(n + \frac{1}{2}\right)\hbar\omega_c + \frac{\hbar^2 k_z^2}{2m} \tag{7.45}$$

Up to now, the confinement of the electron in the volume of the solid has not been taken into account. Let us discuss its effects.

Along the x-direction, as shown by Eq. (7.43), the electron oscillates about the origin $(\hbar k_y/eB)$. Along this direction, it is therefore *confined in the solid* by the magnetic field provided the amplitude of the oscillation and the shift of the origin are not larger than the width of the sample along x. The Born–Von Karman condition is therefore meaningless in this direction.

Along the y- and z-directions the Born–Von Karman condition remains relevant. It determines a standard quantization of the k_i components $(k_i = 2\pi m_i/L_i)$.

Quantum numbers and degeneracy

To summarize the various considered effects, the state of the electron depends on the three quantum numbers n, k_y, and k_z. The first one, n, defines unambiguously the function u in Eq. (7.43). Indeed, as is well known, the energies of the linear harmonic oscillator are not degenerate and thus each n-value will correspond to a single function u_n. However, the knowledge of this function does not specify entirely the probability of presence of the electron along x. The quantized value of $k_y = (2\pi m_y/L_y)$ determines the center of the spatial distribution of this probability. Finally, in the y- and z-directions, the electron is described by a plane wave associated to the quantized value $(k_z = 2\pi m_z/L_z)$. Such a result is in agreement with the classical treatment hereabove which has emphasized the uniform motion of the electron along the magnetic field.

The energy only depends on the two quantum numbers n and k_z. The energy-spacing relative to k_z is very small and, for $L_i \sim 1$mm, is, for instance, of the order of $(4\pi^2\hbar^2/2mL_i^2)^\sim 10^{-12}$eV. The energy spacing corresponding to the n-quantum number is, for $B = 1$ Tesla, of the order of 10^{-4}eV. The structure of the spectrum is therefore complex. It involves on the one hand a quasi-continuous spectrum determined by the $\hbar^2 k_z^2/2m$ levels, and, on the other hand, regularly spaced discrete levels, called *Landau levels,* corresponding to the n quantum number and whose spacing, proportional to $\omega_c = eB/m$, depends on the magnetic field.

Since the energy does not depend of the third quantum number k_y, each energy level is degenerate. Since $\hbar k_y/eB = 2\pi\hbar m_y/eBL_y$ specifies the position of the center of the oscillations along x, the number m_y of possible values of k_y is determined by the width of the sample along x. Hence:

$$\frac{2\pi\hbar m_y}{eBL_y} = L_x \quad \Rightarrow \quad m_y = \frac{eBL_xL_y}{2\pi\hbar} \qquad (7.46)$$

For a sample of 1 cm^2 and a magnetic field of 1 Tesla, the degeneracy is 10^{10}, and is independent of the energy-level considered.

Oscillatory behaviour of the physical quantities

Figure 7.6 shows the resulting density of states $g(E)$ of the electron in a magnetic field. The smooth parabolic variation relative to the free-electron, due to the uniform quantization along the three reciprocal axes, is replaced by a series of peaks superimposed on the free-electron curve. The peaks are due to the fact that the quantum states relative to the k_x- and k_y-directions are no longer uniformly

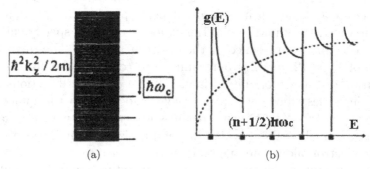

(a)   (b)

Fig. 7.6.   (a) Quantum spectrum of an electron in a magnetic field. (b) Resulting denity of states. The density for zero magnetic field is represented by the dashed line.

distributed on the energy scale but rather concentrated around the values $(n+\frac{1}{2})\hbar\omega_c$. The density of states $g(E)$ diverges for these values of the energy.

As shown by Eq. 6.14, the values of quantities such as the electronic specific heat or the magnetic susceptibility are proportional to the magnitude of the density of states $g(E_F)$ at the Fermi energy E_F. Clearly these values will differ notably, depending if the Fermi level is close to a singularity of $g(E)$, or in the interval between singularities.

Let us consider the electron gas in a metal at $T = 0$ K. Note that the Fermi energy E_F amounts to a few electron-volts while the interval between singularities is $\approx 10^{-4}$ eV. Hence, as apparent on Figure 7.6, the surface below the curve $g(E)$, for $0 < E < E_F$, which is equal to the number of electron states, is the same in the absence or in the presence of the magnetic field. We can therefore assume that the value of the Fermi energy E_F is independent of the field. Since the location of the singularities of $g(E)$ at $E = (n + \frac{1}{2})\hbar eB/m$ can be displaced continuously by modifying the magnitude of the field, one expects an oscillation of the value of $g(E_F)$, and accordingly an oscillation of the values of the quantities which depend critically of $g(E_F)$ (Fig 7.7).

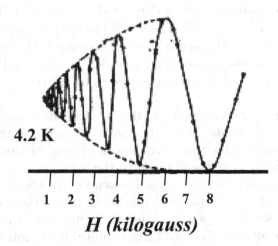

4.2 K

1   2   3   4   5   6   7   8

H (kilogauss)

Fig. 7.7.   Oscillations of the value of the diamagnetic susceptibility as a function of the magnetic field (De Haas–Van Alphen effect).

If the temperature of the metal is $T \neq 0$, the Fermi factor $f(E)$ does not vanish abruptly at E_F. It decreases continuously on an interval of width $\sim k_B T$. The expected oscillations will then be attenuated. Their observation requires that $k_B T \lesssim \hbar \omega_c$, a condition expressing that the thermal energy be smaller than the interval between singularities.

It is worth pointing out that in ordinary metals the orbital diamagnetic effects related to the occurrence of Landau levels are expected to be hidden by the stronger paramagnetic effects related to the spin. Thus the paramagnetic electronic susceptibility is

$$\chi = (2\mu_B^2 / k_B T) \tag{7.47}$$

while the diamagnetic one is shown on Fig. 7.7. However, in certain elements such as bismuth (cf. Section 6.5.2) the effective mass of the electron is exceptionally small, and accordingly the cyclotron frequency is large.

7.3　Collective Dynamics of Bloch-Electrons

We consider now the case of non-interacting electrons submitted to the periodic crystal potential $V(\vec{r})$. In the absence of external field, the quantum state of each electron is described by a Bloch wavefunction $\psi_{n,k}(\vec{r}) = u_{n,k}(\vec{r}) \cdot \exp(i \vec{k} \cdot \vec{r})$ whose precise form is determined by $V(\vec{r})$. In the presence of an external field, one expects a complicated motion of the electron which is submitted simultaneously to the field and to the forces exerted by the constituents of the crystal.

A natural generalization of the method developed for free-electrons would be to write, in the framework of the classical approximation of quantum mechanics, a dynamical equation relating the momentum of the electron to the *total force* resulting from the external fields and the crystal potential. However, although a "wavepacket of Bloch functions" can be defined in the same manner as a wavepacket of planewaves, such a formulation is not possible. Indeed, as denoted by Eq. (7.12), the applied forces must be almost uniform over the size of the wavepacket in direct space, and henceforth, over many unit cells of the crystal. This condition is obviously not fulfilled by the crystal potential $V(\vec{r})$ which varies rapidly at the

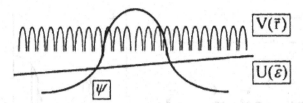

Fig. 7.8. Various scales considered in the semi-classical approximation. The width of the wavepacket is large as compared to the scale of variations of the crystal potential, but small with respect to that of the external electric field.

scale of a single unit cell and is therefore highly non-uniform over the size of a wavepacket. Thus, *with respect to $V(\vec{r})$ the electron cannot be considered as a classical particle.* If a classical dynamical equation can be built, it can only be valid for the external forces whose spatial range can be controlled. One will then necessarily *have a dual description* involving a classical dynamics with respect to the external forces and a quantum dynamics with respect to the crystal potential (Fig. 7.8). It is the latter framework which defines the so-called *semi-classical approximation.*

7.3.1 Origin of the apparent change of mass of the electron

In order to analyze qualitatively the complexity of the electron motion in the semi-classical description of its dynamics, let us consider the one-dimensional model of a quasi-free electron described in Section 5.3 and Section 5.2.2, which we recall schematically on Figure 7.9. As shown, the electron is adequately described by a plane-wave circulating from left to right, if its wavevector k is distant enough from the boundary of the first Brillouin zone. In by contrast, if k is close to this boundary, the quantum state of the electron involves an additional "reflected wave" generated by the constituents of the crystal.

Let us place ourselves in the situation where the value of $k = k_l$ lies at the borderline between the two above regimes, the single plane-wave constituting a good approximation of its state. The application of an external electric field $\vec{\mathcal{E}}$ directed from right to left will tend to accelerate the electron rightwards. Accordingly, the value of k increases to $k' = k_l + \Delta k$, and brings the state of the electron in a

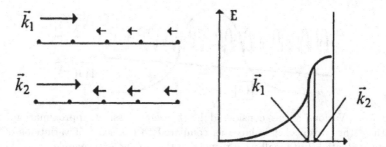

Fig. 7.9. K-vector of the electron close to the limit of the first Brillouin zone: its acceleration is in the direction opposite to that of the applied force.

range of k-values for which the reflected wave has a significant amplitude. Consequently, the application of the field induces a fraction of electron-state circulating from right to left, opposite to the incoming direction of the electron. The resulting variation of the momentum is thus of a sign opposite to that of the applied force. If we put $(d\vec{p}/dt) = m^*\vec{F}$, we have $m^* < 0$. The electron whose state is close to the boundary of the Brillouin zone behaves as if its mass is negative with respect to the effect of the external field.

Clearly, this situation is caused by the quantum diffraction of the electron wavefunction due to the crystal potential, which is increased by the change of k value imposed by the external field. Hence, one can understand that a classical dynamical equation applicable to the external forces, will involve an "effective mass" representing the quantum effects of the crystal potential. One can expect that such a mass will differ from the actual mass of the electron, and will depend on the value and on the direction of $\vec{k}$.

7.3.2 Semi-classical dynamical equation

In a somewhat paradoxical way, despite the complexity of the crystal potential, the semi-classical dynamics of the Bloch-electron will be shown to be governed by the same equation as the free-electron. Namely, the time-dependence of the $\vec{k}$-vector is provided by Eq. (7.11). However, the meaning of this equation is different from that of the free electron. On the one hand, the force $\vec{F}$ appearing in the second member of the equation is not the total force acting on the particle but only the *external force*. On the other hand, the vector $\overrightarrow{k}$ which is the vectorial quantum number attached to the Bloch function, is not proportional to the momentum $\overrightarrow{p}$ of the particle. Actually, it is easy to check that a Bloch function, which is an eigenstate of the translation operators (cf. Section 3.2.2), with the eigenvalues being related to $\overrightarrow{k}$, is not an eigenfunction of the momentum operator $(\hbar/i)\overrightarrow{\nabla}$.

If a classical equation existed, it would relate the momentum to the total force applied to the electron. The semi-classical one relates $\overrightarrow{k} \neq \overrightarrow{p}$ to $\overrightarrow{F}_{external}$.

Derivation of the semi-classical dynamical equation

Let us establish this equation in the case of an external electric field. The proof given below clarifies the evolution of a Bloch function in the presence of such a field. It involves two ingredients. On the one hand, it makes use of the translation operators τ defined in Eq. (3.3). On the other hand, it relies on the fact that the potential associated to the electric field is proportional to the positional quantum operator $\widehat{\overrightarrow{r}}$.

Commutation of translation and field operators. Consider an electric field whose spatial variations are slow enough to be compatible with the classical approximation (cf. Eq. 7.12). Such a field can be considered as *uniform* at the scale of the variations of the wavefunction ψ. Hence, we can write the electron-Hamiltonian as:

$$\widehat{H} = \frac{\widehat{\overrightarrow{p}}^2}{2m} + \widehat{V}(\overrightarrow{r}) + e\overrightarrow{\mathcal{E}}.\widehat{\overrightarrow{r}} = \widehat{H}_0 - \overrightarrow{F}.\widehat{\overrightarrow{r}} \qquad (7.48)$$

where $V(\overrightarrow{r})$ is the crystal potential. The translation operators τ, defined in Eq. (3.3), commute with the "internal", periodic, part $\widehat{H}_0$ of the Hamiltonian. Let us show that:

$$[\widehat{H}, \tau] = [\widehat{H} - \widehat{H}_0, \tau] = [-\overrightarrow{F}.\widehat{\overrightarrow{r}}, \tau] = -(\overrightarrow{F}.\overrightarrow{T})\tau \qquad (7.49)$$

where $\overrightarrow{T}$ is the primitive crystal translation associated to τ. In this view, apply the second member of Eq. (7.49), which is equal to $[-\overrightarrow{F}.\widehat{\overrightarrow{r}}, \tau]$, to an arbitrary wavefunction $\psi(\overrightarrow{r})$. We can write, on the basis of Eqs. (3.2)–(3.3):

$$[-\overrightarrow{F}.\widehat{\overrightarrow{r}}, \tau]\psi(\overrightarrow{r}) = -\overrightarrow{F}.\widehat{\overrightarrow{r}}.\psi(\overrightarrow{r} - \overrightarrow{T}) + \overrightarrow{F}.(\overrightarrow{r} - \overrightarrow{T}).\psi(\overrightarrow{r} - \overrightarrow{T})$$

$$= -(\overrightarrow{F}.\overrightarrow{T}).\tau\psi(\overrightarrow{r}) \qquad\qquad (7.50)$$

Evolution of the quantum average of the translation operator. Let us apply to τ the Ehrenfest equation of quantum mechanics, which governs the evolution of quantum operators. We have, since $\widehat{H}_0$ commute with τ:

$$\frac{d\langle\tau\rangle}{dt} = \frac{i}{\hbar}\langle[\widehat{H},\tau]\rangle = -\frac{i}{\hbar}\overrightarrow{F}.\overrightarrow{T}\langle\tau\rangle \qquad\qquad (7.51)$$

by integration of this equation we obtain:

$$\langle\tau\rangle = \langle\tau\rangle_0.\exp\left(-\frac{i\overrightarrow{T}}{\hbar}\int_0^t \overrightarrow{F}.dt\right) \qquad\qquad (7.52)$$

Evolution of a Bloch function. Assume that the quantum state of the electron is described by $\psi_{k,n}$ in the absence of applied field. The initial value of the translation operator is then

$$\langle\tau\rangle_0 = \langle\psi_{k,n}|\tau|\psi_{k,n}\rangle = \exp(-i\overrightarrow{k}.\overrightarrow{T}) \qquad\qquad (7.53)$$

Applying the field $\overrightarrow{\mathcal{E}}$ at time $t = 0$, the state of the electron at time t is ψ' which we can express as a linear combination of Bloch functions $\psi_{\kappa,\nu}$. Thus:

$$\langle\tau\rangle = \sum_{\kappa,\nu,\kappa',\nu'} A^*_{\kappa',\nu'}A_{\kappa,\nu}\langle\psi_{\kappa',\nu'}|\tau|\psi_{\kappa,\nu}\rangle \qquad\qquad (7.54)$$

Using the fact that the $\psi_{\kappa,\nu}$ are eigenfunctions of τ, and the mutual orthogonality of the Bloch functions, we deduce from Eq. (7.52):

$$\langle\tau\rangle = \sum_{\kappa,\nu} A_{\kappa,\nu}\exp(-i\overrightarrow{\kappa}.\overrightarrow{T}) = \exp\left(-i\overrightarrow{T}.[\overrightarrow{k} + \frac{1}{\hbar}\int_0^t \overrightarrow{F}.dt]\right)$$

$$(7.55)$$

This equation, which is valid for any translation $\vec{T}$, has the consequence that the sum over the vectors $\vec{\kappa}$ in the first member is reduced to a single term corresponding to:

$$\vec{\kappa}(t) = \vec{k} + \frac{1}{\hbar} \int_0^t \vec{F}.dt \tag{7.56}$$

Hence, at any time of its evolution, the electron is in a state defined by the preceding unique translational quantum number. By derivation of $(\vec{\kappa} - \vec{k}) = \Delta\vec{k}$ in Eq. (7.56), we obtain finally the required dynamical equation, identical to (7.11):

$$\hbar \frac{d\vec{k}}{dt} = \vec{F} \tag{7.57}$$

It is worth pointing out that, although its $\vec{k}$-vector is uniquely defined, the electronic state needs not be a Bloch function. As apparent in Eq. (7.55) it can be a sum of Bloch-functions corresponding to several band indices. The evolution has therefore a certain probability of transfering the electron from the initial band to another band. It can be shown, however, that this probability becomes significant only if the energy communicated to the electron by the field is of the same order of magnitude as the energy separation between bands.

On the basis of the identity between Eqs. (7.11) and (7.57) we could infer that Eq. (7.39), governing the evolution of the occupation factor and derived from (7.11) in the case of freeelectrons, also holds for Bloch electrons. This is not strictly true. Indeed, the validity of Eq. (7.39) relies on the fact that the evolution of the electron state can be entirely described by the evolution of the corresponding $\vec{k}$-vector. This requires the electron to remain in the same band and therefore excludes the case of strong applied fields. We will see in Chapter 8 that such a condition is satisfied by the evolutions realized in transport experiments, and that consequently, Eq. (7.39) can be considered valid.

7.3.3   Dynamical effective mass

As already stressed, for Bloch electrons, Eq. (7.11) is not equivalent to the equation $d\vec{p}/dt = \vec{F}$ governing the evolution of the momentum. However, by analogy with the case of freeelectrons we write:

$$\overline{\overline{m^*}} \cdot \frac{d\overrightarrow{v}_g}{dt} = \overrightarrow{F} \qquad (7.58)$$

Such an equation can seem meaningless since it relates the electron velocity to a *small fraction* of the forces acting on the electron. Indeed, the potential of the applied field is generally much smaller than the crystal potential. Let us however consider, within the approximation of quasi-free electrons, an electronic state whose $\overrightarrow{k}$-vector lies far from a Brillouin plane. We know that the crystal potential has a negligible influence on the electron-state. The motion of the electron is then entirely determined by the external forces. Equation (7.58) can be used with a scalar effective mass equal to the real mass of the electron. As already mentioned in Chapter 6, such a situation is realized for conduction electrons in alcaline or alcaline-earth metals. In other situations, the use of Eq. (7.58) has the advantage of allowing a simple study of the dynamics of electrons, the effective mass being a parameter which can be determined experimentally. Since this quantity incorporates the effect of the forces which are not represented by $\overrightarrow{F}$, the deviations from the real mass will reflect the influence of the crystal potential on the considered electron state. In general, $d\overrightarrow{v}_g/dt$ having no reason to be parallel to $\overrightarrow{F}$, the effective mass $\overline{\overline{m^*}}$ is a tensorial quantity which depends of the quantum numbers $\overrightarrow{k}$ and n. Let us derive its expression as a function of the characteristics of the spectrum $E(\overrightarrow{k}, n)$. In this view, we refer to the expression in Eq. (7.2) $\overrightarrow{v}_g = \overrightarrow{\nabla}_k E(\overrightarrow{k}, n)/\hbar$, of the group velocity and write, for each coordinate $v_{g,i}$:

$$\frac{dv_{g,i}}{dt} = \frac{1}{\hbar} \overrightarrow{\nabla}_k \left(\frac{\partial E(\overrightarrow{k}, n)}{\partial k_i} \right) \cdot \frac{d\overrightarrow{k}}{dt} = \frac{1}{\hbar} \sum_j \frac{\partial^2 E}{\partial k_i \partial k_j} \cdot \frac{dk_j}{dt} \qquad (7.59)$$

The vector $\hbar(d\overrightarrow{v}_g/dt)$ is the product of the vector $d\overrightarrow{k}/dt$ by the symmetric second rank tensor formed by the 9 partial derivatives $\partial^2 E/\partial k_i \partial k_j$. Inserting (7.59) into Eq. (7.58) and replacing $(\overrightarrow{F})$ by its expression Eq. (7.57), we obtain:

$$\frac{1}{\hbar}\left\{\overline{\overline{m^*}}.\overline{\nabla_k.\nabla_k}E(\vec{k},n)\right\}\frac{d\vec{k}}{dt} = \hbar\frac{d\vec{k}}{dt} \qquad (7.60)$$

Eliminating $(d\vec{k}/dt)$ we obtain the expression of the symmetric second rank tensor associated to the effective mass:

$$\left(\overline{\overline{m^*}}\right)_{\lambda\mu} = \left(B^{-1}\right)_{\lambda\mu} \quad \text{with} \quad \left(\overline{\overline{B}}\right)_{ij} = \frac{1}{\hbar^2}\left(\frac{\partial^2 E(\vec{k},n)}{\partial k_i \partial k_j}\right) \qquad (7.61)$$

The matrix formed by the components of the effective mass is proportional to the inverse matrix formed by the second derivatives of the energy in $\vec{k}$-space. For freeelectrons, the energy is $E = \hbar^2 \vec{k}^2/2m$. We easily verify that, in this case, Eq. (7.61) yields $m^* = m$. In the general case, whenever the crystal potential $V(\vec{r})$ cannot be neglected, the effective mass is related to the curvature of the dispersion curves $E(\vec{k},n)$ in each direction of the reciprocal space, and its value and sign can be different from those of the real mass of the electron.

Note that this concept will only be of interest if the value of m^* keeps a constant value in the range of $\vec{k}$-values considered. This implies an approximately constant curvature of $E(\vec{k},n)$ which will only be realized at a maximum or a minimum of a band. As shown on Fig. 7.10, for quasi-free electrons in the vicinity of a Brillouin plane the curvature and the effective mass are *negative* in agreement with the qualitative discussion in Section 7.3.1, while in the center of the Brillouin zone it is positive and close to its real value, since the crystal potential has little effect in this k-range. For strongly bound electrons lying in bands well described by the tight binding approximation, large variations of the value of m^* can occur.

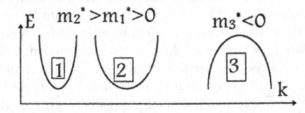

Fig. 7.10. Curvature of $E(\vec{k})$ and effective electron mass.

Finally, it is worth pointing out that the effective mass defined here (Eq. 7.61) on the basis of the dynamical equation coincides with the *static effective mass* defined in Section 6.5.2.

7.3.4   Trajectories of Bloch electrons in a field

The application of the dynamical equation to Bloch electrons has common aspects with its application to freeelectrons, as well as important differences. In the following paragraphs, we will dwell on the situation in metals. The case of semiconductors will be discussed in Chapter 9.

Effect of an electric field in a metal

Two features distinguish the analysis developed hereafter from that applied in Section 7.2.3 to free electrons. On the one hand, the velocity $\overrightarrow{v}_g$ which appears in Eqs. (7.23)–(7.25) is not equal to $(\hbar\overrightarrow{k}/m)$ but to $(\overrightarrow{\nabla}E/\hbar)$. On the other hand, the Fermi surface, which separates in reciprocal space the occupied states from the empty states is not the sphere $|\overrightarrow{k}| = k_F$.

In contrast, a result common to free and Bloch-electrons, since it is exclusively based on the dynamical Eq. (7.57), is that an electric field will induce a uniform drift of $\overrightarrow{k}$-states parallel to the electric field. The generalization of the expression of the current density is obtained by substituting in (7.23) the correct expression of the group velocity:

$$\overrightarrow{j} = -\frac{e}{\hbar}\int_{\overrightarrow{k}}\overrightarrow{\nabla}_k E(\overrightarrow{k},n)[\rho(\overrightarrow{k}) - \rho_0(\overrightarrow{k})].d^3\overrightarrow{k} \qquad (7.62)$$

The non-spherical shape of the Fermi surface modifies the value of the above integral and makes it more difficult to evaluate. However, in the absence of field, it is not necessary to calculate this integral to show that the current density is equal to zero. Such a result derives from the fact that the bands $E(\overrightarrow{k},n)$, and correlatedly, the Fermi surface, are always symmetric with respect to the origin (cf. Section 3.4.1). Consequently, each state occupied by an electron with velocity $\overrightarrow{v}_g \propto \overrightarrow{\nabla}E$ can be put in correspondance with another occupied state with velocity $\overrightarrow{v'}_g \propto -\overrightarrow{\nabla}E$.

Likewise, in the presence of a field, the *asymmetry* of the distribution of occupied states is sufficient to conclude that the current density is non-zero, and that it grows as a function of time. More precisely, we can deduce the group velocity from Eq. (7.61) by integration on the time variable:

$$\overrightarrow{\mathbf{v}}_g(t) = \overrightarrow{\mathbf{v}}_g(0) - e\left(\overline{\overline{m^*}}\right)^{-1}\overrightarrow{\mathcal{E}}.t \qquad (7.63)$$

The contribution of $\overrightarrow{\mathbf{v}}_g(0)$ to the current density being zero, we obtain, by inserting (7.63) into (7.62):

$$\overrightarrow{j} = e^2.t\left\{\int_{\overrightarrow{k}}\left(\overline{\overline{m^*}}\right)^{-1}[\rho(\overrightarrow{k}) - \rho_0(\overrightarrow{k})].d^3\overrightarrow{k}\right\}\overrightarrow{\mathcal{E}} \qquad (7.64)$$

Due to the symmetry of $E(\overrightarrow{k}, n)$, the contribution of ρ_0 vanishes. Finally:

$$\overrightarrow{j} = \frac{Ne^2t}{\Omega}\left\langle(\overline{\overline{m^*}})^{-1}\right\rangle\overrightarrow{\mathcal{E}} \qquad (7.65)$$

where $\langle(\overline{\overline{m^*}})^{-1}\rangle$ is the statistical average over all the occupied states in $\overrightarrow{k}$-space of the inverse of the effective mass matrix. Clearly, this expression generalizes the one (Eq. 7.21) obtained for free electrons. It appears that, in general, *the current density will not be in the direction of the field* and that its magnitude will depend on this direction.

As in Section 7.2.4, we can express the current density in a manner to emphasize the irrelevance of states which do not lie in the neighbourhood of the Fermi surface. The generalization of Eq. (7.25) is

$$\rho - \rho_0 = -\frac{1}{4\pi^3}\frac{df(E)}{dE}\overrightarrow{\nabla}_k E.\Delta\overrightarrow{k} \qquad (7.66)$$

For a metal this expression is non-zero only in the neighbourhood of the Fermi surface, since $(df/dE)^\sim - \delta(E - E_F)$. Inserting this expression in (7.62), and using (7.27):

$$\vec{j} = -\frac{1}{4\pi^3}\frac{e}{\hbar}\int_{\vec{k}} \vec{\nabla}_k E(\vec{k},n).\delta(E-E_F)\vec{\nabla}_k E.\Delta\vec{k}.d^3\vec{k} \qquad (7.67)$$

which yields

$$\vec{j} = \frac{e^2\hbar^2 t}{3\Omega}.v_F^2.n(E_F)\vec{\mathcal{E}} \qquad (7.68)$$

Effect of a magnetic field. For Bloch electrons, Eq. (7.32) is replaced by

$$\frac{d\vec{k}}{dt} = -\frac{e}{\hbar}\vec{\nabla}_k E \times \vec{B} \qquad (7.69)$$

The conclusions drawn from this equation are the same as for free electrons. The reciprocal trajectory, which describes the motion of the endpoint of $\vec{k}(t)$, is the intersection of the surface of constant energy corresponding to $\vec{k}(0)$, with the plane perpendicular to $\vec{B}$, and passing through $\vec{k}(0)$. Indeed, $d\vec{k}/dt$ is normal to $\vec{\nabla}_k E$ and therefore tangent to the surface $E = E_0$. To be compatible with the Pauli principle, the trajectory must pass through unoccupied states. It must therefore lie in the neighbourhood of the Fermi surface.

Unlike the case of free electrons for which the trajectory is necessarily circular, the trajectories for Bloch electrons can have various shapes depending on the form of the Fermi surface and of the direction of the $\vec{k}_0$-vector characterizing the initial quantum state. As illustrated by Fig. 7.11, such a trajectory can be open or closed, while for free electrons trajectories are always closed.

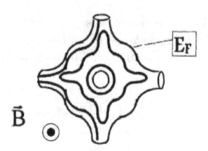

Fig. 7.11.   Shapes of the trajectories obtained by cutting the Fermi surface by a plane perpendicular to $\vec{B}$.

Another difference concerns the rotation frequency on closed trajectories. Whenever the effective mass can be considered as constant along the reciprocal trajectory, Eq. (7.58) yields:

$$m^* \frac{d\overrightarrow{\mathbf{v}}_g}{dt} = -e\overrightarrow{\mathbf{v}}_g \times \overrightarrow{B} \qquad (7.70)$$

which defines a cyclotron frequency $\nu_c = (eB/2\pi m^*)$ dependent on the value of the effective mass. If the mass is variable along the trajectory, ν_c is equal to:

$$\nu_c = \frac{eB}{\hbar^2} \cdot \frac{1}{\partial E/\partial S} \qquad (7.71)$$

where S is the area of the reciprocal trajectory.

A last difference concerns the sense of rotation on closed reciprocal trajectories. This sense is determined by (7.57). It will therefore depend of the fact that the projection of $\overrightarrow{\mathbf{v}}_g = \overrightarrow{\nabla}_k E/\hbar$ on the plane normal to $\overrightarrow{B}$ is directed inwards or outwards the closed trajectory considered. In the first situation, the sense of rotation is the same as for freeelectrons. In the second one, the sense is opposite. Figure 7.12 shows that the two situations correspond respectively to a Fermi surface which "bulges" or "indents" in the direction of the electric field.

For both senses of precession the response of the solid to the application of the magnetic field is of diamagnetic nature as for freeelectrons. This apparently paradoxical result will be clarified in the next paragraph. It is due to the fact that motions occurring in the opposite sense as compared to freeelectrons can be considered as motions of positive particles (holes) occupying the corresponding electron states.

Equation (7.35) which determines the projection of the trajectory in *direct space* for freeelectrons, also holds for Bloch electrons. Hence, this projection is again obtained from the (complicated) reciprocal trajectory by a rotation of angle $(+\pi/2)$ and a scaling by a factor $(\hbar/eB)$ (Fig. 7.12(b)).

The cyclotron frequency ν_c relative to closed trajectories can be measured through a *cyclotron resonance* experiment (Fig. 7.13), in which the considered solid is submitted to an alternating electric field normal to the direction of the magnetic field. Such a field, thus situated in the plane of the reciprocal trajectory can be decomposed

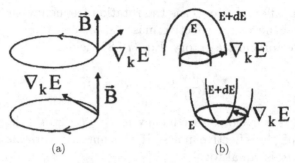

(a)   (b)

Fig. 7.12.   (a) Direction of motion of the electron depending on the direction of
the projection of $\vec{v}$. (b) Corresponding local curvatures of the Fermi surface.

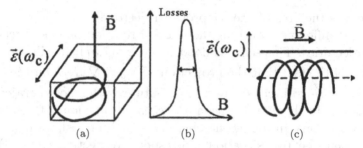

(a)   (b)   (c)

Fig. 7.13.   (a) Schematic organization for measuring the cyclotron resonance. (b)
Shape of the resonance curve. Its width depends on the relaxation time related
to the collisions (see Section 8.2). (c) Measurement in the case of a metal.

into the sum of two rotating fields turning respectively in the same
sense as the electron and in the opposite sense. If the frequency of
the alternating field is equal to the cyclotron frequency the electron will be accelerated by the rotating field on its trajectory. It will
then absorb energy from the source of the electric field. In practice,
the experiment generally consists in applying an electric field of set
frequency, and varying continuously the magnitude of the magnetic
field. One observes a peak of absorption at the resonance. In metals,
a difficulty arises from the fact that the "skin effect" confines the
high frequency electric field $(> 10^{10} Hz)$ to a thin layer close to the
surface of the sample. One is then led to use a setup (Fig 7.13(c))
in which the electric field is parallel to the surface of the solid. In
this configuration, electrons are only accelerated at each period of
their motion in the portion of their trajectory which is close to the

surface. An acceleration of electrons on the entire trajectory can be obtained in the case of semiconductors.

As will be explained in Chapter 8, the motion of electrons is interrupted by collisions. This effect induces a widening of the resonance which increases with the frequency of the collisions. As this frequency decreases on lowering the temperature, one is led to perform the resonance experiments at low temperatures.

7.3.5 Holes

An energy band whose states are all occupied in the absence of field (totally filled band) is "inert" with respect to the application of an electric or a magnetic field. Indeed, due to their symmetry, bands contain pairs of states $\vec{k}$ and $(-\vec{k})$ having the same energy. If these states are both occupied, their contribution to the current density $\vec{j}$ or to the diamagnetic response cancel. Totally filled bands are not involved in the conduction properties of a solid due to the lack of available empty states in the vicinity of the occupied bands (Fig. 7.14).[1]

If a band is *almost filled,* the only occupied $\vec{k}$-states which can contribute to the dynamical physical properties will be those for which $(-\vec{k})$ is not occupied. This criterion does not result from the dynamical equation. *It is a quantum effect of the crystal potential* which structures the set of the $\vec{k}$-states into bands containing a finite number of states. Hence, in a similar manner as the concept of effective mass, the concept of *hole* is introduced in the view of complementing the semi-classical dynamical equation. It allows a more convenient study of the collective dynamics of the electrons in a filled band without having to consider all the states in the band whose dynamics compensate mutually.

We define a hole as an empty state within a band. As we now proceed to show, such a state can be considered as a particle possessing a positive charge $+e_h$, a $\vec{k}_h$-vector specifying its quantum state, a velocity $\vec{v}_h$, and an effective mass m_h^*.

---

[1]The same limitation, invoked in Chapter 6, and pertaining to the fact that the field must not be too large, holds here. Indeed, we have stressed above that a strong field could induce a transfer of electrons from one band to another.

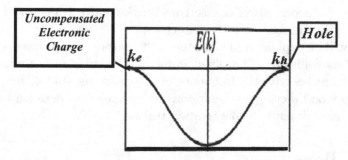

Fig. 7.14. Energy band containing one empty state at the upper limit of the band. The non-compensated electron occupies the symmetric state with the same energy.

Consider a band containing a single hole, i.e. a single unoccupied state $\overrightarrow{k}_0$. This state is associated to a value of the electron velocity $\overrightarrow{v} = \overrightarrow{\nabla} E_k(\overrightarrow{k}_0)/\hbar$, through Eq. (7.2).

(i) The inertia of a totally filled band with respect to the application of an electric field leads to consider this band as *electrically neutral*. The negative charge of the $2N_M$ electrons contained in this band is compensated by the positive charge of the crystal ions. Hence, the presence of a hole, consisting in the withdrawal from the band of one electron without change of the positive environment, determines a positive non-compensated charge $+e$. This *charge is assigned to the hole*.

(ii) Each band being symmetric, in reciprocal space, towards the transformation $\overrightarrow{k} \rightarrow (-\overrightarrow{k})$, i.e. $E(\overrightarrow{k}, n) = E(-\overrightarrow{k}, n)$, the sum of all the $\overrightarrow{k}$-vectors in a given band vanishes. Hence, the sum of $\overrightarrow{k}$-vectors of the occupied states of a band involving one hole is therefore equal to $(-\overrightarrow{k}_0)$. *This sum is the vectorial quantum number assigned to the hole.*

(iii) The current density generated by the band is equal to the current density associated to the single electron occupying the non-compensated state $(-\overrightarrow{k}_0)$ (all the other occupied states have cancelling contributions). Thus:

$$\overrightarrow{j} = -\frac{e}{\hbar}\overrightarrow{\nabla}_k E(-\overrightarrow{k}_0, n) = +\frac{e}{\hbar}\overrightarrow{\nabla}_k(\overrightarrow{k}_0, n) \qquad (7.72)$$

where the last term is determined by the symmetry of the band (the energies at $\vec{k}$ and $-\vec{k}$ are equal but their gradients are opposite). The charge of the hole being $+e$, and the current density (7.72) being assigned to the hole, and being also equal to $+e\vec{v}_h$, we can deduce $\vec{v}_h = \vec{\nabla}_k(\vec{k}_0, n)/\hbar = \vec{v}$. The velocity of the hole is the same as that associated to the unoccupied state of the band.

(iv) Let us assume the validity for holes of Eq. (7.61) defining the effective mass, and compare its application to a hole and to the corresponding unoccupied electron state. We obtain:

$$m_h^* \frac{d\vec{v}_h}{dt} = +e\vec{E} + e\vec{v}_h \times \vec{B} \tag{7.73}$$

$$m^* \frac{d\vec{v}}{dt} = -e\vec{E} - e\vec{v} \times \vec{B} \tag{7.74}$$

This comparison shows that the effective mass which must be assigned to the hole is $m_h^* = -m^*$. The frame below summarizes the correspondance between the characteristics of the hole and that of the unoccupied state:

$$q = +e \quad \vec{k}_h = -\vec{k} \quad \vec{v}_h = \vec{v} \quad m_h^* = -m^* \tag{7.75}$$

(v) Finally, based on the definition of the occupation factor, we can infer that a hole satisfies, as does an electron, the Fermi statistics. However, the occupation factor of a "hole state", i.e. the probability for a state to be empty is $(1-f)$ while this factor is f for an electron.

The concept of *hole* provides an easy interpretation of the experimental fact, revealed by certain Hall measurements (cf. Chapter 1), that in some solids the conduction seems due to positive charges. In these solids, the conduction band is more than half filled. The Fermi level is then located in an energy range where the band curvature is negative, and thus the *effective electron mass* is negative. The force exerted by a magnetic field on the current will have the same direction as if this current is due to the motion of positive charges. The study of the action of the electric and magnetic fields on the entire set of conduction electrons can then be replaced by the study of the motion of the holes present in the band.

Chapter 8

Electronic Transport Properties of Solids

Main ideas: Drude model and its shortcomings (Section 8.2). Collisions of electrons with collective oscillations and defects (Section 8.2.4). Origins of electrical and heat conductivities (Section 8.4)

8.1 Introduction

For Bloch electrons as for free-electrons, Eqs. (7.21)–(7.65) in the preceding chapter show that the application of an electric field $\overrightarrow{\mathcal{E}}$ induces a current $\overrightarrow{j}$ whose magnitude is proportional to the duration t of application of the field. Such a result means that there is no steady-state regime. This is in contrast with Ohm's law $\overrightarrow{j} = \sigma \overrightarrow{\mathcal{E}}$ where σ is time-independent quantity characteristic of the nature of the solid. The numerical values of the conductivity determined by Eqs. (7.62)–(7.65) correspond, even for short durations t, to an irrealistically large value of σ. Thus, for copper ($N/\Omega \sim 10^{29} \text{cm}^{-3}$), submitted to a field of 1V/cm during a time t corresponding to a drift of the Fermi–surface of the order of 1% of its size, the *resistivity* $(1/\sigma)$, calculated through Eq. (7.21), is $\sim 10^{-10} \text{ ohm}^{-1} \text{cm}^{-1}$, which is four orders of magnitude smaller than the experimental value at room temperature.

Hence, in the framework of the dynamical description developed in Chapter 7, the electric current carried by the conduction electrons of a solid can be considered to flow without loss through the solid.

Such a result is easily understandable for a free-electron gas in which there exist no obstacles to the collective motion of electrons. It can also be understood in the case of Bloch electrons although these interact, through the crystal potential $V(\vec{r})$, with the ions of the structure, and are scattered by them. In this case, the absence of loss is due to the more subtle fact that *the scattering of electrons by the periodic potential is already taken into account, in the form of the Bloch functions* describing the stationary states of electrons.

In order to explain the existence of conduction losses of a solid, and, more generally, the dissipative character of all its *transport properties*, it is therefore not sufficient to consider that electrons are only submitted to the periodic crystal potential $V(\vec{r})$. In this chapter, we examine the additional elements which must be included into the electronic description of a solid, in order to clarify the physical origin of Ohm's law, and of the finite electrical conductivity of solids. This clarification will allow us to analyze, more generally, other transport properties of solids such as, for instance, their thermal conductivity.

8.2 Physical Origin of the Finite Conductivity

8.2.1 Drude model of collision with the fixed ions

A first explanation of the finite conductivity of metals (Drude 1900) recurred, as already mentioned in Chapter 1, to the assumption that the global drift of the electrons induced by application of a field is perturbed by the presence of the ions in the structure. The conduction electrons collide with these obstacles situated at fixed positions in the structure, and part of the energy which is acquired from the field is transfered to the ions, thus justifying the existence of a dissipation.

Although the quantum argument, relative to Bloch functions, given in the above introduction, infirms the validity of this theory, it is worth summarizing its content, because part of it, i.e. the occurrence of certain collisions, remains valid in a realistic description of solids.

In Drude's theory, the conduction electrons form a classical gas in thermal equilibrium, whose velocity distribution is determined by the classical (Maxwell–Boltzmann) statistics (Fig. 8.1). In addition, there is a limited mean-free-path of the electrons, corresponding to a time interval τ between collisions. The larger the interionic distance

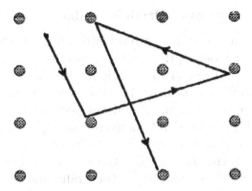

Fig. 8.1. Drude model for conduction. The electrons constituting a gas obeying the Maxwell statistics, collide with the atomic nuclei forming the crystalline structure.

in the structure, the smaller the electron–ion collision cross-sections, and accordingly the larger the value of τ.

In the absence of external field, the distribution of electron-velocities, governed by the Maxwell-statistics, is isotropic and there is no global drift of the electrons: the average velocity is $\langle \vec{v} \rangle = 0$.

The application of an electric field accelerates each electron in the time interval between collisions. In a simplified description of the effect of the field, we assume that the *average drift velocity* $\langle \vec{v}_d \rangle = -(e\tau/m)\vec{\mathcal{E}}$ resulting from the field is added to the thermal equilibrium velocity $\vec{v}$, with $\langle \vec{v} \rangle = 0$. The global drift of the electrons determines a current density:

$$\vec{j} = -\frac{N}{\Omega}e\langle \vec{v}_d \rangle = \frac{N}{\Omega}\frac{e^2\tau}{m}\vec{\mathcal{E}} = \sigma\vec{\mathcal{E}} \qquad (8.1)$$

Hence, in agreement with Ohm's law, a finite conductivity is determined. It is proportional to the average time-interval τ separating successive collisions. In metals, conductivities are of the order of $10^7 - 10^8 \mathrm{ohm}^{-1}.\mathrm{m}^{-1}$ at room-temperature. Taking $(N/\Omega) \approx 10^{29}\mathrm{m}^{-3}$, we obtain from Eq. (8.1) $\tau \approx 10^{-13} - 10^{-14}\mathrm{s}$. On the other hand, in a classical electron gas at room temperature, the velocity $\mathbf{v}$ of electrons is of the order of $10^4\mathrm{m/s}$. The *mean-free-path* $\Lambda_e = \mathbf{v}.\tau$ is therefore in the range $1 - 10\text{Å}$, consistent with the interatomic distances in a crystal.

8.2.2 Shortcomings of Drude's model

We know that in a metal, electron-states are of a quantum nature and that a classical description of their collective thermal equilibrium is inadequate. However, in itself, this inadequacy is unsufficient to invalidate Drude's model. Indeed, within the semi-classical approximation of the electron dynamics, developed in Chapter 7, Drude's result (Eq. (8.1)) can be simply derived from Eq. (7.21) by introducing a finite time τ of free-electron path assigned to the occurrence of collisions with the fixed ions of the structure. The actual irrelevance of the model is revealed by the evaluation of the numerical value of the mean-free-path for a quantum gas. We know that for such a gas, the number of active electrons in the conduction process are a small fraction of the total number of electrons. These active electrons whose states are close to the Fermi surface, possess a high average velocity $\langle \mathbf{v} \rangle \approx 10^6$ m/s, two orders of magnitude higher than the average velocity in a classical gas. Accordingly, their mean-free-path $\Lambda_e = \langle \mathbf{v} \rangle \tau$ is also two orders of magnitude higher and is in the range $\Lambda_e \approx 10^2 - 10^3$ Å. Such a value is much larger than the interatomic distances in a solid and it denotes that the obstacles set to the free displacements of electrons in a solid are not the fixed ions of the structure.

Another shortcoming of Drude's model is its inability to explain the very large temperature-dependence of the conductivity of a metal. Thus, the conductivity of copper increases ten-fold on cooling from 300 K to 77 K. Such a variation can only result from changes in the density (N/Ω) or in the collision-time τ. In a metal, the number N of conduction-electrons is equal to the number of electrons in the conduction band. This number can be considered as independent of the temperature since the interval between the top of the conduction band and the Fermi-level is generally large as compared to $k_B T$ (Fig. 8.2). The variation of the density therefore only results from the thermal expansion of the solid which is 3–4 orders of magnitude too small to account for the observed increase in the conduction. On the other hand, the temperature-dependence of the interatomic distances being also related to the thermal expansion of the solid, the collision time τ should be almost temperature independent.

In conclusion, the physical origin of the conduction losses in a solid must be looked-for in another mechanism than in the scattering of electrons by the fixed ions of the structure.

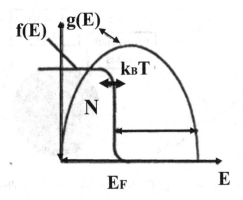

Fig. 8.2. In a metal, the electronic density, which is determined by the number of electrons in the conduction band does not vary because $(E_m - E_F) \gg k_B T$.

8.2.3 Irrelevance of collisions between electrons

In the kinetic theory of molecular gases, which accounts, on a microscopic basis, for the properties of these systems, the transport properties, such as the thermal conductivity or the viscosity, are explained convincingly by the existence of collisions between the molecules constituting the gas. One could therefore imagine that the simpler way of accounting for the transport properties of an electron gas would be to derive them from the existence of mutual collisions *between the electrons*. We would then be led to take into account the interactions between electrons which were assumed to be inexistent in the preceding chapters. Let us show that even if such interactions were significant (cf. Chapter 10), they only have a marginal effect on the mean-free-path of the electron, owing to the submission of the electron gas states to the Pauli principle. Hence, as we argue now, collisions between electrons cannot be the main ingredient of a theory of the conduction losses.

In this view, let us consider the collision between two electrons of a quantum electron-gas. The electrons are assumed to have no other interactions apart from their collisions. Let us first assume that the electrons are almost-free electrons. Then, the occupied quantum states fill, in reciprocal space, the Fermi sphere of radius k_F. A collision involves two *incident* electron-states defined by the vectors $\vec{k}_1$ and $\vec{k}_2$ and two scattered electron-states defined by $\vec{k}_3$ and $\vec{k'}_4$ (Fig. 8.3). The incident states are necessarily *occupied states*

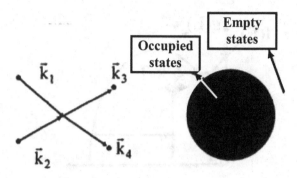

Fig. 8.3. Restriction imposed by Pauli principle to the collisions between electrons.

with $|\vec{k}_i| \leq k_F$, while, owing to the Pauli principle, the scattered states are necessarily empty states with $|\vec{k}_s| \geq k_F$. Hence, the conservation of energy $(\hbar^2/2m)(\vec{k}_1^2 + \vec{k}_2^2) = (\hbar^2/2m)(\vec{k}_3^2 + \vec{k}_4^2)$ imposes to the four moduli to be approximately equal to k_F. For free-electrons, the conservation of momentum is equivalent to the equality $\vec{k}_1 + \vec{k}_2 = \vec{k}_3 + \vec{k}_4$. The two former conditions imply that *the two scattered states are identical to the incident states*. Thus *a collision does not modify the distribution of occupied states* and henceforth the properties of the quantum gas. For Bloch electrons this is also the more general situation, although, strictly speaking, the conservation of momentum only implies the conservation of the sum of $\vec{k}$-vectors up to a reciprocal lattice vector.

8.2.4   Interaction with collective oscillations

As pointed out in the introductory paragraph, there are no conduction losses for a gas of independent electrons in a periodic structure. Within the description of electrons in solids as a set of independent particles, the only origin of a finite conductivity has therefore to be searched in the fact that the electron is submitted to a *non-periodic potential*. In this line, two possible mechanisms of the conduction losses are relevant. In metals, the dominant mechanism at room temperature, which determines an instantaneous deviation from periodicity, is the thermal oscillation of atomic nuclei (or ions) around their equilibrium periodic positions in the structure.

The other mechanism, which can be important at low temperatures, is the presence of defects of the crystal structure such as impurities. As already underlined in Chapters 1 and 2, the periodicity of a crystal only characterizes the time-average of the atomic positions. The atoms of a solid undergoes a variety of types of *collective motions*, the amplitudes of which increase as a function of the temperature (cf. Chapter 10, Section 10.3.2). Consequently, at any time, the configuration of atoms is not strictly periodic. This instantaneous deviation from periodicity is small at room temperature, the typical amplitude of oscillation of an atom about its average position being less than a tenth of an Angström. Besides, the different types of collective oscillations occur randomly. We can therefore consider that an electron moving through a crystal is submitted to a *small random non-periodic potential* $\tilde{v}(\overrightarrow{r})$ in addition to the periodic potential $V(\overrightarrow{r})$ discussed in the preceding chapters. Such a random potential has the effect of deviating the electron from their trajectories (Fig. 8.4). They will thus limit the free-path of the electron in the same manner as the collisions considered in Drude's model.

The effect of thermal oscillations is strongly temperature-dependent. Indeed, let us, for instance, consider a temperature range in which the collective oscillations can be considered as classical (i.e. non-quantical). This range corresponds to temperatures such as $T \gg \Theta_D$, where Θ_D is the so-called Debye temperature (cf. Fig. 1.3 in Chapter 1). The equipartition theorem in classical statistics indicates that the average value of the oscillation energy of a microscopic oscillator (e.g. an atom) is of the order of $k_B T$. The energy being proportional to the square of the oscillation amplitude A, this amplitude varies as $\sqrt{T}$. The "effective surface of the atom" which constitutes an obstacle to the motion of an electron is proportional to $A^2 \propto T$.

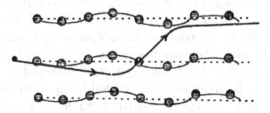

Fig. 8.4. Scattering of an electron by the collective oscillations of atoms of a crystal.

Table 8.1.

Metal	σ $(10^7\Omega^{-1}\text{m}^{-1})$	$\frac{N}{\Omega}$ (10^{28}m^{-3})	τ_c (10^{-14}s)
Copper	6.4	8.5	2.7
Silver	6.6	5.9	4
Lead	0.53	13	0.15

Accordingly, the mean time τ between collisions, and the conductivity (cf. Eq. (8.1)) are expected to be proportional to $(1/T)$ in qualitative agreement with the experimental observations in metals.

At low temperatures $(T \ll \Theta_D)$, the effect of the collective oscillations must be discussed in the framework of quantum mechanics. As will be shown in Chapter 10, the equilibrium state of collective excitations in a solid can be described as equivalent to the presence in the solid of particles called *phonons,* and whose energies are quantized. The interaction of an electron with the collective atomic oscillations can then be represented as consisting of collisions between the electron and the phonons. It can be shown that the number of phonons present in the solid decreases as T^3. Hence, at low temperatures, the conductivity has a stronger temperature dependence, which can be shown to follow a $\approx T^5$ law.

Table 8.1 reproduces the values of the conductivities and of the electronic densities of three usual metals. By reference to Eq. (8.1) one can observe that the large differences between the conductivities of "good conductors" such as copper or silver, and a "bad conductor" such as lead, correspond to collision times τ differing by more than one order of magnitude.

These differences depend on a variety of specific details of the materials, such as the nature of the collective oscillations of the atoms, the strength of the interaction between the electrons and the collective oscillations, etc. It is therefore difficult to account simply for the values of the collision times appearing in the table. However, a general qualitative argument can be pointed out. Thus, as is well known, lead is a "mechanically softer" material than copper or silver, i.e. it can be more easily deformed by application of mechanical stresses. Such a property can be thought to have its origin in a relative weakness of the magnitude of certain interatomic forces, which will determine smaller "restoring forces" for the collective oscillations. Accordingly, the frequencies of these oscillations, which are proportional to the square-root of the restoring forces, are also expected to be lower in lead than in copper or silver. At given temperature $(T \gg \Theta_D)$, the mean oscillation energy $(M\omega^2 < r^2 > \propto k_B T)$ is the same for any collective mode and is proportional to the mean square of the oscillation amplitude $\langle r^2 \rangle$. Consequently, the amplitude of oscillation will be larger in lead, thus determining a larger cross-section of the collision between an electron and the collective motion of atoms, and, henceforth, a smaller collision time, in qualitative agreement with Table 8.1.

8.2.5 Interaction between electrons and structural defects

A large variety of defects can exist in a crystalline structure (Fig. 8.5), the simpler of which are vacancies, substitutional, or interstitial impurities, etc. They have the common effect to determine a partial destruction of the perfect periodicity of the solid. Such defects will be centers of random scattering of the electrons, and will limit the free-path of the electron. A finite conductivity, related to a specific collision time τ_d, will thus be determined.

The number of certain types of defects, such as vacancies depends on the temperature. Although the creation of such a defect in a solid involves a positive energy ϵ, thermal equilibrium corresponds to the presence of a large number $n_d \propto \exp(-n_d \epsilon / k_B T)$ of defects in the solid. This number will rapidly decrease on lowering the temperature and vanish at $T = 0K$. As for collective oscillations such defects therefore determine a rapidly decreasing conductivity at low temperatures.

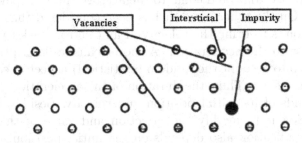

Fig. 8.5.   Vacancies, interstitial defects and impurities.

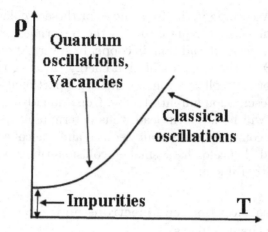

Fig. 8.6.   Schematic temperature dependence of the resistivity of metals, with the dominant mechanism in each temperature range.

In contrast the number of impurities and their mutual distances do not depend on the temperature. It is therefore these types of defects which are at the origin of the "residual" resistivity of a conducting solid at low temperatures, when the collisions due to collective oscillations and vacancies disappear. Figure 8.6 shows the expected schematic variation of the conductivity for a metal.

8.3   Electron Dynamics in the Presence of Collisions

The occurrence of collisions has been considered in the framework of two simplifying assumptions. In the first place, it has been assumed that each electron is accelerated for a given definite duration τ, identical for all electrons, and equal to the average time τ separating two collisions. On the other hand, it has been considered that the effect of a collision is to bring the velocity of an electron back to its initial value, prior to the acceleration. A more realistic description would be to take into account the random (statistical) aspect of each of the two former effects. Thus, the duration of the acceleration of an electron depends of its initial position and velocity. Besides, the effect of a collision is to modify the direction and value of the velocity. Such a modification also depends on the initial position and velocity of the electron. Since these characteristics are random, the whole collective evolution has to be considered on a statistical basis. An additional cause of randomness resides in the objects at the origin of the collisions. Hence, in the case of collisions with the collective atomic oscillations, the probability of a collision depends of the phase of the oscillation-wave which is random.

In order to include the collisions in a statistical description of the collective evolution of the electrons in a solid (cf. Eq. (7.38)), one can examine the effect of the collisions on the occupation factor $\mathbf{f}(t, \overrightarrow{r}, \overrightarrow{k})$.

8.3.1 Boltzmann equation

For an electrons occupying a volume Ω, the density of states is $\Omega/4\pi^3$. If they are localized in an infinitesimal volume $d^3\overrightarrow{r}$, this density of states will be $d^3\overrightarrow{r}/4\pi^3$. On the other hand, the occupation factor of these states is $\mathbf{f}$. Hence, the infinitesimal quantity $(\mathbf{f}/4\pi^3)d^3\overrightarrow{r}d^3\overrightarrow{k}$ is equal to the number of electrons which are located at time t in the volume $d^3\overrightarrow{r}$, and whose quantum number $\overrightarrow{k}$ belongs to the volume $d^3\overrightarrow{k}$ of the reciprocal space. A collision does not change the position $\overrightarrow{r}$ of an electron, since collisions are *local* , but it modifies its $\overrightarrow{k}$-state. Hence, $\mathbf{f}$ will generally change. Writing this change in a formal manner $(\partial \mathbf{f}/\partial t)_{\text{coll}}$, and including it in the evolution equation (7.39), we obtain:

$$\frac{\partial \mathbf{f}}{\partial t} + \overrightarrow{\mathbf{v}}_g.\overrightarrow{\nabla}_r\mathbf{f} + \frac{\overrightarrow{F}}{\hbar}.\overrightarrow{\nabla}_k\mathbf{f} = \left(\frac{\partial \mathbf{f}}{\partial t}\right)_{\text{coll}} \tag{8.2}$$

This is the *Boltzmann equation* which governs the evolution of the statistical occupation factor in the presence of collisions.

8.3.2 Relaxation time and local equilibrium

Equation (8.2) is obviously incompletely formulated. Indeed, the form of the second member should be specified for each of the physical processes at the origin of collisions (collective oscillations, point defects, etc.). However, it is possible to infer on the basis of a qualitative discussion, the general characteristics of the collective evolution determined by the preceding equation.

Note first, that in the absence of collisions $((\partial \mathbf{f}/\partial t)_{\text{coll}} = 0)$, and of external forces $(\overrightarrow{F} = 0)$, and for a spatially uniform state of the system $(\overrightarrow{\nabla}_r \mathbf{f} = 0)$, Eq. (8.2) reduces to $(\partial \mathbf{f}/\partial t) = 0$. It thus determines a time-independent occupation factor. Consequently, if an arbitrary distribution of electrons over the $\overrightarrow{k}$-vectors of the reciprocal space is initially realized, this distribution will be preserved at any time. Such a result clearly differs from the actual situation in a solid. Indeed, one expects that starting from an arbitrary distribution, the system will evolve progressively towards a state characterized by the thermodynamic equilibrium distribution (i.e. the Fermi factor) $\mathbf{f}_0$.

This argument suggests that the collisions are the "driving force" underlying the time-evolution of the occupation factor towards its thermodynamic equilibrium form $\mathbf{f}_0$. In order to express in a simple, but still formal, manner this role of the collision term, let us write:

$$\left(\frac{\partial \mathbf{f}}{\partial t}\right)_{\text{coll}} = -\frac{\mathbf{f}(t, \overrightarrow{r}, \overrightarrow{k}) - \mathbf{f}_0}{\tau(\overrightarrow{r}, \overrightarrow{k})} \tag{8.3}$$

Such an expression vanishes when the equilibrium is reached, consistently with the further preservation of this equilibrium.

Assume first, that $\overrightarrow{F} = 0$, and that the system is spatially uniform (no dependence of $\mathbf{f}$ and τ on the coordinate $\overrightarrow{r}$). An integration with respect to t then leads to:

$$\frac{\mathbf{f}(t, \overrightarrow{k}) - \mathbf{f}_0}{\mathbf{f}_0} = \exp\left(-\frac{t}{\tau}\right) \tag{8.4}$$

The occupation factor tends towards $\mathbf{f}_0$ with a characteristic time τ, which is called the *relaxation time* of the electron-system towards equilibrium. Since the changes in the occupation factor of the states is due to the collisions, one can expect that the relaxation time τ will have the same order of magnitude as the average time between collisions considered in Eq. (8.1).

Let us now examine qualitatively the evolution towards thermal equilibrium in the more complex case of a *spatially non-uniform* occupation factor. Assuming again the absence of external forces $(\overrightarrow{F} = 0)$. Let us, in addition, assume that the scale λ of the spatial variations of $\mathbf{f}$ is large $(\nabla_r \mathbf{f} \approx \Delta \mathbf{f}/\lambda)$ compared to the free-path $\Lambda = \mathbf{v}.\tau$ of the electron $(\Lambda/\lambda \ll 1)$. For instance, the mean-free-path Λ being of the order of 10^3Å, we consider temperatures, and electronic densities which are almost uniform within regions *several microns in size*. Multiplying the two members of Eq. (8.2) by (τ/f_0), and *replacing the derivatives by finite differences* $(\partial f/\partial t \to \Delta f/\Delta t)$, Eq. (8.2) can be written as

$$\left(\frac{\Delta f}{f_0}\right)\left(\frac{\tau}{\Delta t} + \frac{\Lambda}{\lambda}\right) = -\frac{f - f_0}{f_0} \qquad (8.5)$$

Consider an initial state f far from equilibrium. f and f_0 will have similar magnitudes. The second member of Eq. (8.5) is then *of order unity*. The condition $(\Lambda/\lambda) \ll 1$ then implies that, *in a first stage of the evolution*, the temperature of the occupation factor is essentially determined by the term $(\tau/\Delta t)$. This evolution is therefore governed by Eq. (8.4): at each point $\overrightarrow{r}$ of the system. The difference $(f - f_0)$ tends to zero in a time $\tau(\overrightarrow{r})$. The system approaches an equilibrium distribution *locally*. At this first stage, no significant evolution towards spatial uniformity is achieved, since the "driving force" of such a uniformity is the small term $\overrightarrow{v}.\overrightarrow{\nabla}_r f \ \tilde{} \ (\Lambda/\lambda)(\Delta f/f_0)$.

When this local thermodynamic equilibrium is almost realized, we have $(f - f_0)/f_0 \ll 1$. All the terms in Eq. (8.5) then become relevant. The evolution will tend to achieve simultaneously spatial uniformity and completion of the local thermodynamic equilibrium, eventually reaching a global equilibrium. During this stage, we can write:

$$\frac{\Delta t}{\tau} \cong \frac{\lambda}{\Lambda} \gg 1 \qquad (8.6)$$

Hence, the characteristic time of this second stage of the evolution is much larger than τ.

8.4 Electronic Transport Properties

Let us show, on the basis of the preceding qualitative discussion, that Eqs. (8.2)–(8.3) allow to derive the electronic transport properties of a solid. In this section, we first examine the transport of electric charges under application of an electric field to the solid, in the view of specifying the conduction properties for Bloch electrons. We then consider the transport of heat by electrons when a temperature gradient is imposed to the solid in the view of expressing the form of the thermal conductivity.

8.4.1 Evolution in local equilibrium

We have seen that, once local equilibrium is achieved in a system in the absence of external forces (Fig. 8.7), the occupation factor $f(t, \overrightarrow{r}, \overrightarrow{k})$ is close to the local Fermi factor $f_0(\overrightarrow{r}, \overrightarrow{k})$. This conclusion holds even if an external force exists. Consider, for instance, the application of an external electric field. The application of equations $\overrightarrow{j} = \sigma\overrightarrow{\mathcal{E}}$ and $\overrightarrow{j} = -e(N/\Omega)\overrightarrow{v}$ (Eqs. (7.15), (7.16)) with $\sigma = 10^8 \Omega^{-1}.m^{-1}$ shows that for a field of $10^4 V/m$, the drift velocity v is of the order of $10^2 m/s$. It is therefore four orders of magnitude smaller than the individual thermal velocities of the electrons.

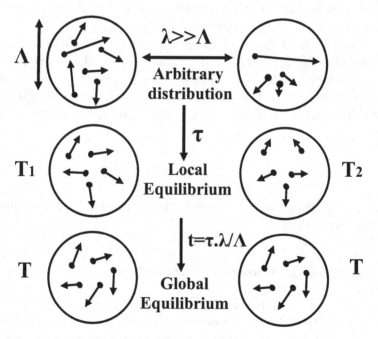

Fig. 8.7. Evolution towards equilibrium in two steps: Establishment of a local equilibrium followed by a spatial uniform distribution of the electrons energy. Λ is the free path of electrons between collisions, and λ measures the scale of variations of the occupation fact f.

The flow of current through a solid thus corresponds to a small deviation from equilibrium. This remark entitles us to search the solutions of (8.2)–(8.3) in the form:

$$\mathbf{f}(\vec{r}, \vec{k}) = \mathbf{f}_0(\vec{r}, \vec{k}) + \phi(\vec{r}, \vec{k}) \qquad (8.7)$$

where ϕ is a small correction to $\mathbf{f}_0$. In the subsequent calculations, we can then retain the non-zero terms of lowest degrees in ϕ, which are generally the linear terms.

8.4.2 Electrical conductivity

Let us consider the situation in which a uniform and time-independent electric field $\vec{\mathcal{E}}$ induces the occurrence of a uniform and time-independent current density $\vec{j}$ in the solid (Fig. 8.8). In the evolution equations (8.2)–(8.3), we must set $\vec{\nabla}_r \mathbf{f} = 0$. Moreover, the local equilibrium distribution factor $\mathbf{f}_0(\vec{r}, \vec{k})$ reduces to the global equilibrium Fermi-factor $\mathbf{f}_0(\vec{k})$. Using the linear approximation (8.7) we obtain, at lowest order:

$$\frac{\hbar}{e\tau(\vec{k})}\phi(\vec{k}) = \vec{\mathcal{E}}.\vec{\nabla}_k\mathbf{f} \cong \vec{\mathcal{E}}.\vec{\nabla}_k\mathbf{f}_0 = (\vec{\mathcal{E}}.\vec{\nabla}_k E)\frac{d\mathbf{f}_0}{dE} \qquad (8.8)$$

This expression shows that the deviation ϕ from the equilibrium factor is proportional to the electric field. The expression (7.18) of the density of current (Chapter 7) leads to:

$$\vec{j} = -\frac{e}{4\pi^3}\int \vec{v}.\mathbf{f}.d^3\vec{k} = -\frac{e}{4\hbar\pi^3}\int \phi(\vec{k}).\vec{\nabla}_k E.d^3\vec{k} \qquad (8.9)$$

In writing the last member, we have taken into account the fact that *the equilibrium distribution f_0 determines no current* through the solid. Combining the two last equations we finally obtain:

$$\vec{j} = \left\{\left(\frac{e^2}{4\hbar^2\pi^3}\right)\int \tau(\vec{k}).(\overline{\vec{\nabla}_k E \times \vec{\nabla}_k E})\left(-\frac{d\mathbf{f}_0}{dE}\right)d^3\vec{k}\right\}.\vec{\mathcal{E}} = \overline{\overline{\sigma}}.\vec{\mathcal{E}} \qquad (8.10)$$

The conductivity is represented by a tensor of rank 2 whose expression depends of the collision time $\tau(\vec{k})$ and of the specific energy dispersion function $E(\overrightarrow{k})$ of the solid through the tensor:

$$\overline{\overline{(\overrightarrow{\nabla}_k E \times \overrightarrow{\nabla}_k E)}} \Rightarrow \left[\frac{\partial E}{\partial k_i} \cdot \frac{\partial E}{\partial k_j}\right] \tag{8.11}$$

As already pointed out in Section 7.2.4 (Eqs. (7.26)–(7.28)), for a metal, the derivative of the Fermi factor can be approximated as $(-d\mathbf{f}_0/dE) \approx \delta(E - E_F)$. Hence, we verify again that the current is only determined by the *active electron states* close to the Fermi surface. Noting the similarity with Eq. (7.67), we deduce from Eqs. (7.65)–(7.68) two equivalent but useful expressions of the current:

$$\overrightarrow{j} = \frac{Ne^2\tau}{\Omega} \left\langle \overline{\overline{\left(\frac{\partial^2 E}{\partial k_i \partial k_j}\right)}} \right\rangle \cdot \overrightarrow{\mathcal{E}} = \frac{Ne^2\langle\tau\rangle}{\Omega} \left\langle \overline{\overline{\left(\frac{1}{m^*}\right)}} \right\rangle \cdot \overrightarrow{\mathcal{E}} \tag{8.12}$$

where $\overline{\overline{m^*}}$ is the anisotropic (tensorial) effective mass, and where the brackets $\langle\rangle$ represent the *statistical average* over all occupied states $\overrightarrow{k}$. On the other hand we also have:

$$\overrightarrow{j} = \frac{e^2\hbar^2\langle\tau\rangle\mathbf{v}_F^2 \cdot g(E_F)}{3\Omega} \overrightarrow{\mathcal{E}} \tag{8.13}$$

where $\mathbf{v}_F$ is the velocity of electrons, and $g(E_F)$ the density of states at the Fermi level. These various expressions of the conductivity are the microscopic formulations of Ohm's law in an anisotropic solid.

Note that Eq. (8.12)–(8.13) are closely similar to the expression (8.1) of the conductivity derived within the framework of Drude's model. The latter expression would be obtained if the relaxation time τ is independent of $\overrightarrow{k}$, and if the effective mass is equal to the real mass of the electron. These equations can be given a common interpretation: namely, the semi-classical trajectory of the Bloch electrons is interrupted after a time $\langle\tau\rangle$, and, consequently, the drift of the Fermi surface in reciprocal space (Fig. 7.3) is limited by the collisions in such a manner that after a time τ, each electron is brought back to its initial state.

Also note that this simplified description of the collective evolution of the electrons in which the drift of the Fermi surface is limited to the duration τ, can also be obtained by using an "individual" dynamic equation, relying on an additional term to Eq. (7.57):

$$\hbar\frac{d\vec{k}}{dt} = \vec{F} - \frac{\hbar\vec{k}}{\tau}$$ (8.14)

Indeed, it is easy to verify that, in the absence of external forces, $\vec{k}$ is brought back to its initial value after the time τ, and also that in a steady state regime, the drift of the occupied states induced by the application of an electric field $\vec{\mathcal{E}}$, is limited to $(-e\tau\vec{\mathcal{E}}/\hbar)$.

8.4.3 Electronic heat conduction

Let us now consider a situation in which the external forces are absent, but there exists a temperature gradient $\vec{\nabla}_r T$ which induces a heat transfer through the solid (Fig. 8.8). In the same manner as the transfer of electric charges induced by an electric field, the heat transfer can be characterized by a vector $\vec{\delta}$ representing the density of the heat flux. The contribution to this flux of the electrons possessing a given velocity is provided by an expression similar to Eq. (7.15):

$$d\vec{\delta} = E.\vec{v}.\rho(\vec{k})d^3\vec{k}$$ (8.15)

the total density having the form:

$$\vec{\delta} = \left(\frac{1}{4\pi^3}\right)\int E(\vec{k}).\vec{v}.f.d^3\vec{k}$$ (8.16)

The analog of Ohm's law is the Fourier law which defines the *electronic heat conduction* κ:

$$\vec{\delta} = -\kappa.\vec{\nabla}_r T$$ (8.17)

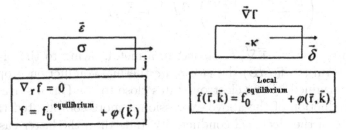

Fig. 8.8. Compared schemes of definition of the electrical conductivity (left) and of the thermal conductivity (right).

This conductivity can be calculated, in the same manner as σ, by using the linearized evolution equations (8.2)–(8.3)–(8.5).

In this view, note first that the local equilibrium occupation factor $f_0(\vec{r}, \vec{k})$ varies with the position $\vec{r}$ in the solid. This is due to the fact that the temperature gradient imposes a non-uniform state in the system. Hence, we can write:

$$f_0(\vec{r}, \vec{k}) = f_0[E, E_F(T), T] = \left(1 + \frac{E(\vec{k}) - E_F(T(\vec{r}))}{k_B T(\vec{r})}\right)^{-1}$$

(8.18)

The $\vec{r}$ dependence results in particular from the dependence of the Fermi energy on the temperature. We derive from this expression:

$$\vec{\nabla}_r f_0 = \vec{\nabla}_r T. \left(-\frac{df_0}{dE}\right) \left(\frac{E - E_F}{T} + \frac{dE_F}{dT}\right)$$

(8.19)

On the other hand, taking into account the linearization of the evolution equations (8.2)–(8.5), we obtain:

$$\phi = -\frac{\tau(\vec{r}, \vec{k})}{\hbar} (\vec{\nabla}_k E).(\vec{\nabla}_r f_0)$$

(8.20)

Finally, reporting (8.19)–(8.20) into (8.16) we derive an expression of the heat flow density similar to that (8.10) of the electric current-density $\vec{\delta}$:

$$\left\{\left(\frac{1}{4\hbar^2\pi^3}\right) \int \tau(\vec{r}, \vec{k}).\overline{(\vec{\nabla}_k E \times \vec{\nabla}_k E)} \left(-\frac{df_0}{dE}\right)\right.$$
$$\left.\left(\frac{E - E_F}{T} + \frac{dE_F}{dT}\right) d^3\vec{k}\right\}.(-\vec{\nabla}_r T)$$

(8.21)

Thus $\vec{\delta} = = -\overline{\overline{\kappa}}\vec{\nabla}_r T$. Hence, in a metal, owing to the presence of the term $(-df_0/dE)$ the *electronic* thermal conduction properties are determined by the electron states close to the Fermi surface.

A reduction of the above expression is more complicated than in the case of the electrical conductivity. Indeed, in the latter case, one could neglect the effect of the local variation of the Fermi energy. In the present case, this effect is an essential contribution to the thermal conduction. It is implicated both in the form of $(-df_0/dE)$ and in the expression of $\{(E - E_F)/T + dE_F/dT\}$.

The variations of E_F for Bloch electrons is not necessarily simple. Let us consider, as an approximation the variations calculated for a free electron gas which is provided by the Sommerfeld expansion (6.13). Assuming that τ does not vary, we obtain:

$$\vec{\delta} = \left\{ \left(\frac{k_B^2 T \langle \tau \rangle}{12\pi\hbar^2} \right) \int \overline{(\vec{\nabla}_k E \times \vec{\nabla}_k E)} \left(-\frac{df_0}{dE} \right) d^3\vec{k} \right\} .(-\vec{\nabla}_r T)$$

(8.22)

By comparison with Eqs. (8.10)–(8.12) we derive finally:

$$\kappa = \frac{\pi^2 k_B^2 T \langle \tau \rangle}{9\Omega} \mathbf{v}_F^2 . g(E_F) = \frac{\pi^2 k_B^2 T}{3e^2} \sigma$$

(8.23)

There should exist a proportionality between the electric conductivity of a metal and the electronic part of its thermal conductivity, the proportionality factor $(\pi^2 k_B^2 . T/3e^2)$ being independent of the nature of the substance. The experimental manifestation of this result constitutes *Wiedemann and Franz law*. The table hereunder indicates the values measured for the ratio $L = \kappa'/\sigma T$ between the *total thermal conductivity* and the electric conductivity in various metals at two different temperatures. It appears that the constancy of the ratio $L = (\kappa/\sigma T)$ is verified within a precision of 10%, and agrees, within the same precision, with the theoretical value hereabove: $L = 2.4 \times 10^{-8} \Omega.K^{-2}$.

Metal	$L = \frac{\kappa}{\sigma T} (10^{-8} W\Omega/K^2)$ (273 K)	L (373 K)
Copper	2.20	2.29
Silver	2.31	2.38
Lead	2.64	2.53

Such an agreement shows that *in a metal, heat is essentially carried by the conduction electrons*. As mentioned in Chapter 1, heat can also been transfered by the collective oscillations of atoms. This mechanism is secondary in metals, but is dominant in insulators. The fact that the thermal conductivity of metals is much larger than that of insulators, is due to the greater efficiency of the electron transport-mechanism.

Bibliography

Electrical Conductivity in Solid Materials, *J.P. Suchet, and B.R. Pamplin*, Pergamon Press (1975).

Thermal Conduction in Solids, *R. Berman and R. Berman*, Clarendon Press (1976).

Chapter 9

Intrinsic and Doped Semiconductors

Main ideas: Fermi level in insulators and intrinsic semiconductors (Section 9.2). Band structures of silicon and gallium arsenide (Section 9.2.3). Electron and holes at equilibrium in doped semiconductors (Section 9.3). Principles of the diode and the transistor (Section 9.4).

9.1 Introduction

As explained in Chapter 6, insulators are solids which, at 0K, possess a *fully occupied valence band* separated by a finite energy interval V_G (a *gap*) from the next unoccupied energy band (the so-called conduction band). *Intrinsic semiconductors* are insulators, which are distinguished quantitatively from other insulators, by the magnitude of their gap in the range $\gtrsim 2$ eV.

The most common ones are not characterized by specific crystal structures. The diamond structure of carbon (C) is classified as an insulator, with a gap ≈ 6 eV, while silicon (Si), the prominent example of semiconductor, most used in technology, possesses the same crystal structure, with a gap equal to ≈ 1.1 eV. The gaps of some other intrinsic semiconductors are indicated on Table 9.1.

The semiconductors used in technology are generally "doped" semiconductors in which additional atoms (termed *dopants*), of various types, introduced deliberately in the composition of the solid constitute defects of the perfect crystal structure of the intrinsic insulator. Such impurities determine additional energy levels for the electrons (or, as will be seen, the holes) forming additional bands

Table 9.1. Gaps (in electron-volts), atomic diameter, and cubic lattice parameter of a few common intrinsic semiconductors.

Semiconductor	V_G (eV)	Atomic diameters	Lattice parameter
Silicon	1.12	2.22 Å	5.43 Å
Germanium	0.67	2.5 Å	5.66 Å
GaAs	1.43	2.6 Å/ 2.30 Å	5.63 Å
InAs	0.33	3.1 Å/ 2.30 Å	6.05 Å
InP	1.25	3.1 Å/ 2 Å	5.87 Å

(the impurity bands) generally located between the valence band and the conduction band of the intrinsic semiconductor.

9.2 Properties of Intrinsic Semiconductors

9.2.1 Location of the Fermi level

The density of states $g(E)$ in the valence and conduction band of an insulator (and thus of an intrinsic semiconductor) are schematically reproduced on Fig. 9.1. This figure shows, as well, the shape of the occupation factor $f(E)$ (Fermi factor), and the schematic variations of $E(\vec{k})$ (the dispersion curve) for the two bands. Near the top of the valence band, and the bottom of the conduction band, the $E(\vec{k})$ dependence is approximately parabolic.[1] Let us show that at 0 K the Fermi level (cf. Section 6.2.2) $\mu = E_F$ is situated in the middle of the gap, and that it remains in the vicinity of this value as long as $k_B T \ll E_F$. Indeed, let us reproduce here Eq. (6.5) of Chapter 6, in which N is total number of electrons in the valence and conduction bands per unit volume:

$$N = \int_{E_{\vec{k},n}} g(E).f(E \mid E_F, T).dE \qquad (9.1)$$

This equation contains the number of states $g(E)$ in the energy interval dE, and the occupation factor $f(E)$ of each state at the temperature T. Their product, integrated on the entire energy spectrum

---

[1] An anisotropic shape of these surfaces will be considered in Eq. (9.10).

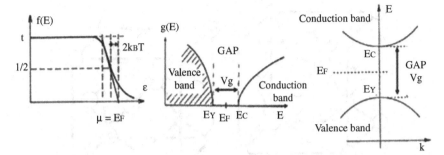

Fig. 9.1. Left: Shape of the Fermi factor (occupation of the quantum states. Middle: density of states of an intrinsic semiconductor with an almost fully occupied valence band and an empty conduction band. Right: Band structure in reciprocal space.

determines the total number of electrons. The occupation factor f is equal to (cf. Eq. (6.8)):

$$f(E_{\vec{k},n} \mid E_F, T) = \frac{1}{1 + \exp\{(E_{\vec{k},n} - E_F)/k_B T\}} \qquad (9.2)$$

Let us write Eq. (9.1) on the one hand, for $T = 0$ K, and, on the other hand, for $T > 0$. The first integral in Eq. (9.3) hereunder expresses that at 0 K all the states of the valence band are occupied and that the conduction band is empty, while the second and third integrals, relative to $T > 0$, express that a certain proportion of electrons occupies the states of the conduction band:

$$N = \int_0^{E_V} g(E)dE = \int_0^{E_V} g(E)f(E - E_F)dE$$
$$+ \int_{E_C}^{\infty} g(E)f(E - E_F)dE \qquad (9.3)$$

Substracting the second integral from the first one and using (cf. Eq. (9.2)) the symmetry of the Fermi factor $[1 - f(E - E_F)] = f(E_F - E)$, leads to:

$$\int_0^{E_V} g(E)f(E_F - E)dE = \int_{E_C}^{\infty} g(E)f(E - E_F)dE \qquad (9.4)$$

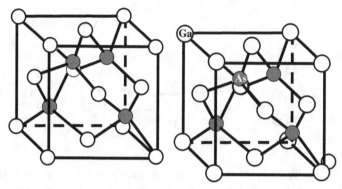

Fig. 9.2.   Crystal structures, of the cubic FCC type, of silicon and gallium arsenide (GaAs).

We consider T small: $(k_B T \ll (E_F - E_V)$ and $k_B T \ll (E_C - E_F))$. Hence:

$$f(\pm[E_F - E]) \propto \exp\{\mp[E_F - E]/k_B T\} \qquad (9.5)$$

E_V is the top of the valence band $E(\vec{k})$. Likewise, E_C is the bottom of the conduction band $E(\vec{k})$. Assuming that in the vicinity of E_v and E_C the bands are isotropic in $\vec{k}$ space, we can express their respective parabolic shapes as (Fig. 9.2).

$$g(E) = K\sqrt{E_V - E} \; ; \; g(E) = K'\sqrt{E - E_C} \qquad (9.6)$$

Replacing in Eq. (9.4) expressions (9.5) and (9.6), we obtain, by putting $x = (E_V - E)/k_B T$ in the first member of Eq. (9.4) and $y = (E - E_C)/k_B T$ in the second member:

$$K(k_B T)^{3/2} e^{(E_V - E_F)/k_B T} \int_0^{E_V/k_B T} \sqrt{x} e^{-x} . dx$$

$$= K'(k_B T)^{3/2} e^{(E_F - E_C)/k_B T} \int_0^{\infty} \sqrt{y} e^{-y} dy \qquad (9.7)$$

For $k_B T \ll E_V$ we can take $(E_V/k_B T) \approx \infty$, the two integrals in Eq. (9.7) are then equal to $(\sqrt{\pi}/2)$ and thus:

$$E_F = \frac{E_V + E_C}{2} + \frac{k_B T}{2} \text{Ln}\left(\frac{K}{K'}\right) \qquad (9.8)$$

showing that for $T = 0$ K the Fermi energy is in the middle of the gap V_G. At $T \neq 0$ it remains approximately at this location when the temperature is increased, as long as $k_B T \ll (E_C - E_V)$. The coefficients K and K' in Eq. (9.6) are related to the effective masses of the electron in the corresponding bands. We have seen in Chapter 5, Eq. (5.7) that, for free electrons in a uniform potential, the density of states is, allowing for the two spin orientations, $g(\varepsilon) = (4\pi\Omega/h^3)(2m)^{3/2}\sqrt{\varepsilon} = K_0\sqrt{\varepsilon}$. For Bloch electrons, the electron mass is replaced by the *effective mass* in the considered band. Hence, taking $\Omega = 1$ (a volume equal to the unit volume) (cf. Eq. (7.61)):

$$K = \frac{4\pi}{h^3}(2m_V^*)^{3/2} \qquad K' = \frac{4\pi}{h^3}(2m_C^*)^{3/2} \qquad \rightarrow \frac{K}{K'} = \left(\frac{m_V^*}{m_C^*}\right)^{3/2}$$

$$(9.9)$$

Whenever the bands are not isotropic K and K' depend on the direction in reciprocal space, since the effective masses are then tensorial quantities (cf. (7.61)):

$$(\overline{\overline{m^*}})_{\lambda\mu} = (B^{-1})_{\lambda\mu} \quad \text{with} \quad (\overline{\overline{B}})_{ij} = \frac{1}{\hbar^2}\left(\frac{\partial^2 E(\vec{k}, n)}{\partial k_i \partial k_j}\right) \qquad (9.10)$$

Let us consider an alternate derivation of Eq. (9.8) by analyzing separately the situation in the valence and conduction band. This derivation has the advantage of distinguishing the contributions of electrons and of holes. In this view, if, at temperature T, there are n_C electrons in the conduction band, the same number of electrons have left the valence band and n_C states are empty. We have seen, in Chapter 7 (Section 7.3.5), that these states can be considered as occupied by "holes" in number $n_h = n_C$. These speculative particles are assigned (cf. Eq. (7.75)) the positive charge $+e$, the effective mass $-m_{V^*}$, where m_{V^*} is the effective mass of an electron in the valence band (Eq. 9.9), and a wavevector $(-\vec{k})$ where $(\vec{k})$ is the state occupied by the hole.

In the conduction band, we have, using Eqs. (9.6) and (9.5), and putting as in Eq. (9.7), $y = (E - E_C)/k_B T$, the number of electrons

n_c *per unit volume* in the conduction band

$$n_c = \frac{4\pi}{h^3}(2m_C^* k_B T)^{3/2} \exp\frac{E_F - E_C}{k_B T} \int_0^\infty \sqrt{y}.\exp(-y).dy$$

$$= N_C.\exp\frac{E_F - E_C}{k_B T} \tag{9.11}$$

in which the integral is, again, equal to $\sqrt{\pi}/2$. Likewise, the number of holes per unit volume in the valence band is also $n_h = n_c$. However, the occupation factor for holes is, for $k_B T \ll V_G$:

$$f_h = (1 - f) = \frac{1}{1 + \exp(E_F - E)/k_B T} \approx \exp(E - E_F)/k_B T$$

$$\tag{9.12}$$

Thus, the number n_h of holes in the valence band is:

$$n_h \approx \frac{4\pi}{h^3}(2m_V^* k_B T)^{3/2} \exp\frac{E_V - E_F}{k_B T}(\sqrt{\pi}/2)$$

$$= N_V.\exp\frac{E_V - E_F}{k_B T} \tag{9.13}$$

As $n_h = n_e$ the equality of (9.11) and (9.13) is equivalent to (9.7), thus establishing again (9.8)–(9.9).

Note that:

$$n_c.n_h = N_C.N_V \exp\left(\frac{-E_G}{k_B T}\right) = n_i^2 \tag{9.14}$$

is independent of the location of the Fermi level. Also note that n_i^2 only depends of the densities N_C and N_V, of the gap E_G, and of the temperature T. Hence, if, as will be considered in doped materials hereunder, additional quantum states are present in the gap, the product of the number of electrons per unit volume will remain equal to n_i^2.

9.2.2   Number of carriers, conductivity, and mobility

While in an insulator, as diamond, the factor $(E_C - E_F)/k_B T$ relative to the bottom of the conduction band is 120, at room temperature, in silicon and germanium it is respectively ≈ 22, and ≈ 15. The occupation factor of theses states will therefore be, in diamond, $f \lesssim 10^{-52}$, totally negligible. In silicon and germanium it will be of the order of $f \approx 10^{-6} - 10^{-10}$, in a range which determines a detectable number of electrons, at room temperature, in the states of the conduction band. In silicon, this number, provided by Eq. (9.11), is of the order of $n_e \simeq 10^8/\text{cm}^3$ while in a metal, this density is of the order of $10^{22}/\text{cm}^3$. Note that, due to the small value of the free-electron density in the conduction band and of the free-hole density in the valence band, the Fermi–Dirac statistics for these particles *reduces*, as already expressed in Eq. (9.5), *to the Boltzmann classical statistics*.

In a metal, the density of current is $\overrightarrow{j} = -e\langle\overrightarrow{v}\rangle$ where $\langle\overrightarrow{v}\rangle$ is the drift current generated by the field $\overrightarrow{\varepsilon}$. The conductivity σ, defined by $\overrightarrow{j} = \sigma\overrightarrow{\varepsilon}$ and is equal to (cf. Chapter 8):

$$\sigma = n_c \cdot \frac{e^2\langle\tau\rangle}{m*} \qquad (9.15)$$

On the other hand, the mobility is defined by $\overrightarrow{v} = \mu_0 \overrightarrow{\varepsilon}$. Thus $\mu_0 = e\langle\tau\rangle/m^*$.

In Eq. (9.15), n_c is the density of electrons in the conduction band, $\langle\tau\rangle$ the average of the time between collisions ($\tau \approx 10^{-14}$s cf. Table 8.1), and m^* the effective mass of the electron. In a metal, the carriers determining the conductivity are the electrons of the conduction band. However, in a semiconductor, there are two contributions to the electric current induced by an electric field. besides the "free electrons" of the conduction band, there is a current carried by the "free holes" of the valence band. Hence, one must write:

$$\sigma = n_c \frac{e^2\langle\tau\rangle}{m_e*} + n_h \cdot \frac{e^2\langle\tau\rangle}{m_h*} \qquad (9.16)$$

Let us estimate qualitatively the electronic conductivity σ of silicon. Taking as a first approximation (the actual values are reproduced on Table 9.2) $m_e^* \approx m_h^* \approx m \approx 9 \times 10^{-31}$ kg and $e \approx 1.6 \times 10^{-19}$ Cb, we find for silicon, at room temperature, $\sigma \approx 10^{-7}$ Ohm^{-1}.m^{-1} (or a resistivity $\rho \approx 10^5$ Ohm.cm) thus many orders of magnitude lower than in good metals for which $\sigma \approx 10^7$ Ohm^{-1}.m^{-1}. The different values of the effective masses in silicon and gallium arsenide will be further discussed in the next paragraph.

The most striking *qualitative* difference between a semiconductor and a metal resides in the *temperature dependence of the conductivity*, on heating above room temperature. As mentioned in Chapter 8, Section 8.2.4, the electron density in the conduction band of a metal does not vary significantly, while the average collision time $\langle \tau \rangle$ decreases slowly as $\propto (1/T)$. Thus, the conductivity of metals has a slow decrease on heating (cf. Fig. 8.6). In contrast, in an intrinsic semiconductor, there is a steep increase of the electron density in the conduction band and, relatedly of the holes in the valence band, governed by the temperature factor (cf. Eq. (9.11)) $T^{3/2} \exp(E_F - E_C)/k_B T$ or $T^{3/2} \exp(E_V - E_F)/k_B T$. In silicon $(E_F - E_C) \approx 0.56\,\text{eV}$. Hence, between $T = 300$ K $(k_B T = 0.025$ eV) and $T = 350$ K $(k_B T = 0.029$ eV) there is an increase of conductivity by a factor $(350/300)^{3/2}.\exp(3.1) \approx 30$.

9.2.3 Real band structures of intrinsic semiconductors

The schematic shape in $\overrightarrow{k}$ space of an intrinsic semiconductor has been recalled on Fig. 9.1, with the parabolic upward curvature of the conduction band, determining a positive effective mass for the electrons and the downward curvature of the valence band determining a negative effective mass for the electron-states (and thus a positive effective mass for the holes). Let us examine the real, and more complex, shapes of these bands in the density of states and in $\overrightarrow{k}$ space for two prominent examples of intrinsic semiconductors, namely *silicon* and *gallium arsenide*.[2] These band structures are determined by the crystal structures of these solids.

Crystal structures of silicon and gallium arsenide

These two semiconductors have the same type of structure as diamond (Fig. 1.5). They both possess a cubic Bravais lattice of the FCC type (cf. Figs. 2.15–2.22). An Si atom occupies each corner and face center of the conventional cubic unit cell (empty circles on Fig. 9.2). However, this FCC structure *is not a close packing one* (cf. Section 2.4.1) in which the Si atoms of the corners and face

---

[2]As will be briefly recalled in the next section, silicon is the basis of electronic technology, while gallium arsenide and its related compounds are the basis of optoelectronic technology.

centers would be in contact. This is due to the presence, in the tetrahedral voids (cf. Fig. 2.21) of the cubic cell, of four other Si atoms (shaded circles) too large to occupy the tetrahedral voids between the preceding atoms. The elementary unit cell of the FCC lattice thus contains two atoms, with each Si atom surrounded by four other Si atoms forming a tetrahedron. A similar configuration holds for gallium arsenide: the gallium atoms (Fig. 9.2) occupy the vertices and face centers of the conventional cubic cell while the arsenic atoms are in the tetrahedral sites.

Qualitative characteristics of the band structure

These two crystals are *covalent*, as diamond (cf. Chapter 5, Section 5.3) i.e. the bonding between neighbouring atoms involves *shared electron bonds*. Let us focus on the case of silicon. Each Si atom possesses 14 electrons. Taking into account the spin, the atomic states of an Si atom are $(1s)^2(2s)^2(2p)^6(3s)^2(3p)^2$. Among the 28 electrons of the two silicon atoms in each elementary unit cell 20 fill the lowest energy bands associated to the (1s), and (2s)(2p) states. The 3s and 3p atomic states give rise to the valence band of the solid. They generate $8N$ electron states. As in the case of diamond (Fig. 6.10), the s, p_x, p_y, p_z orbitals of two Si atoms are linearly combined to give 3sp orbitals pointing towards the four neighbouring Si atoms of the surrounding tetrahedron. As shown on Fig. 9.3, the cohesion of the crystal is ensured by bonds between neighbouring Si atoms, constituted by two shared valence electrons occupying the oriented 3sp orbitals. Hence, in each Si atom, there are $4 \times 2 = 8$ electrons occupying the eight 3sp electron states. The valence band (Fig. 9.3) generated by these atomic states, which contains $8N$ electron-states will be fully occupied at 0 K. This full occupation of the valence band

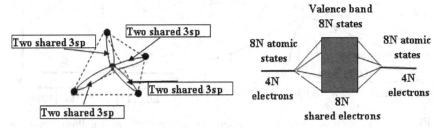

Fig. 9.3. Schematic generation of the valence band of silicon by 3sp orbitals.

explains the insulating character of silicon. Note that the density of electrons in the valence band is $\approx 2 \times 10^{29}/\mathrm{m}^3$. More excited atomic levels contribute to the formation of the conduction band. A similar scheme is valid for gallium arsenide.

Actual band structures of silicon and gallium arsenide

We have seen in Chapter 5, how the band dispersion curves $E(\overrightarrow{k})$ of a simple cubic crystal, with a monoatomic basis, can be calculated by the tight-binding approximation. For more complex structures such as silicon and gallium arsenide, one has to use numerical methods involving linear combinations of atomic orbitals. The results of such calculations are represented on Fig. 9.4, for the $E(\overrightarrow{k})$ dispersion curves, in two directions of k-space. Since the reciprocal Bravais lattices of those two cubic crystal are also cubic. The chosen directions in k-space are along the edge and the diagonal of the cubic cell.

Let us first describe, *for silicon*, the part of the $E(k)$ spectrum with $E \leqslant 0$ corresponding to the valence band. Three dispersion curves (L'_3 and below) are reproduced. A fourth band lying at lower energies is not represented. The qualitative analysis, described above, of the 3s3p orbitals determining the valence band, has stressed that these four bands should contain a total of $8N$ states. Hence, each

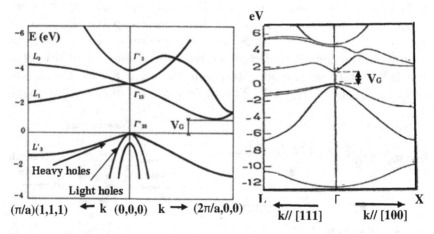

Fig. 9.4.   Valence and conduction bands of silicon (left) and gallium arsenide (right).

Table 9.2. Heavy and light effective masses in Si and GaAs.

Semiconductor band	m_H^*/m_0	m_L^*/m_0
Si-valence	0.49	0.16
Si-conduction	1	0.33
GaAs-valence	0.5	0.082
GaAs-conduction	–	0.07

of these bands contains $2N$ states, taking into account the spin-degeneracy. Besides, since two of the bands are tangent at $k = 0$, there is an additional degeneracy at this k-value (4 states instead of two). Calculations show that the third valence band, which lies below, should also be tangent at $k = 0$, if it were not for an interaction between the spin of the electron and its orbital momentum, which lifts the degeneracy and lowers the energy of the third band. The three curves show a downward curvature at $k = 0$. Hence, the corresponding *electron effective masses*, which are related to $(\partial^2 E/\partial k_i^2)$ are negative, and, accordingly, the *hole effective masses* are positive. Besides, at $k = 0$, the higher energy curve (L_3') is "flatter" than the lower ones. Since the effective mass at this k value is determined by $m^* = \hbar^2(1/\partial^2 E/\partial k_i^2)$, the dispersion curve (L_3') describes a hole with a "heavy" effective mass $m_{V,H}^*$, while the lower one describes a hole with a "light" mass $m_{V,L}^*$. The values of these masses are indicated in Table 9.2.

The overlapping bands forming the conduction band are separated from the valence band by a gap of $V_G \approx 1.1$ eV. A peculiarity of this gap is that the bottom of the conduction band is not situated at $k = 0$. Silicon has a, so-called, *indirect gap*. It is situated, *inside the first Brillouin zone*, in the (100) direction of the reciprocal cubic cell (Fig. 9.4). Thus, there are six symmetry related minima of the conduction band, and this implies a six-fold additional degeneracy at the bottom of this band. In Eq. (9.11), the effective degeneracy N_C (Eq. (9.14)) must include this factor of 6.[3] Also note that the effective mass of this band is a heavy mass.

---

[3] The bottom of the conduction band is also anisotropic. The "scalar" effective mass m_c^* must also be modified to take in account this feature (cf. Eq. (9.10)).

The band structure of gallium arsenide has different characteristics: the top of the valence band and the bottom of the conduction band are both situated at $k = 0$, with $V_G \approx 1.43$ eV. This gap is *direct*.

Optical implication of a direct or indirect gap

The fact that an intrinsic semiconductor possesses a direct or indirect gap has a consequence on its absorption or emission of light. A beam of light with frequency ν such as $h\nu > V_G$, can be absorbed by an electron of the valence band, and transfered to the conduction band. The light can also induce the inverse mechanism (stimulated emission) by "assisting" the transfer of an electron of the conduction band to an unoccupied state of the valence band. The latter process is a *recombination* of an electron and a hole. Both the absorption and stimulated emission of light can be considered as due to a collision between an electron and a photon of the light beam which results in an electron of different energy. For instance in the case of absorption of light, such a collision is submitted to two conditions:

$$E_{\text{valence electron}} + E_{\text{photon}} = E_V + h\nu = E'_{\text{conduction}} \qquad (9.17)$$

$$\overrightarrow{k}_V + \overrightarrow{K}_{\text{photon}} = \overrightarrow{k}_C \qquad (9.18)$$

In an indirect gap semiconductor, such as silicon, although the light beam has a frequency ν corresponding to $h\nu \gtrsim V_G$ the second equation cannot be satisfied. Indeed, as seen on Fig. 9.4, the difference $(\overrightarrow{k}_C - \overrightarrow{k}_V) \approx \overrightarrow{k}_C$ is of the order of $0.8.2\pi/a \approx 0.9 \times 10^{10}$ m^{-1}. In contrast, the photon wavevector is

$$\overrightarrow{K}_{\text{photon}} = \frac{2\pi h\nu}{hc} = \frac{2\pi . V_G}{hc} \approx 0.6 \times 10^7 \qquad (9.19)$$

thus one thousand times too small to satisfy Eq. (9.18). The situation is different in GaAs in which the gap is direct and $(\overrightarrow{k}_C - \overrightarrow{k}_V) \approx 0$. The implication of this difference is that in silicon, light with an energy close to the gap will not be absorbed by the above simple mechanism, while in GaAs the absorption (or emission) will be very efficient, hence its use in optical generation of light.

9.3 Doped Semiconductors

The technology based on semiconductors (and evoked briefly in Table 1.1 of Chapter 1) is not directly related to the small residual conductivity of intrinsic semiconductors evaluated with Eq. (9.15): ($\rho \approx 10^5 \, \Omega \, Ohm$.cm in silicon at 300 K). It relies on a much higher conductivity obtained by substituting deliberately "foreign" atoms (termed *dopants*), of various types, to the intrinsic components of the solid. Such impurities, which constitute defects of the crystal structure, determine for the electrons (or as will be seen, the holes) additional energy levels or bands (the impurity bands) located between the valence band and the conduction band of the intrinsic semiconductor. Depending on the chemical nature of the doping atoms, the additional band can be an empty band (at 0 K) located close in energy to the valence band of the intrinsic semiconductor (*acceptor band*). It plays a role similar to the conduction band, i.e. it can host electrons from the valence band and thus add holes to this band, but with a reduced gap. It can also be an occupied band (*donor band*) close to the conduction band. This impurity band will play a role similar to a valence band, a reservoir of electrons, separated from the conduction band by a small gap. More complex situations involving several additional bands can also occur. The resulting temperature factors $\exp[-(\varepsilon - E_F)/k_B T]$ or $\exp[-(E_F - \varepsilon)/k_B T]$ are then much larger than that of the intrinsic semiconductor. The resulting conductivity at finite temperatures is also higher and can be controlled, not only by the temperature, but also by the nature, and the concentration (the number per unit volume) of the doping atoms.

9.3.1 Donor impurity in silicon (n-doping)

Phosphorus (P) has a similar atomic size as the Si atom and can be chemically substituted to an Si atom in the crystal structure of silicon. However, the substituted P atom has five valence electrons in atomic states $(3s)^2(3p)^3$, four of which will participate to the four sp^3 bonding with the neighbouring four Si atoms (Fig. 9.5). The additional electron $(-e)$ is *loosely* bound to the positive ion P^+. Indeed, in the first place, this electron is not part of a "strong" covalent bond. More significantly, the electrostatic attraction between P^+ and e is much weakened in comparison with their attraction "in vacuum".

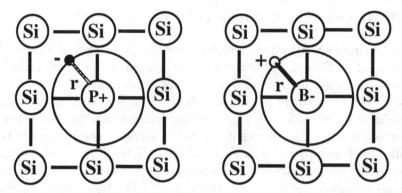

Fig. 9.5. Left: Substitution of a silicon atom by a phosphorus donor atom. Right: Substitution by a boron acceptor atom.

Indeed, this attraction is exerted in a medium with high dielectric constant ε_r (of the order of 12 in silicon). Instead of an electrostatic potential of attraction $V = e/4\pi\varepsilon_0 r$, equivalent to the potential in a hydrogen atom, the potential has the form $V = e/4\pi\varepsilon_0\varepsilon_r r$ 12 times weaker. This effect is due to the polarization of the particles of the crystal by the electric field, generating an antagonist electric field, which "screens" the applied field. The binding energy of this "hydrogenoid" (hydrogen-like) atom will be:

$$E_{\text{Donor}} = E_0 \cdot \frac{m^*}{m_0} \cdot \frac{1}{\varepsilon_r^2} \qquad (9.20)$$

where $E_0 = 13.5$ eV is the ground-state/ionization energy of the hydrogen atom. Hence, taking $(m^*/m_0) \approx 0.33$ (Table 9.2), and $\varepsilon_r \approx 12$, the binding energy of the additional electron is $E_{\text{Donor}} \approx 30$ meV. Supplying this energy to the electron will promote it in a "free" state, i.e. in a state of the conduction band. Hence, the donor state is an electron state situated 30meV below the conduction band (cf. Fig. 9.6). P is termed a *shallow donor*. Note that a donor state, which is analog to a hydrogen atom, *has a spin degeneracy of two*, although it contains at most one electron. This feature will affect the statistical occupation factor.

On the other hand, the radius of the hydrogenoid $e - P^+$ atom is $(r_D = r_0 \cdot \varepsilon_r \cdot m_0/m^*) \approx 36r_0 \approx 20$ Å. Its volume is $\approx 3 \times 10^{-20}$ cm^3. In order that the donor level remains a well-defined energy level, and does not enlarge into a band, by maintaining the various pairs

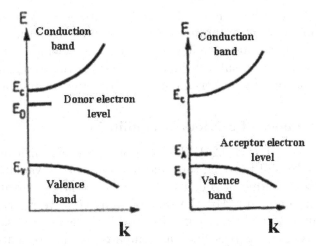

Fig. 9.6. Donor and acceptor level positions with respect to the conduction and valence bands.

$e - P^+$, well separated, the concentration of donors must be such that the radius of the various additional electrons do not overlap, i.e. that the concentration of P atoms be $\ll 10^{22}/\text{m}^3$, as compared to the $\approx 10^{29}/\text{m}^3$ of the "intrinsic" electrons.

9.3.2 Acceptor impurity in silicon (p-doping)

As in the preceding example of substitution, boron (B) has a similar atomic size as the Si atom and can be chemically substituted to an Si atom in the crystal structure of silicon. In this case, the substituted B atom has three valence electrons in atomic states $(3s)^2(3p)^1$. To form the four sp^3 bonding with the neighbouring Si atoms (Fig. 9.5), boron will use an electron of the valence band. It is an *acceptor*. We then have, on the one hand, a boron ion B^-, and, on the other hand, an empty state (a positive hole) in the valence band. Their mutual attraction will form, as in the preceding paragraph, an "hydrogenoid" atom, the binding energy of which can be evaluated, as in Eq. (9.19). Taking into account, again, the dielectric "screening", by ε_r, of the attraction, and the effective mass, m_h^* of the hole, one determines a binding energy of the neutral pair $hole - B^- \approx 45$ meV. Supplying this energy to the pair will ionize it and the hole will be restored as a

free-hole of the valence band. The bound state of the pair $hole - B^-$
is therefore ≈ 45 meV above the valence band (Fig. 9.6).

Similar dopings are used in gallium arsenide in which, for instance
atoms of silicon Si or sulfur S (donors) are substituted to gallium and
in which magnesium or beryllium are substituted to arsenic.

9.3.3   Number of carriers at equilibrium

In Sections 9.2.1 and 9.2.2, the location of the Fermi level as well
as the numbers of free-electrons and holes has been evaluated for an
intrinsic semiconductor. Let us consider here these quantities for a
doped semiconductor. For the sake of simplicity, we restrict to the
case of an *n-type doping*, e.g. Si doped by arsenic, whose concentration of dopants N_D is small as compared to the concentration of
silicon atoms (e.g. $10^{20}/m^3$ as compared to $10^{29}/m^3$).

Qualitative analysis

At 0 K all the electrons are either in the valence band or in the
donor band. The highest occupied level being in the impurity band,
and the conduction band being empty, the Fermi level is located
between the impurity band and the conduction band (Fig. 9.7). For
$0 < k_B T \lesssim (E_C - E_D)$, the first electrons transfered to the conduction band are those of the donor-impurity band. Hence, the "tail"
of the Fermi function covers the energy range between E_D and E_C,
and the Fermi level is progressively lowered below the donor band.
For higher temperatures, the donors are completely ionized, and a

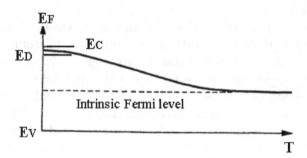

Fig. 9.7.   Evolution of the Fermi level as a function of temperature for an n-doped
semiconductor.

certain proportion of the electrons of the valence band is transfered to the conduction band. Their number becomes dominant at higher temperatures, and the behaviour of the semiconductor becomes similar to that of the intrinsic material. Hence, the Fermi level progressively approaches its location in the intrinsic semiconductor, i.e. close to the middle of the gap (cf. Fig. 9.7).

Quantitative analysis

In presence of a donor level possessing N_D states *per unit volume*, the electrons in the conduction band have their origin either in the donor band (n_D electrons per unit volume) or in the valence band (in number equal to the number p of holes in this band). Their number n_c is thus

$$n_c = p + n_D \tag{9.21}$$

On the other hand n_c is related to the location of the Fermi level E_F through

$$n_c = N_C \exp(E_F - E_C)/k_B T \tag{9.22}$$

The former Boltzmann form of the occupation factor is justified since (n_c/N_C) is small. In contrast, the occupation factor to be used in evaluating (n_D/N_D), which is not necessarily small, is the standard Fermi factor, also taking into account (as mentioned in Section 9.3.1) the spin degeneracy of the donor band.

$$f_D = \frac{1}{1 + \frac{1}{2}\exp(E_D - E_F)/k_B T} \tag{9.23}$$

Thus, the number of electrons present in the donor band at a temperature T is:

$$N_D - n_D = N_D.f_D \tag{9.24}$$

We can combine the Boltzmann factor in n_c, and the Fermi factor f_D to express the Fermi factor as function of the other quantities.

Thus

$$\exp(E_F/k_BT) = (n_c/N_C)\exp(E_C/k_BT) \qquad (9.25)$$

and (Eq. (9.24))

$$\exp(E_F/k_BT) = \frac{N_D - n_D}{2N_D}\exp(E_D/k_BT) \qquad (9.26)$$

By equalling the two expressions, the position of the Fermi level is determined as

$$E_F = \frac{E_C + E_D}{2} - k_BT.\mathrm{Ln}\left(\frac{2N_C}{N_D}\right) \qquad (9.27)$$

As shown on Fig. 9.7, at $T = 0$ K, the Fermi level is at half distance between the donor band and the conduction band. Besides, as $N_C \gg N_D$ its position decreases on increasing the temperature. For instance, for a moderate doping, $(2N_C/N_D) \approx 10^5$, which determines a position of the Fermi level $\approx 10k_BT \approx 250$ meV, hence significantly below the donor level, which, for arsenic in silicon, is 30 meV below the conduction band.

We now use Eq. (9.24) to eliminate n_D and establish a relation between the number of electrons n_c and of holes p as a function of temperature:

$$n_c(n_c - p) = \frac{N_C.}{2}(N_D - N_C + p)\exp(E_D - E_C)/k_BT \qquad (9.28)$$

For low temperatures, the Fermi level is sufficiently distant from the valence band for the number p of holes to be negligible ($p = 0$). The above equation then reduces to:

$$n_c \approx \sqrt{\frac{N_C.N_D}{2}}\exp[(E_D - E_C)/2k_BT] \qquad (9.29)$$

On Fig. 9.8, this behaviour is represented in a $\mathrm{Ln}(n_c)$ versus $(1/T)$ diagram by a straight line, of slope $(E_C - E_D)/2k_B$.

For a certain range of higher temperatures, the Fermi level lies sufficiently below E_D to ensure that the donor level has transfered all its electrons to the conduction band. In this range, it is also sufficiently distant from the valence band, and this will result in the fact that the number of holes in the valence band will remain very small. In this

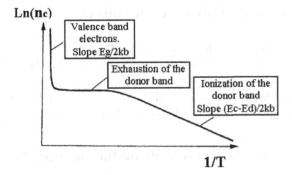

Fig. 9.8. Three regimes of electrons transfer to the conduction band in an n-doped semiconductor.

range, the number of electrons in the conduction band remains almost constant. It is the *saturation regime* (exhaustion of the population of the donor band). Finally, at still higher temperatures, electrons of the valence band, in number larger than N_D, will become dominant in supplying in electrons the conduction band. As mentioned above, the doped semiconductor has a behaviour resembling the intrinsic regime. Equation (9.14) applies with $n_c \approx p$:

$$n_c \approx \sqrt{N_C.N_V} \exp\left(\frac{-E_G}{2k_BT}\right) \qquad (9.30)$$

This regime is represented on Fig. 9.8.

A similar calculation can be performed in the case of a p-doped semiconductor. By analogy with the case of n-type semiconductor, the Fermi level at 0 K will be situated between the acceptor level and the valence band. On heating, it will evolve upward in energy and reach at higher temperatures, the location close to the middle of the gap, as in an intrinsic semiconductor. Likewise, on heating progressively from 0 K, electrons from the valence band will be transferred to the acceptor band until a "saturation" regime occurs in which the number of holes in the valence band is almost constant and equal to N_A (number of acceptor states in the acceptor band). Finally, on further heating, valence band electrons will be transferred to the conduction band as in the intrinsic material.

In a multiple doped (n and p) semiconductor, the situation can be more complicated, depending of the respective amounts of n and p-dopings.

9.4 Principles of Two Semiconductor Devices

In this section, we outline the basic principles of structures which landmarked the use of semiconductors in industrial technology, namely the p–n junction (semiconducting diode), and the transistor. In these devices, one uses the spatial contact between layers of semiconductors (or insulator) having different types of doping.

9.4.1 P–n junction and semiconducting rectifier

A p–n junction is the superposition of one layer of p-doped semiconductor (e.g. silicon doped with boron) and of a layer of n-doped semiconductor (e.g. silicon doped with arsenic), which can exchange electrons and holes through their common surface.

Consider, first, these two layers as spatially separated. Their energy scheme within each layer is represented on Fig. 9.9. As noted in Section 9.3 and on Fig. 9.7 the Fermi level E_F in the n-doped region, *at low enough temperature*, is situated slightly above or slightly below the donor level E_D. Likewise, in the p-doped region it is situated in the vicinity of the acceptor level E_A. As described in the preceding section, the n-doped layer contains free-electrons in the conduction band, and ionized positive donor atoms (e.g. As^+ ions), hosts of electron states in the donor band. In the p-doped

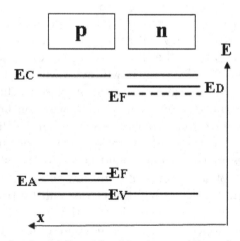

Fig. 9.9. Energy levels and Fermi level in the two (distant) p-doped (left) and n-doped (right) semiconductor.

layer, there are free-holes in the valence band, and ionized negative ions (e.g. B^-). *Each layer is locally neutral.*

Equilibrium in the absence of external field

Qualitative description. Now consider the two layers in contact in such a way as to permit the passage of carriers (electrons, holes) from one layer to the other.[4] Due to the difference in concentration of free-electrons and free-holes in the two regions there will be a *migration* of free-electrons electrons from the n-region to the p-region and a migration of holes from the p-region to the n-region.

The driving force of this transfer is the fact that, as stressed in Chapter 6 (Section 6.1), when two systems in contact can exchange particles, the Fermi energy has the same value *at equilibrium*, in the two systems. Indeed, the Fermi energy has a similar property with respect to the exchange of particles, as the temperature with respect to the exchange of heat. Thus, two systems in thermal contact will exchange heat until equilibrium is reached and their temperatures equalize. Likewise, in two systems in contact which can exchange particles, equilibrium will correspond to an equal Fermi energy E_F in the two systems. Before equilibrium is reached, the Fermi level in the p-region is close to the valence band. Hence, the occupation factor of the conduction band of the p-region is almost zero. As a consequence, the free-electrons transfered from the n-region will recombine with holes and occupy states of the valence band. Conversely, free-holes transfered to the n-region, will combine with electrons, originating from the conduction band of the n-region. Three main effects result from the transfer of charges.

(a) The transfer of charges from one layer to the other has the effect that the electric neutrality in each layer is no longer preserved. The p-layer in which electrons migrate from the n-layer, becomes negatively charged. Likewise the n-layer, in which holes from the p-layer migrate, becomes positively charged. As a consequence, an electric field $\vec{\mathcal{E}}$ is generated by this unbalance of charges,

---

[4]The fabrication of the p–n junction and of other devices using "wafers" of silicon, has an interesting historical and technological background which can be found in the references at the end of this chapter.

directed, across the contact surface, *from the n-region (positively charged) to the p-region (negatively charged).* This field has the effect of acting on the electrons $(-e)$ to flow from the p-layer to the n-layer. Hence, *the field* has an *antagonist effect* to the inverse flow induced by *the diffusion* of charges. Similarly, its action on the holes will be antagonist to that of the diffusion, i.e. it induces a drift of the holes towards the p-region. When these two effects (field induced drift and diffusion) compensate, an equilibrium is reached, and the Fermi level is the same in the entire system.

(b) Since the diffusion process is a local effect, the spatial range of the n- and p-layers most concerned by the migration of charges is a layer (more or less extended) surrounding the contact surface of the two regions.The recombination of electrons with holes in this layer has the effect of leaving few free charges. This *depleted region* is also the region in which the electric field is generated. It is called the *space-charge zone,* in which the electric charges present are predominantly the ionized atoms (donors and acceptors) cf. Fig. 9.10.

(c) The electric field $\vec{\varepsilon}$ derives from an electrostatic potential Φ. This potential is exerted on the particles (electrons, holes) of the crystal and will therefore modify the band structure of the semiconductor which is determined by the crystal interactions.

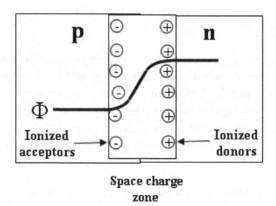

**Space charge
zone**

Fig. 9.10.   Transfer of charges between the p and n layers in contact.

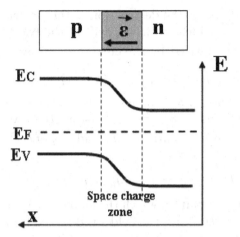

Fig. 9.11. p–n junction at equilibrium in the absence of external field. The presence of an electric field and of the related potential Φ, in the space-charge zone, has the effect of changing the shape of the conduction and valence bands.

Hence, the bands will have a different energy, mainly in the space-charge region. This is particularly the case of the conduction band and the valence band as shown on Fig. 9.11.

Quantitative description. Let x be the coordinate perpendicular to the contact surface of the two layers, the axis being oriented from the n-region to the p-region. Denote $\overrightarrow{\delta}_e$ and $\overrightarrow{\delta}_h$ the diffusion current densities of electrons and of holes, i.e. the respective numbers of these particles crossing the unit surface per unit time, and induced by the variable concentration of electrons and holes. If $n(x)$ and $p(x)$ are the local concentrations of electrons and holes, the diffusion equations can be written as

$$\overrightarrow{\delta}_e = -D_e \overrightarrow{\nabla}(n) \Rightarrow -D_e \frac{\partial n(x)}{\partial x} \tag{9.31}$$

and

$$\overrightarrow{\delta}_h = -D_h \overrightarrow{\nabla}(p) \Rightarrow -D_h \frac{\partial p(x)}{\partial x} \tag{9.32}$$

Each of these diffusions of charged particles ($n \Rightarrow -e$) and ($p \Rightarrow +e$) carries an *electric* diffusion current, respectively $J_{D,e}(x) = eD_e \partial n/\partial x$ and $J_{D,h}(x) = -eD_h \partial p/\partial x$.

As explained above, the destruction of the electric neutrality generates the field $\vec{\varepsilon}$ which induces a drift of the two types of particles respectively characterized by the current densities $\vec{j}_e$ and $\vec{j}_h$:

$$\vec{j}_e = ne\mu_e\,\vec{\varepsilon} \qquad \vec{j}_h = pe\mu_h\,\vec{\varepsilon} \tag{9.33}$$

In which μ_e and μ_h are the mobilities of the electrons and of the holes. The field being in the positive x-direction, we can write that, at equilibrium, the total currents $\vec{J}_e$ and $\vec{J}_h$ of each type of particle is equal to zero:

$$J_e = j_e + eD_e\partial n/\partial x = 0 \qquad J_h = j_h - eD_h\partial p/\partial x = 0 \tag{9.34}$$

The potential $\Phi(x)$ from which the field ε derives satisfies the relation $\varepsilon(x) = -\partial\Phi/\partial x$. Since the field ε is directed from the n-region to the p-region, the potential $\Phi(x)$ increases in the opposite direction, since $\varepsilon(x) = -\partial\Phi/\partial x$. There is a barrier potential, in the space-charge zone, which opposes, at equilibrium, the drift of holes from the p-region to the n-region, and the drift of the electrons from the n-region to the p-region (cf. Fig. 9.10).

On the other hand, the potential energy of an electron in the potential Φ is $(-e\Phi)$. For an electron in the conduction band this (negative) energy is added to the energy E_C of the conduction band. Hence, the occupation of the conduction band will be modified with respect to the value in Eq. (9.22) and become:

$$n_c(x) = N_C \exp \frac{E_F - E_C + e\Phi(x)}{k_B T} \tag{9.35}$$

from which we obtain, since only the potential Φ depends on x in the second member,

$$\frac{1}{n_c}\frac{\partial n_c}{\partial x} = \frac{e}{k_B T}\cdot\frac{\partial\Phi}{\partial x} = -\frac{e}{k_B T}\cdot\varepsilon \tag{9.36}$$

Using (9.33), (9.34) and (9.36) leads to a relation between the diffusion coefficient and the mobility of electrons:

$$D_e = \mu_e\cdot\frac{k_B T}{e} \tag{9.37}$$

A similar relation holds for holes:

$$D_h = \mu_h\cdot\frac{k_B T}{e} \tag{9.38}$$

The two latter relations are known as *Einstein relations*.

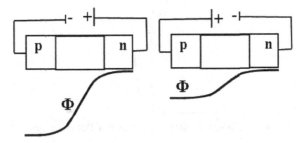

Fig. 9.12.   p–n junction polarized by an "inverse" external electric field directed from region n to region p (left) or by a "direct" field directed from region p to region n (right).

Equilibrium with an applied external field

If an external electric field ε' is applied to the p–n junction with the positive voltage connected to the n-region (cf. Fig. 9.12, left), inside the junction, this internal field will be in the same direction as the field ε generated by the unbalance of charges induced by the transfer of electrons and holes. It will increase the value of the field $(\varepsilon + \varepsilon')$ and that of the barrier potential Φ in the space-charge zone. As stressed above, this barrier will oppose more efficiently the flow of carriers from one layer to the other. Besides, the space-charge zone, being depleted of carriers, the current induced by the external field will be very small.

If the positive voltage is applied to the p-region (Fig. 9.12, right) the field ε' is applied in the opposite direction and the resulting field $(\varepsilon - \varepsilon')$ will be associated to a potential Φ having a smaller amplitude. Hence, free-carriers will flow more efficiently from the p-region to the n-region, and a large current can be generated by application of the voltage.

The resulting current versus voltage characteristics of the polarized p–n junction is represented on Fig. 9.13. This curve explains the "rectifying" property of the p–n junction. Indeed, apply to the junction an alternating (AC) voltage (Fig. 9.14) to the junction, at each negative half-period of the voltage a very small current will pass in a resistor inserted in the circuit, which will generate a negligible voltage in the resistor. During the positive half-period a much larger current flows through the resistor. Hence, the average voltage thus generated is positive.

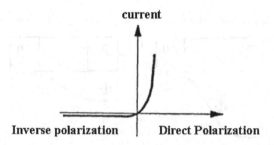

Fig. 9.13.   Current versus voltage in a p–n junction.

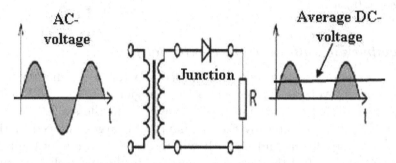

Fig. 9.14.   Schematic configuration of a voltage rectifier using a p–n junction. The incoming voltage alternately polarizes the junction from p to n and from n to p.

9.4.2   P–n–p transistor

A transistor is the superposition of *three* doped layers of semiconductors, either a n–p–n sequence, or a p–n–p sequence of layers. The intermediate layer is designed to be much thinner than the two surrounding layers. e.g. 50 microns for the external layers and 5 microns for the inner layer. Besides there is a small concentration of dopants in the n-region, with the result that a small number of electrons are available to recombine with holes. Such a structure can either be used as an amplifier of electric signal or as a fast electronic switch. Let us examine here the principle of operation of a p–n–p transistor.

In the absence of external voltages (Fig. 9.15 (a)), the p–n–p transistor is a succession of a p–n junction and of an n–p junction. Following the transfer of electrons and holes between layers, as described in the preceding section (see Fig. 9.10), the equilibrium generates in an n–p-junction the onset of a space-charge zone, depleted from

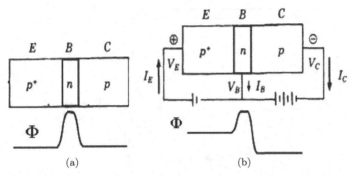

(a)   (b)

Fig. 9.15. (a) P–n–p transistor at equilibrium in the absence of polarization. (b) Polarized transistor.

carriers, and of an upward potential barrier in the $p \to n$ junction. A symmetric barrier is set in the $n \to p$ junction. In principle this barrier will oppose the drift of free-holes from p to n. However, *due to the small width of the n-layer as compared to the diffusion length of holes*, a diffusion-driven current of holes will enter deeply enough in the n-layer to reach the neighborhood of the p-region. The downward $n \to p$ potential then allows the hole to reach the other p-layer. At equilibrium, the symmetry of the double junction ensures that no global current crosses the device.

Let us now insert the p–n–p transistor in an electric circuit (Fig. 9.15 (b)) in which a positive polarization is applied to the p+ layer (called the *emitter*) with respect to the n-layer (called the *base*). On the other hand, a negative polarization is applied between the other p-layer (called the *collector*) and the n-layer. As shown on Fig 9.12, the first polarization will *decrease* the height of the barrier between the emitter p+ and n, while the second *increases* the magnitude of the barrier between the collector p and the basis n.

In the p+ region, free-holes drift towards the n-region under the action of the positive voltage between the emitter and the basis. The barrier (although diminished) opposes their crossing to the n-region. However, as mentioned above, the diffusion-assisted drift of holes will allow a positive current to penetrate the n-layer. When a hole penetrates the n-layer, the large downward potential n–p carries it to the collector. This "evacuation" towards the collector permits again a substantial diffusion of holes from the p+ region to the n-region, thus repeating the process. As a result, a relatively large current

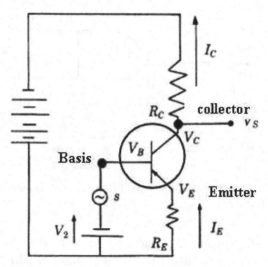

Fig. 9.16.   Schematic configuration of an amplifier of AC current using a p–n–p transistor.

can cross from the emitter to the collector. Besides, few holes are deviated towards the base since the concentration of electrons which can recombine with the holes is small. Thus a small fraction of the charges which penetrate the n-region are carried by the electric field towards the basis, inducing a small current in the basis-emitter external circuit. If the voltage between the base and the emitter is slightly increased, more holes will be injected by the emitter, the majority of which will flow to the collector. More electrons will then be injected in the base which will recombine with a fraction of the holes. Thus a relationship exists between the current in the base and the current in the collector which is a proportionality with a large ratio β. Inserting the transistor into a circuit with resistors and a source of AC voltage (cf. Fig. 9.16) will allow to take advantage this proportionality and amplify the voltage of the source s into the collector voltage V_s. The coefficient of amplification will be equal to β which can be of the order of 100.

Bibliography

Electrons and Holes in Semiconductors with Applications to Transistor Electronics, *W. Shockley*, D. Van Nostrand (1950).

Semiconductor Physics, *K. Seeger*, Springer (1973).

Semiconductors, *R.A. Smith*, Springer (1978).

Introduction to Solid State Electronics, *F.F.Y. Wang*, North-Holland (1989).

Physique des Semiconducteurs (In French), *B. Sapoval and C. Hermann*, Ellipses (1990).

Quantum Theory of the Optical and Electronic Properties of Semiconductors, *H.H. Haug*, World Scientific (2009).

Chapter 10

Solids as Systems of Particles in Interaction

Main ideas: Basis of the model of independent electrons: Born–Oppenheimer and Hartree approximations (Section 10.2). Some structural properties of solids, phonons (Section 10.3)

10.1 Introduction

In the preceding chapters, the electron in a solid has been considered in the framework of the so-called *one-electron approximation*. In this approximation, the electrons are isolated particles only submitted to a potential $V(\vec{r})$ representing, in a simplified manner, the interactions existing between an electron and the other particles of the solid (nuclei and other electrons). In the present chapter, we examine the physical basis of the former description and we specify its underlying approximations. We also briefly examine the physical phenomena related to the interactions between nuclei.

10.2 Justification of the Independent Electrons Approximation

A more complete quantum description of a solid consists in considering the electrostatic interactions between all its constituting particles[1] namely the various nuclei, each defined by position $\vec{R}_j$, momentum $\vec{P}_j$ mass M_j and electric charge $Z_j e$, and the electrons defined by the sets ($\vec{r}_i, \vec{p}_i, m, -e$). These interactions determine the following form for the Hamiltonian:

$$\widehat{H} = \widehat{H}_{NN} + \widehat{H}_{eN} + \widehat{H}_{ee} \tag{10.1}$$

in which the terms of the second member are:

$$\widehat{H}_{NN} = \left\{ \sum_{\mu} \frac{\vec{P}_{\mu}^2}{2M_{\mu}} \right\} + \left\{ \sum_{\mu,\nu} \frac{Z_{\mu} Z_{\nu} e^2}{|\vec{R}_{\mu} - \vec{R}_{\nu}|} \right\} = \{K_N\} + \{V_{NN}\}$$

$$\tag{10.2}$$

relative to the momentum and coordinates of the various nuclei,

$$\widehat{H}_{eN} = \sum_{i} \left\{ \frac{\vec{p}_i^2}{2m} - \sum_{\mu} \frac{Z_{\mu} e^2}{|\vec{r}_i - \vec{R}_{\mu}|} \right\} = K_e + V_{eN} = \sum_{i} \widehat{h}(i)$$

$$\tag{10.3}$$

relative to the electron-momentum and to the interaction between each electron and all the nuclei, and

$$\widehat{H}_{ee} = \frac{1}{2} \sum_{i \neq j} \frac{e^2}{|\vec{r}_i - \vec{r}_j|} \tag{10.4}$$

which is the interaction-potential between electrons. *The electrostatic units have been chosen as to make* $4\pi\epsilon_0 = 1$.

Note that such a Hamiltonian is not specific of a solid. It describes any system formed by the same constituents (nuclei and electrons) interacting electrostatically, such as, for instance, the liquid and vapour phases having the same chemical composition. Its adaptation to the situation in a crystalline solid results from some of the approximations which we now examine.

---

[1]The nature of the arguments developed in order to justify the single electron approximation would not be different if magnetic interactions between the electrons or between the electrons and the nuclei were also considered.

10.2.1 Born–Oppenheimer approximation

The term H_{eN} contains both the coordinates $\vec{r}_i$ of the electrons and $\vec{R}_\mu$ of the nuclei. Such a term has the consequence that, in principle, one cannot determine separately the quantum state of the electrons and that of the nuclei. A first approximation aims at realizing such a separation. It consists in assuming that the state of the electrons can be determined by considering that the nuclei occupy fixed positions in space, and are not subject to motion.

This assumption is consistent with the fact that the ratio of the masses of the nuclei and of the electrons being of the order of 10^4, the velocities of nuclei are orders of magnitude less than the velocities of electrons. At room temperature, the average kinetic energy of nuclei is $(P^2/2M) = 3k_BT/2 \cong 4.10^{-2} eV$ while, as noted in Chapter 6, the average kinetic energy of electrons is of the order of several eV. The ratio of velocities is thus $\sim 10^3$.

The practical implementation of this assumption consists in searching the electronic wavefunction of the N electrons of the solid in the form $\Psi_{el}(\vec{r}_1, \ldots, \vec{r}_N | \vec{R}_\mu)$ in which the coordinates $\vec{R}_\mu$ of the nuclei are *fixed parameters*, set to values $\vec{R}_\mu^0$. Ψ_{el} is then determined as a solution of the *electronic Schrödinger equation*:

$$[\widehat{H}_{eN}(\vec{R}_\mu^0) + \widehat{H}_{ee}] \quad \Psi_{el}(\vec{r}_1, \ldots, \vec{r}_N | \vec{R}_\mu^0)$$

$$= E_{el}(\vec{R}_\mu^0).\Psi_{el}(\vec{r}_1, \ldots, \vec{r}_N | \vec{R}_\mu^0) \qquad (10.5)$$

If it can be solved, this equation will also provide the values of the electronic energies $E_{el}(\vec{R}_\mu^0)$ as functions of the $\vec{R}_\mu^0$ parameters. We can now go back to the complete Schrödinger equation of the solid $\widehat{H}\Phi = E_T\Phi$, postulating that the wavefunction is of the form $\Phi = \Psi_N(\vec{R}_\mu)\Psi_e(\vec{r}_1, \ldots, \vec{r}_N | \vec{R}_\mu)$. In the operator $\widehat{H_{NN}}$, only $\sum(\vec{P}_\mu^2/2M_\mu)$ involves second derivatives with respect to the R_μ. The other terms are mere multiplicative terms. Hence, by noting that:

$$\frac{\partial^2}{\partial^2 R_\mu^2}\Psi_N(R_\mu)\Psi_e(r_j \mid R_\mu) = 2\frac{\partial \Psi_N(R_\mu)}{\partial R_\mu}.\frac{\partial \Psi_e(r_j \mid R_\mu)}{\partial R_\mu}$$

$$+ \Psi_e\frac{\partial^2 \Psi_N}{\partial^2 R_\mu^2} + \Psi_N\frac{\partial^2 \Psi_e}{\partial^2 R_\mu^2} \qquad (10.6)$$

and inserting (10.5) into $\widehat{H}\Phi = E_T\Phi$, we obtain:

$$\widehat{H}\Phi = E_T\Phi = \Psi_e \left[\widehat{H}_{NN} + E_{el}(\overrightarrow{R}_\mu) \right] \Psi_N - A \qquad (10.7)$$

with

$$A = \sum_\mu \frac{\hbar^2}{M_\mu}(\partial\Psi_N/\partial R_\mu)(\partial\Psi_e/\partial R_\mu) + \sum_\mu \frac{\hbar^2}{2M_\mu}\Psi_N(\partial^2\Psi_e/\partial R_\mu^2)$$

$$(10.8)$$

The A term is small and can be neglected. Indeed, compare these two terms to E_{el} which is of the same order of magnitude as the kinetic energy of the electrons $\approx (\hbar^2/2m)(\partial^2\Psi_e/\partial r_i^2)$. Assuming that all derivatives have the same order of magnitude, we can see that these terms are such as $(|A|/E_{el}) \approx (m/M_\mu)$, in agreement with the qualitative inference stated in the beginning of the paragraph.

Hence, we find that the "nuclear" wavefunction Ψ_N is a solution of the equation:

$$\widehat{H}_{\text{Nuclei}}\Psi_N = \left[\widehat{H}_{NN} + E_{el}(\overrightarrow{R}_\mu) \right] \Psi_N = E_T\Psi_N \qquad (10.9)$$

which also determines the total eigen-energy E_T of the solid (which, in contrast with $E_{el}(\overrightarrow{R}_\mu)$, is a mere constant). We can see that in Eq. (10.9) the potential energy is $[V_{NN} + E_{el}(\overrightarrow{R}_\mu)]$. Since the nuclei are in a collective *bound state* (they remain at close distance from each other), hence of *negative total energy* E_T, and that V_{NN} is a positive repulsive potential, this implies that E_{el} plays the role of an *attractive* potential between the nuclei. This is consistent with the fact that, as in molecules, electrons form the bonds between the constituents nuclei of the solid. Conversely, in Eq. (10.5) the negative, attracting, part of the potential, is the interaction V_{eN} between the electrons and the nuclei. This reveals the complex nature of the separation of the solid into two subsystems (electrons, and nuclei) since the stability of each of these subsystems is ensured by the configuration of the other subsystem.

The principle of the determination of the overall configuration of the solid, deriving from this mutual dependence, is discussed in Section 10.3.1. Let us first assume, in the two following paragraphs, that the nuclear positions $\overrightarrow{R}_\mu$ are set to the values $\overrightarrow{R}_\mu^0$ *experimentally determined* for the crystal structure of the solid (e.g. by X-ray diffraction cf. Chapter 4). Equation (10.5) is then entirely specified, and the determination of the collective electronic wavefunction is a well defined problem. However, its solution requires further approximations.

10.2.2 Hartree solution of the electronic equation

As emphasized by Eq. (10.3), H_{eN} is a sum of terms $\widehat{h}(i)$, which can be qualified as "monoelectronic" since each such term depends of the coordinates and momentum of a single electron:

$$\widehat{h}_i = \frac{\overrightarrow{p}_i^2}{2m} - \sum_\mu \frac{Z_\mu e^2}{|\overrightarrow{r}_i - \overrightarrow{R}_\mu|} \qquad (10.10)$$

If the interaction between electrons $\widehat{H}_{ee}$ could be neglected, one would be able to reduce the determination of the global electronic wavefunction Ψ_{el} to that of single-electron wavefunctions $\phi_i(\overrightarrow{r}_i)$, which are solutions of the one-electron Schrödinger equation $\widehat{h}(\overrightarrow{r}_i)\phi_i(\overrightarrow{r}_i) = \epsilon_i\phi_i(\overrightarrow{r}_i)$. Indeed, if we write

$$\Psi_{el}(\overrightarrow{r}_1, \overrightarrow{r}_2, \ldots, \overrightarrow{r}_N) = \prod_{1,N} \phi_i(\overrightarrow{r}_i) \qquad (10.11)$$

we obtain

$$\widehat{H}_{eN}\Psi_{el} = \left(\sum_{1,N} \epsilon_i \right) \Psi_{el} \qquad (10.12)$$

since each $\widehat{h}(i) = \widehat{h}(\overrightarrow{r}_i)$ only acts non-trivially on the ith single-electron wavefunction. In such a situation, we can see that not only the single-electron model considered in the preceding chapters is validated in a simple manner, but also that the periodic potential $V(\overrightarrow{r})$ introduced in this model is merely equal to the sum of the attractive Coulomb potentials of the various nuclei:

$$V(\overrightarrow{r}) = -\sum_\mu \frac{Z_\mu e^2}{|\overrightarrow{r} - \overrightarrow{R}_\mu|} \qquad (10.13)$$

However, neglecting $\widehat{H}_{ee}$ is obviously not correct. Indeed, the electrostatic repulsion $\widehat{H}_{ee}$ between electrons is of the same order of magnitude as the attraction exerted by the nuclei, since the shortest distances between electrons and those between electrons and nuclei are comparable. Accordingly, the exact resolution of the Schrödinger equation (10.5) cannot, in general, be achieved. It is then of interest to examine methods of determination of an approximate form of the electronic states Ψ_{el}.

With this purpose, several such methods of increased sophistication have been described.

The simplest one is the *Hartree method* in which the presence of the electrons other than the one[2] occupying the considered electron-state is replaced by a mean charge density $-e\delta(\overrightarrow{r})$. The potential energy resulting from this density, which replaces the contribution of the ith electron to $\widehat{H}_{ee}$ is:

$$h'(i) = h'(\overrightarrow{r}_i) = e^2 \int_{\overrightarrow{r}} \frac{\delta(\overrightarrow{r})}{|\overrightarrow{r}_i - \overrightarrow{r}|} d\overrightarrow{r} \qquad (10.14)$$

This replacement has the advantage of reducing the electronic Hamiltonian to a sum of single-electron terms $[\widehat{h}(i) + \widehat{h}'(i)]$ and allowing the determination of the electronic wavefunction as in Eqs. (10.11)–(10.12), provided that the density $\delta(\overrightarrow{r})$ is known.

In the framework of this approximation, we find again a *validation of the independent electron model* developed in the preceding chapters, with a potential $V(\overrightarrow{r})$ of the form:

$$V(\overrightarrow{r}) = -\sum_{\mu} \frac{Z_\mu e^2}{|\overrightarrow{r} - \overrightarrow{R}_\mu|} + h'(\overrightarrow{r}) \qquad (10.15)$$

Physically, the effect of the positive term $h'(\overrightarrow{r})$ is to "screen" partly the attraction of an electron by the nuclei.

The difficulty of this approach is that the density $\delta(\overrightarrow{r})$ is not known beforehand. It is taken equal to the sum of the probability

---

[2]For the sake of simplifying the description we distinguish one specific electron in spite of the indiscernibility of electrons. Strictly speaking, the distinction bears on a given occupied electron-state.

densities determined by the $(N-1)$ occupied electron-states other than the ith one considered[3]:

$$\delta(\overrightarrow{r}) = \sum_{j \neq i} |\phi_j|^2 \qquad (10.16)$$

Its knowledge therefore relies on the determination of the functions ϕ_j which itself relies, through solving of the one-electron Schrödinger equation, on the expression of $\delta(\overrightarrow{r})$. Hence, the determination of, both the single-electron wavefunctions ϕ_j, and the density $\delta(\overrightarrow{r})$ (and henceforth of the independent electron potential $V(\overrightarrow{r})$), result from an *iterative procedure*:

$$\{\phi_i^0\} \to \delta_0(\overrightarrow{r}_i) \to \{\phi_i^{(1)}\} \to \delta_1 \to \cdots \to \{\phi_i^{(m)}\} \to \delta_m \to \cdots . \qquad (10.17)$$

The $\{\phi_i^0\}$ are a set of initial try-functions, from which the density $\delta_0(\overrightarrow{r}_i)$ is determined through Eq. (10.16). The $\{\phi_i^{(m)}\}$ are the monoelectronic wavefunctions determined at the mth stage of the iteration through:

$$\{\widehat{h}(i) + \widehat{h}'_{m-1}(i)\}\phi_i^{(m)} = \epsilon_i^{(m)} \phi_i^{(m)} \qquad (10.18)$$

Note that, since the $\phi_i^{(m)}$ are single-electron wavefunctions in a crystal, they necessarily possess the form of Bloch functions : $\phi_i^{(m)} = \exp(i\overrightarrow{k}.\overrightarrow{r}_i).u_{\overrightarrow{k},n}^{(m)}(\overrightarrow{r}_i)$. Equation (10.18) is then equivalent to Eq. (3.16) satisfied by $u_{\overrightarrow{k},n}^{(m)}$:

$$\left\{\left(-\frac{\hbar^2}{2m}\right)(\overrightarrow{\nabla} + i\overrightarrow{k})^2 + V^{(m)}(\overrightarrow{r})\right\} u_{\overrightarrow{k},n}^{(m)} = \epsilon^{(m)} u_{\overrightarrow{k},n}^{(m)} \qquad (10.19)$$

with $V^{(m)}$ provided by Eq. (10.14) as a function of $h'^{(m)}$.

The joint determination of the periodic functions $u_{\overrightarrow{k},n}$ defining the Bloch states, and of the crystal potential $V(\overrightarrow{r})$ can be considered

---

[3]This formulation of the density is only exact if the electrons are independent, otherwise the probability of presence of an electron at a given point would depend of the probability of presence of the other electrons. Hence, this expression is consistent with the assignment of an individual wavefunction to each electron.

as achieved, when, within a predefined precision, one has a "self-consistent" fulfilment of the three equations:

$$\left\{ \left(-\frac{\hbar^2}{2m} \right) (\vec{\nabla} + i\vec{k})^2 + V(\vec{r}_i) \right\} u_{\vec{k},n}(\vec{r}_i) = \epsilon_i u_{\vec{k},n}(\vec{r}_i) \quad (10.20)$$

$$\delta(\vec{r}_i) = \sum_{\vec{k}',n',j \neq i} |u_{\vec{k}',n'}(\vec{r}_j)|^2 \quad (10.21)$$

$$V(\vec{r_i}) = -\sum_{\mu} \frac{Z_\mu e^2}{|\vec{r}_i - \vec{R}_\mu|} + e^2 \int_{\vec{r}} \frac{\delta(\vec{r})d\vec{r}}{|\vec{r}_i - \vec{r}|} \quad (10.22)$$

Thus, Eqs. (10.20)–(10.22) justify the model assumed in Chapters 3 and 5 of a free electron in a crystal potential. Note that the former formula defines *a different crystal potential* for each electron state ϕ_i owing to the presence in the expression of $\delta(\vec{r}_i)$ of a summation over $j \neq i$. It is also worth noting that, unlike the case of non-interacting electrons, Eq. (10.11) determining the total electronic energy has to be written

$$\hat{H}_{eN} \Psi_{el} = E\Psi_{el} \quad (10.23)$$

where $E \neq \sum \epsilon_i$. Indeed, in summing the ϵ_i in Eq. (10.19) over the N electrons, one takes into account *twice* the energy of interaction between electrons.

10.2.3 Shortcoming of the Hartree method

A flaw of the Hartree wave functions is their non-compliance with the Pauli principle. The *Hartree–Fock* method will remediate to this flaw by incorporating this principle. In this method, the idea of replacing the interactions between electrons by an average potential is preserved. To implement it, let us first show that the Hartree approximation, which has been introduced above through an "averaging" of the electron positions, can also be introduced as the result of a "variational" formulation.

Variational formulation of the Hartree approximation

For any quantum mechanical system of Hamiltonian $\hat{H}$, the ground-eigenstate can be obtained by means of a variational procedure. If Ψ

is an arbitrary state belonging to the state-space of the system, and $J = \langle \Psi | \widehat{H} | \Psi \rangle$ the corresponding matrix element of $\widehat{H}$, it can be shown that the ground state Ψ_0 of the system is the uniquely defined state for which J is minimum. This condition is implemented for *normalized* states Ψ, by the condition[4]:

$$\langle \delta\Psi | \widehat{H} - \lambda | \Psi \rangle = \int \delta\Psi(H - \lambda)\Psi d\tau = 0 \qquad (10.24)$$

where λ is the Lagrange multiplier determined afterhand by the normalization condition $\langle \Psi | \Psi \rangle = 1$ of $|\Psi\rangle$. If the variations $\delta\Psi$ explore a limited subspace of quantum states, the state Ψ_0 determined in this manner is an approximation of the true ground state. Within the considered limited subspace, it corresponds to the *best approximation* to the ground state energy $E = J_0$.

The Hartree approximation is the implementation of this general variational method in the special case where $|\Psi\rangle$ is constrained to be of the form (10.11), each ϕ_i being a normalized function of the coordinates $\vec{r}_i$ of the ith electron. In this case, since there are N constraints relative to the normalization of the wavefunctions of the N electrons, one has to use N Lagrange multipliers which we denote ϵ_i (as λ, they have the dimension of an energy). Equation (10.24) takes the form:

$$\left(h_i + \sum_{j \neq i} \langle \phi_j | h_{ij} | \phi_j \rangle \right) |\phi_i\rangle = \epsilon_i |\phi_i\rangle \qquad (10.25)$$

which is identical to the form obtained in the Hartree approximation:

$$\left\{ h_i + e^2 \int_{\vec{r}} \frac{\sum_{j \neq i} |\phi_j|^2}{|\vec{r}_i - \vec{r}|} d\vec{r} \right\} \phi_i$$

$$= \left\{ -\frac{\hbar^2}{2m}\Delta_i - \sum_{\mu} \frac{Z_\mu e^2}{|\vec{r}_i - \vec{R}_\mu|} + e^2 \int_{\vec{r}} \frac{\sum_{j \neq i} |\phi_j|^2}{|\vec{r}_i - \vec{r}|} d\vec{r} \right\} \phi_i \qquad (10.26)$$

---

[4]In fact the condition is obviously written as $\langle \delta\Psi | \widehat{H} - \lambda | \Psi \rangle + \langle \Psi | \widehat{H} - \lambda | \delta\Psi \rangle = 0$ but it can be shown that the simpler condition is equivalent.

Hartree–Fock approximation

To improve the Hartree approximation and take into account the
Pauli principle, one has to express the condition that two electrons
cannot occupy the same quantum state. In other terms, the wavefunctions of all the N electrons must be different. Since the Pauli
"exclusion" principle is valid for particles possessing, in addition to
their space coordinates, a spin (with two possible states), the wavefunction of a single electron has to be written $\phi_j = \varphi_j(\vec{r_j}).\alpha_j$ where
α_j has one of the two possible values of the spin state. Instead of
writing the wavefunction of the N electrons as $\Psi_e = \Pi_1^N \prod_j$ as in
the Hartree formulation, we write Ψ_e in the form of a determinant:

$$\sqrt{\frac{1}{N!}} \sum_{perm} (-1)^{perm} \phi_{\nu_1}(1)\phi_{\nu_2}(2)...\phi_{\nu_N}(N) \qquad (10.27)$$

where the ϕ_{ν_i} are N electron states, and where the sum is over all
the permutations of these states among the N electrons. This determinant vanishes if two of the functions in the products are identical,
i.e. if two electrons occupy the same state. The Pauli principle is
thus automatically taken into account by this expression. The same
variational method (Eq. (10.26)), used in the Hartree approximation, applied to this form of determinantal wavefunction, leads to
the expression:

$$\left\{ -\frac{\hbar^2}{2m}\Delta_i - \sum_{\mu} \frac{Z_\mu e^2}{|\vec{r}_i - \vec{R}_\mu|} + e^2 \int_{\vec{r}} \frac{\sum_{j \neq i} |\phi_j|^2}{|\vec{r}_i - \vec{r}|} d\vec{r} \right\} \phi_i - B = \varepsilon_i \phi_i$$

$$(10.28)$$

with

$$B = \sum_{j \neq i} \left[\int \phi_j(\vec{r}) \frac{e^2}{|\vec{r} - \vec{r_i}|} \phi_i(\vec{r}) d\vec{r} \right] \phi_j \qquad (10.29)$$

This expression differs from the Hartree expression (10.26) by
the B term which is the *exchange term*. Note that in its expression
the $\phi_i = \varphi_i.\alpha_i$ and $\phi_j = \varphi_j.\alpha_j$ the values of the spin variables α_i
and α_j are identical. *Indeed, the integrations are only over the space
variables* $\vec{r}$. Thus, the spin variables are factorized and therefore
their scalar products is equal to zero if they are distinct. *It can be
shown that $B > 0$.* Its contribution $(-B)$ to the electronic energy

ε_i, (10.29), is therefore negative.While the interaction between electrons is expected to be positive (since electrons repel each other), the exchange energy, which is also of electronic origin, increases the stability of the system by lowering this energy. This effect is due to the Pauli principle. Electron with the same spin will not be in the same "space state" $\varphi(\overrightarrow{r})$ and be in the vicinity of each other. Around an electron there will be a "hole" of other electrons which therefore decreases the electron-electron repulsion.

Another consideration of the exchange energy will be of interest in Chapter 11 as it is involved in the mechanism of occurrence of ferromagnetism.

10.3 Structural Properties of Solids

The independent electrons approximation, which has been justified in the preceding section, has allowed, as shown in Chapters 5–9, to account for a large variety of electronic properties of solids. In this section, we consider briefly properties pertaining to the states of the ions constituting the structure of a solid. These properties derive from Eq. (10.9). Figure 10.1 shows the relationship between the various approximations and results.

10.3.1 Ground state of the atomic configuration

Let Ψ_0 be the ground state, solution of Eq. (10.9). At 0 K the thermodynamic stable state of the set of nuclei of the solid is described by this function. Its determination defines their average coordinates $\overrightarrow{R}_j$. In principle, from the knowledge of Ψ_0, one could check that the system of nuclei and electrons forming the solid, is a crystal of a specific configuration (Bravais lattice, unit-cell dimension, configuration of the atomic basis). However, this knowledge is not accessible in practice except in very simple cases. Indeed, as emphasized about Eq. (10.9), it implies that, beforehand, the electronic energy $E_l(\overrightarrow{R}_\mu)$ corresponding to the nuclei configuration has been determined, and that an iterative procedure (e.g. by the Hartree method) has been implemented. The experimental determination (e.g. by diffraction experiments) of the structure of a crystalline solid thus remains an indispensable step of the study of its physical properties.

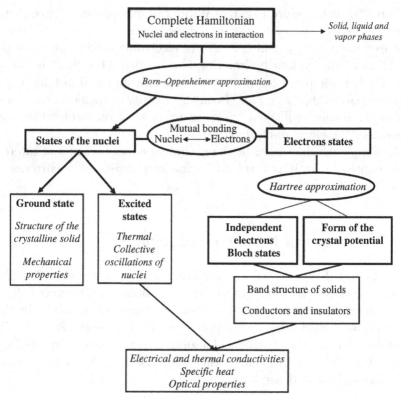

Fig. 10.1. Relationship between the main approximations assumed, and the microscopic properties of solids which they allow to determine.

The availability of powerful numerical methods has recently permitted to address limited aspects of the *ab initio* determination of the structure of solids. For instance, one can, for instance, compare, at 0 K, the cubic structure of sodium chloride NaCl, to a similar structure of Na Cl with another type of structure. This is a less difficult task since it is not necessary to determine the $3N$ coordinates od the $3N$ nuclei of the crystal. The structures are defined, by the relative nuclei positions inside a unit cell, of given shape, but of unknown $|\vec{a}_i|$ dimensions, The iterative procedure concerns only three unknown parameters. It then allows to determine the "equilibrium dimensions of the unit cell at 0 K. These values determine the minimum" of the energy for each considered type of structure. The procedure also supplies the value of the energy at its minimum. The lowest energy will correspond to the most stable structure among those compared.

The energy of the ground state Ψ_0, if it can be determined for a given configuration of the nuclei, defines the *mechanical properties* of the solid. Indeed, the increase of the total energy of the solid (nuclei + electrons) resulting from a given deformation, of the crystal unit cell, i.e. a change of the moduli of the $|\overrightarrow{a}_i|$ and of the angles between the $\overrightarrow{a}_i$ vectors, corresponds to the *elastic energy* of the solid, associated to a deformation. Besides, the second derivatives of this energy with respect to the various components of deformation, also determine the elasticity moduli of the solid. These moduli are an important mechanical characteristics of the solid, e.g. its hardness.

Cohesion and compressibility of metals

Let us examine how the stability of the structure of a metal is ensured by the conduction electrons. For the sake of simplicity, we consider a monoatomic metal. In a metal, the conduction electrons exert an attractive electrostatic potential on the positive ions which maintains them at their equilibrium positions in the crystal structure. We can understand the mechanism of this interaction in the framework of a simple model. As justified hereunder, when the distance d between atoms is reduced, the value of the (negative) electrostatic energy of the electrons increases proportionally to $(1/d)$. In contrast, their (positive) kinetic energy increases as $(1/d^2)$. The equilibrium distance d corresponds to the minimum of the total energy.

Thus, in a monoatomic metal as sodium, the electrons can be considered to a good approximation, as free-electrons. Their density is spatially uniform and is equal to one electron per unit cell of volume ω. A sphere of radius d and of volume ω contains the charge of a single electron $(-e)$, uniformly distributed. This sphere is centered on a positive ion of charge $(+e)$. Each sphere being globally neutral, we assume that there is no interaction between adjacent spheres. As deduced from Eqs. (6.6) and (6.8) in Chapter 6, the average kinetic energy of a single free electron at 0 K is:

$$U_{\text{kinetic}} = \frac{3}{5}E_F = \frac{3}{5}\frac{\hbar^2}{2m}\left(\frac{3\pi^2 N}{4\pi d^3/3}\right)^{2/3}$$

$$= \left\{\frac{3}{10}\left(\frac{9\pi}{4}\right)^{2/3}\frac{e^2\hbar^2}{me^2}\right\}\frac{1}{d^2} = \frac{A}{d^2} \tag{10.30}$$

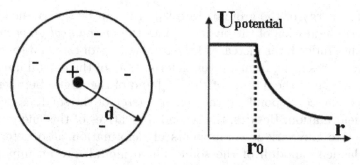

Fig. 10.2.   Equilibrium distance between atoms in a metal as determined by the electrostatic mutual interaction between ions and the free electron gas.

Let us now evaluate the electrostatic energy of the negatively charged sphere interacting with the positive ion. Note that the attracting potential created by the ion does not have the Coulomb form $\propto (1/r)$. Indeed, the ion comprises, besides the positive nucleus exerting an attraction on the free-electron, "core" electrons which exert a repulsion at short distance. The schematic variation of the potential can be taken as in Fig. 10.2: a uniform potential in the internal region of radius r_0, and a Coulomb $(1/r)$ law at larger distance. One can then write, denoting η the density of charge within the sphere of radius d, the total charge in this sphere being $(-e)$:

$$U_{\text{potential}} = -\eta e^2 \left\{ \int_0^{r_0} \frac{4\pi r^2 dr}{r_0} + \int_{r_0}^{d} \frac{4\pi r^2 dr}{r} \right\} = \frac{e^2 r_0^2}{2d^3} - \frac{3e^2}{2d}$$

(10.31)

The core radius of the Na^+ ion is $r_0 \approx 3.2a_0$ (where a_0 is the radius of the hydrogen atom). The total electronic energy, per unit cell, of the crystal is then:

$$U_{\text{Total}} = U_{\text{kinetic}} + U_{\text{potential}} = \left\{ \frac{3e^2 a_0}{2} \right\} \left\{ \frac{1}{5} \left(\frac{9\pi}{4} \right)^{2/3} \left(\frac{1}{d^2} \right) \right.$$

$$\left. + \frac{(3.2)^2 a_0}{3d^3} - a_0 \left(\frac{1}{d} \right)^2 \right\}$$

(10.32)

Minimizing this expression with respect to d leads to the value $d \approx 4.1.a_0 \approx 2.16$ Å. Table 6.1 in Chapter 6 indicates that the lattice parameter of the conventional cell of sodium, which contains two atoms, is $a = 4.2$ Å. It can be deduced that the ions, in contact with each other, have a radius of ≈ 2.1 Å, in good agreement with the above calculation.

This type of calculation provides a good approximation of the distance between atoms in other metals. This shows that the conduction electrons are indeed at the origin of the cohesion of metals. As emphasized in Chapter 2, it is the quasi-isotropic charge-distribution of the electron gas in metals which explains that the packing of adjacent spheres represents satisfactorily the crystal structures of many metals.

Another consequence of the role of the free-electron gas in the properties of a metal is the determination of its mechanical compressibility. The bulk compressibility is defined as the ratio between the relative *decrease* of volume and the external pressure δp applied to the solid:

$$\chi = -\frac{1}{\Omega}\frac{\delta\Omega}{\delta p} \tag{10.33}$$

Assume that the variation of the external pressure is equal to the sole variation of the pressure of the electron gas of the metal when the volume undergoes a variation $\delta\Omega$. The internal pressure of the electron gas is $p = -(\delta U_{\text{kinetic}}/\delta\Omega)$ in which U_{kinetic} is provided by Eq. (10.30). Hence:

$$\chi = \frac{1}{\Omega}\frac{1}{\partial^2 U_{\text{kinetic}}/\partial\Omega^2} \tag{10.34}$$

Adapting expression (Eq. (10.30)) of U_{kinetic} to the case of an atom with z conduction electrons, and having the volume ω, we deduce:

$$\chi = \frac{9\Omega}{10 U_{\text{kinetic}}} = \frac{3\omega}{2z E_F} \tag{10.35}$$

Table 10.1 shows the good qualitative agreement between the values of compressibilities thus calculated, for some metals, with the experimentally measured values (in $10^{-11} \text{N}^{-1}.\text{m}^2$).

10.3.2 Collective oscillations of the atoms: Phonons

The interaction energy between nuclei (Eq. (10.8)) is *minimum* for the equilibrium structure, defined by the coordinates R_μ^0 of the atoms

Table 10.1. Comparison between calculated electronic
compressibilities and measured compressibilities of some
metals.

Metal	$\chi(calc)$	$\chi(exper)$
Potassium	31.3	35.6
Silver	2.9	1.0
Cesium	64.9	69.9

in the crystal structure, as discussed in the preceding paragraph. If small deviations $\vec{\rho}_\mu$ are imposed to these coordinates, the potential energy of the crystal will thus increase as a *positive quadratic form* of the deviations $\vec{\rho}_\mu$. Such a quadratic form can always be transformed into its "diagonal" form, which is expressed as functions of coordinates X_j, linear combinations of the ρ_μ. While the ρ_μ are deviations of the individual positions of the atoms, the X_j describe *collective displacements* of the nuclei around their equilibrium positions. As a function of the X_j and of their "conjugate" coordinates P_j the Hamiltonian of the nuclei can be written as a sum of square terms:

$$\widehat{H}_{\text{Nuclei}} = \widehat{H}_0 + \Delta\widehat{H} = \widehat{H}_{\text{Nuclei}}(\vec{R}^0_\mu) + \sum_j^{3N} \frac{\widehat{P^2_j}}{2M'_j} + \frac{M'_j.\omega^2_j}{2}.\widehat{X^2_j}$$

(10.36)

One recognizes in the expression of $\Delta\widehat{H}$ the Hamiltonian (energy) of a set of $3N$ independent harmonic oscillators, associated to the collective displacements of nuclei, N being the number of atoms in the solid. The possible eigenvalues of the Hamiltonian of a single one-dimensional oscillator is standardly determined in quantum mechanics as $\varepsilon = (n + \frac{1}{2})\hbar\omega$. Hence, the possible energy eigenvalues of $\Delta\widehat{H}$ are:

$$E = \left(n_1 + \frac{1}{2}\right)\hbar\omega_1 + \cdots + (n_j + \frac{1}{2})\hbar\omega_j + \cdots + \left(n_{3N} + \frac{1}{2}\right)\hbar\omega_{3N}$$

(10.37)

Up to a shift by $(-\sum \hbar\omega_j/2)$ of the origin of the total energy value, this expression can be expressed as:

$$E = \sum_{j=1}^{3N} n_j \hbar\omega_j \qquad (10.38)$$

Each term $n_j \hbar\omega_j$ can be considered as the sum of the energies of n_j particles of energy $\hbar\omega_j$. Each of these hypothetical particles carries a *quantum of oscillation energy* $\hbar\omega_j$. They are termed *phonons* by analogy with the term *photon* used for particles which carry quantas of electromagnetic energy. The collective oscillation state of the atoms of a crystal is equivalent to the presence of a certain number n_j of each type of phonon $\hbar\omega_j$. Since the number of these particles is not fixed, their number $\sum n_j$ is arbitrary. Moreover, there can be an arbitrary number of these particles in a given ω_j state, the equilibrium state of this set of particles is then governed by the *Bose–Einstein statistics*.

Phonons energy bands, acoustic and optical phonons

The Hamiltonian (Eq. 10.36) depends, on the one hand, on the equilibrium positions $\overrightarrow{R}_\mu^0$ of the atoms of the crystal, and, on the other hand, on the deviations ρ_μ of these atoms. Both these quantities are *invariant by the lattice translations of the solid* : substituting the deviation of an atom by that of an identical atom will not change the spectrum of the crystal oscillations.

Referring to Chapter 3, we can therefore conclude that the eigenfunctions of $\Delta\widehat{H}$, similarly to the electronic states of the solid, are Bloch functions of the form $u_n(\overrightarrow{k}).\exp(i\,\overrightarrow{k}.\overrightarrow{r})$ (cf. Chapter 3, Section 3.2) with energies $E_n(\overrightarrow{k})$ in which $\overrightarrow{k}$ is a vector of the first Brillouin zone. This analogy entitles us to infer that the $E_n(\overrightarrow{k}) = \hbar\omega_n(\overrightarrow{k})$ form quasi-continuous bands which are termed *phonon branches*. Thus, the index j in ω_j, used hereabove to enumerate the various collective modes of oscillation, is actually a composite index associated to a branch index n and to the quantized components of a $\overrightarrow{k}$ vector.

Recall that the expression (10.36) has its origin in the diagonalization of the quadratic form, function of the deviation positions $\overrightarrow{\rho}_j$

in the potential energy of oscillation of the atoms. The invariance of this Hamiltonian by the lattice translations of the solid, suggests that the linear combinations $\vec{p}_j$ which correspond to this diagonalization are, in fact, Bloch functions, since these functions are the eigenfunctions of the Hamiltonian.

As underlined in Chapter 3, a given energy band comprises N_M states, in which N_M is the number of primitive unit cells of the crystal. If there are p atoms per primitive unit cell, the total number of atoms is $N = pN_M$. Thus, the $3N = 3pN_M$ oscillator frequencies appearing in Eq. (10.38) are grouped into $3p$ phonon branches, some of which can be degenerate. As an example, Fig. 10.3 shows the phonon dispersion curves for silicon in two prominent directions of the reciprocal space. One notes on this figure that among the bands, two have a peculiarity: the value of $\omega(\vec{k}) = 0$ for $\vec{k} = 0$. These branches are the *acoustic branches* of phonons. The others being termed *optical branches of phonons*. These terms have simple physical explanations.

A first manner to distinguish those two types of branches refers to a standard distinction in the study of the motion of a system composed of several particles. Hence, one distinguishes the motion of the *center of gravity* of the system from the *relative motion* of the particles *with respect to this gravity center*. The first motion corresponds

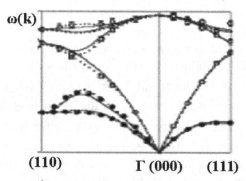

Fig. 10.3. Phonon $\omega(\vec{k})$ dispersion curves of silicon in two directions of the reciprocal space. In an arbitrary direction, there should be three branches of acoustic phonons In the (111) direction there is symmetry degeneracy of one branch. There should also be six optic branches (two atoms in the unit cell). One has represented the lower ones also degenerate in the (111) direction.

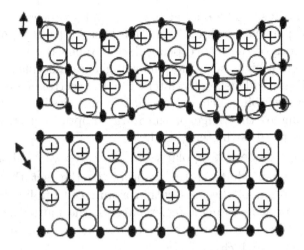

Fig. 10.4. Up: Schematic motion in an acoustic mode. A relative motion of the nodes of the lattice but no relative motion of the atoms within the cell. Down: Optical mode: Relative motion of the atoms within the cell but no deformation of the lattice.

to the *acoustic modes*, and the second to the *optical modes*. Alternately, one can invoke the distinction between the *Bravais lattice* of the crystal structure and the *atomic basis* within a lattice unit cell.

Acoustic modes. The acoustic modes concern the relative displacements of the *different nodes of the Bravais lattice* during the oscillation. Each node is assigned the total mass of the atoms in the basis (or the primitive unit cell). One has in this case, for each collective oscillation, a density wave whose phase $i\vec{k}.\vec{r}$ is determined by the $\vec{k}$ vector. When $\vec{k} \to 0$, the phase is equal to zero, which means that all the nodes are displaced *in phase*. This overall displacement is equivalent to *a global translation of the crystal*. For such a displacement, there is *no restoring force* and therefore no energy involved. This justifies the cancelling of $\omega(\vec{k})$ for $\vec{k} = 0$. Figure 10.4 shows that, if the wavelength corresponding to $\vec{k} \to \lambda = 2\pi/|\vec{k}|$ is much larger than the size of the lattice, the atomic basis will hardly be deformed. The acoustic-mode oscillation is equivalent to the propagation of a local mechanical perturbation of the nodes. This type of motion occurs during the *propagation of sound*, which thus justifies the term "acoustic".

Optical modes. The optical phonon modes concern the relative motions of particles within the atomic basis of each primitive-unit cell, the nodes of the lattice being fixed.[5] When the basis contains ions of different signs such motions will generate oscillating electric dipoles. These will generally interact with electromagnetic waves, emitting or absorbing such waves. Their frequencies is such that these waves belong to the range of the infrared-optical electromagnetic waves. This justifies the term used.

An optical mode is a dipolar wave $\overrightarrow{p}.\exp(-i\overrightarrow{k}.\overrightarrow{r})$. For such an oscillation mode, whatever the value of $\overrightarrow{k}$, the fact that there is a relative motion of the ions within a unit cell implies that there will always be a restoring force, related to the mutual attraction–repulsion of the ions. Hence the optical branches of phonons do not possess points at which $\omega(\overrightarrow{k}) = 0$.[6]

Bibliography

Electronic Structure of Solids, *J.C. Slater*, MIT Press (1954).
Physical Basis of Plasticity in Solids, *J.C. Tolédano*, World Scientific (2012).

---

[5]The two types of motions can interact. This is in particular the case of piezo-electric crystals in which a mechanical stress, which induces an "acoustic" type of motion of the lattice, induces an electric voltage (which is associated to an "optical" type of motion).

[6]An exception is the case of crystals undergoing certain type of transitions from a crystal structure to another. One can then have, on approaching the transition temperature, a so-called *soft optical mode* whose frequency tends to zero at a certain value of the $\overrightarrow{k}$-vector.

Chapter 11

Ferromagnetism and Superconductivity

11.1 Introduction

In Chapters 3–9, it has been assumed that, in a solid, the electrons did not interact with each other. Each electron was submitted to a potential which represented the influence of all the other particles composing the solid, including the other electrons. In Chapter 10, this assumption was substantiated by two theoretical justifications: the Born–Oppenheimer approximation, and the Hartree method of iterative determination of the electronic states. In this chapter, we outline the principles of two properties of solids, *ferromagnetism* and *superconductivity*, whose basic mechanism puts into play an interaction between electrons. These properties have in common another feature. They vanish above a certain temperature T_c characteristic of the material. Ferromagnetism, a property which underlies the attraction effects of magnets, already known in the antiquity, and which has many modern industrial applications, has its explanation both in the existence of the spin of the electron, and to the electrostatic repulsion between electrons. Their combination into the so-called *exchange interaction*, has already been invoked when introducing the Hartree–Fock method in Chapter 10. Other magnetic behaviours had been briefly considered in Chapters 6 (Section 6.3.2) and 7 (Sections 7.2.5 and 7.2.7). Superconductivity, more recently discovered (at the beginning of the 20th century), consists in a complete disappearance of electrical resistivity below a certain temperature. Its mechanism, in the case of the so-called *conventional* superconductors involves another type of interaction between electrons, namely an *attraction* mediated by the phonons of the crystal lattice. For more recently discovered so-called *high-temperature* superconductors, the explanation of their properties is still a matter of current research.

11.2 Magnetic Properties of Solids

A great variety of magnetic properties are observed in solids, which can be of electronic, or of nuclear origin. In this chapter, we focus on the properties of the former type. At macroscopic level, the characteristic quantity is the magnetization $\overrightarrow{M}$. It is the sum of microscopic magnetic dipoles $\overrightarrow{M} = \sum \overrightarrow{\mu}_i$ over the unit volume Ω of the solid. In the atoms, the electrons generate two types of dipoles. On the one hand, a dipole arises from the motion of the electron within the atom on its closed trajectory. It is related to the orbital momentum $\overrightarrow{l}_i = \overrightarrow{r}_i \times m \overrightarrow{v}_i$ of the electron by the relation $\overrightarrow{\mu}_i^l = -(e\hbar/2m)\,\overrightarrow{l}_i$. A second type of dipole is related to the spin $\overrightarrow{s}_i$ of the electron, $\overrightarrow{\mu}_i^s = -(e\hbar/m)\overrightarrow{s}_i$. The total magnetic dipole of the electron is the sum of those two dipoles. The occurrence of a macroscopic magnetization $\overrightarrow{M}$ corresponds to the fact that the various electronic dipoles are not oriented randomly, but that they have a dominant orientation, and that they add up to determine $\overrightarrow{M}$.

As shown briefly by a statistical argument in Chapter 1, the existence of a magnetization in a solid can only be explained by recurring to quantum mechanics. Let us give an additional proof that classical physics excludes the occurrence of a magnetization.

11.2.1 Bohr–Van Leuwen theorem

A formal proof of this shortcoming of classical physics can be given which constitutes the so-called Bohr–Van Leeuwen theorem. The *classical Hamiltonian* function of N electrons in the presence of a magnetic field is

$$H = \sum_{1}^{N} \frac{1}{2m} \left\{ \overrightarrow{p_i} - e\overrightarrow{A}(\overrightarrow{r_i}) \right\}^2 + V(\overrightarrow{r_1}, \overrightarrow{r_2}, \dots, \overrightarrow{r_N}) \qquad (11.1)$$

where $\vec{A}$ is the vector potential of the magnetic field $\vec{B}$, and V is the scalar potential internal to the system to which the electrons are submitted. The statistical average $\langle \vec{M} \rangle$ of the value of the magnetization, at thermal equilibrium is

$$\langle \vec{M} \rangle = -\frac{\partial[k_B T \mathrm{Log}\, Z]}{\partial \vec{B}} \tag{11.2}$$

in which Z is the partition function (cf. Chapter 6, Section 6.2.2), T the temperature, and k_B the Boltzmann constant. The expression of the classical partition function is:

$$Z = \int \exp(-\beta H) \cdot \vec{dr_1} \cdots \vec{dr_N} \cdot \vec{dp_1} \cdots \vec{dp_N} \tag{11.3}$$

Making the change of variables $\vec{P_i} = \{\vec{p_i} - e\vec{A}(\vec{r_i})\}$ it is easy to check that the determinant of this change of variables is equal to unity. This has the consequence that the former sum is independent of $\vec{A}$ (since the variables $\vec{P_i}$ are eliminated by the integration). Thus, it is also independent of the value of the magnetic field $\vec{B}$. Hence, deriving $\mathrm{Log}\, Z$ with respect to $\vec{B}$ yields a zero value for $\langle \vec{M} \rangle$.

11.3 Ferromagnetism

Ferromagnetism consists in the occurrence of a non-zero magnetization $\vec{M}$ in the absence of an external magnetic field $\vec{B}$ applied to the solid. In a given substance, this property exists up to a certain temperature T_c (the Curie temperature) above which it vanishes. T_c is a "phase transition" separating two phases with different magnetic properties, the magnetic phase, stable below T_c, characterized by $\vec{M} \neq 0$, and a phase, stable above T_c in which $\vec{M} = 0$, in the absence of external magnetic field. In the most common ferromagnetic substances (e.g. Iron, Nickel, Cobalt) the value of T_c is higher than $1000\,\mathrm{K}$. Hence, the thermal disorder will suppress this coherent orientation, only for $k_B T \approx 50\text{--}100\,\mathrm{meV}$.

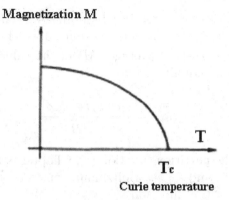

Fig. 11.1.   Schematic temperature dependence of the magnetization in a ferro-magnetic substance.

11.3.1   Relation between the signatures of the magnetic transition

Several specific effects are observed when the temperature is varied across the range of the Curie temperature. Hence, below T_c, the magnitude of the magnetization $\overrightarrow{M}(T)$ decreases progressively on heating towards T_c and vanishes above this temperature (Fig. 11.1). Likewise the magnetic susceptibility $\chi = (\partial M / \partial B)$ tends to diverge to a very large positive value on approaching T_c from above or below. Finally, one observes a magnetic "hysteresis" below T_c, i.e. a non-reversible behaviour of the dependence $M(B)$ when an external magnetic field of sufficient value, is varied in cycles from positive to negative values. Before analyzing the microscopic origin of ferromagnetism, let us show that a simple model can relate those different manifestations. Assume that magnetic dipoles in a solid can only have two possible opposite orientations $(\pm M)$ in a given direction. The resulting magnetization, sum of these dipoles, will have also those two possible orientations. The magnetic equilibrium properties of the system is determined by the minimum of a "magnetic" free energy function of the magnetization and the temperature $\mathcal{F}(M,T)$. Since the two possible orientations of M are equally possible, the free energy must have the same value by reversing M. Hence, $\mathcal{F}$ must only depend on even powers of M. We write

$$\mathcal{F}(M) = \mathcal{F}_0 + \frac{\alpha(T)}{2}M^2 + \frac{\beta}{4}M^4 \qquad (11.4)$$

with β a positive constant.[1] The equilibrium state corresponding to the minimum of $\mathcal{F}$ is determined by $\partial \mathcal{F}/\partial M = 0$, and $\partial^2 \mathcal{F}/\partial M^2 > 0$.

$$\partial \mathcal{F}/\partial M = M(\alpha + \beta M^2) = 0 \qquad \partial^2 \mathcal{F}/\partial M^2 = \alpha + 3\beta M^2 > 0$$
$$(11.5)$$

This state is such that $M_{eq} = 0$ for $T > T_c$ and $M_{eq} \neq 0$ for $T \leq T_c$. Applying these conditions above T_c ($M_{eq} = 0$) results in $\alpha(T > T_c) > 0$. Below T_c ($M_{eq} \neq 0$) we obtain

$$\alpha(T) + \beta M^2 = 0 \qquad (11.6)$$

Hence, α is necessarily negative below T_c. The simplest function which changes sign at T_c has the form $\alpha(T) = \alpha_0(T - T_c)$. We deduce from Eq. (11.6) the temperature dependence of M:

$$|M| = \sqrt{\frac{\alpha_0(T_c - T)}{\beta}} \qquad (11.7)$$

A temperature dependence which specifies the behaviour of the magnetization, qualitatively in agreement with the experimental results on Fig. 11.1. In the presence of a external magnetic field B, the free energy becomes:

$$\mathcal{F}(M, B) = \mathcal{F}(M) - M.B \qquad (11.8)$$

At equilibrium, we have $\partial \mathcal{F}(M, B)/\partial M = 0$ which provides $M[\alpha_0(T - T_c) + \beta M^2] = B$. The magnetic susceptibility χ is defined by $\chi = \partial M/\partial B$. Since $M_{eq} = 0$ above T_c we obtain

$$\chi(T > T_c) = \frac{1}{\alpha_0(T - T_c)} \qquad (11.9)$$

While, below T_c ($M_{eq}^2 = -\alpha/\beta$),

$$\chi(T < T_c) = \frac{1}{2\alpha_0(T_c - T)} \qquad (11.10)$$

the linear dependence of $(1/\chi)$ corresponds to the behaviour represented on Fig. (11.2).

---

[1]When β is negative the theory must take into account terms of higher degrees. The variation of the considered quantities is then discontinuous a T_c. We consider here the simplest situation.

Physics of Electrons in Solids

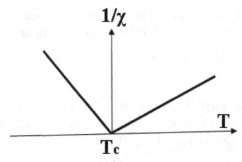

Fig. 11.2.   Linear dependencies of the inverse susceptibility on either sides of T_c.

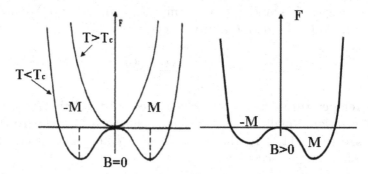

Fig. 11.3.   Magnetic free energy above and below T_c in the absence of external field (left). Change of shape in an applied magnetic field oriented towards positive B (right).

Finally, consider the shape of the free energy for $T < T_c$ (Fig. 11.3). It has two minima, and a maximum (at $M = 0$). Assume that the system is in the minimum $M < 0$. Applying an external field in the opposite direction will tend to drive the system towards the other minimum $M > 0$. The free energy becomes asymmetric (Fig. 11.3, right). However, there is a free energy barrier to overcome and it is only for a high enough value of the field that the system can occupy the opposite minimum ($M > 0$). Reversing the field determines the same sequence of events. Hence, instead of switching from one stable state to the other by application of a small field, one has to apply a large enough field to obtain this switching. This situation generates the $M(B)$, *hysteretic* curve represented on Fig. 11.4.

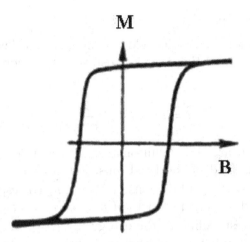

Fig. 11.4. Hysteresis curve $M(B)$ on applying an alternate external magnetic field.

11.3.2 Microscopic origin of ferromagnetism

Irrelevance of the magnetic interaction between dipoles

One could think that this "spontaneous" coherent orientation of the various electronic dipoles would result from the magnetic interaction between the dipoles of the different electrons, since a magnetic dipole will generate a magnetic field which can act on the neighbouring electrons. This is not the case. Indeed, let us consider two parallel dipoles μ_1 and μ_2 of magnitude $\mu_B = (e\hbar/m) = 9.27 \times 10^{-24}\,\text{J.T}^{-1}$ having their origins on two neighbouring atoms of a crystal lattice, at a distance $a \approx 4\,\text{Å} = 4 \times 10^{-10}\,\text{m}$. Each dipole produces a magnetic field which acts on the other dipole. The resulting force of mutual attraction is:

$$F = \frac{3\mu_0}{4\pi} \frac{\mu_B^2}{a^4} \tag{11.11}$$

By integrating the work $F.da$ of this force from a to infinity, we obtain an energy of interaction $E = (\mu_0\mu_B^2)/4\pi a^3 \approx 10^{-3}\,\text{meV}$ $\ll k_B T_c (= 50\text{--}100\,\text{meV})$. Hence, the interaction energy is five orders magnitude smaller than the thermal energy required to offset the ferromagnetic ordering of dipoles.

Exchange energy

Consider two electrons in the field of two fixed ions. The Hamiltonian of the system is (omitting the electrostatic interaction between fixed ions):

$$\widehat{H} = \widehat{h}_1 + \widehat{h}_2 + \frac{1}{4\pi\varepsilon_0}\frac{e^2}{r_{12}} \qquad (11.12)$$

in which the $\widehat{h}_i$ are the Hamiltonians, with identical form, of each electron in the field of the fixed ions. Let $\phi_1(\vec{r}_1)$ and $\phi_2(\vec{r}_2)$ be normalized *orbital* eigenfunctions of $\widehat{h}_1$ and $\widehat{h}_2$ corresponding to the energies ε_1 and ε_2. The product $\phi_1(\vec{r}_1) \cdot \phi(\vec{r}_2)$ is an eigenfunction of $\widehat{h}_1 + \widehat{h}_2$ corresponding to the energy $(\varepsilon_1 + \varepsilon_2)$. Since the $\widehat{h}_i$ have the same form, other normalized eigenfunctions, corresponding to the same energy $(\varepsilon_1 + \varepsilon_2)$, are:

$$\Phi_A = \frac{1}{\sqrt{2}}\{\phi_1(\vec{r}_1)\phi_2(\vec{r}_2) - \phi_1(\vec{r}_2)\phi_2(\vec{r}_1)\} \qquad (11.13)$$

and

$$\Phi_S = \frac{1}{\sqrt{2}}\{\phi_1(\vec{r}_1)\phi_2(\vec{r}_2) + \phi_1(\vec{r}_2)\phi_2(\vec{r}_1)\} \qquad (11.14)$$

which are respectively antisymmetric and symmetric with respect to the permutation of particles (or of wavefunctions) 1 and 2. However, these orbital wavefunction do not describe the complete state of the particles. Their spin state must be taken into account. Since the Hamiltonian equation (11.12) does not depend explicitly on the spin. The above wave functions must be multiplied by an eigenfunction of the *total spin* of the two particles. The spin of each particle of magnitude $(\frac{\hbar}{2})$ can be denoted by the value of its projection $s_i^z = \pm\frac{\hbar}{2} = |\pm\rangle$. The expression of the eigenvalues of the total spin are then, as a function of the spins of the two particles:

$$|S = 1, S_z = \pm 1\rangle = |\pm, \pm\rangle$$

$$|S = 1, S_z = 0\rangle = \frac{1}{\sqrt{2}}\{|+, -\rangle + |-, +\rangle\} \qquad (11.15)$$

and

$$|S = 0, S_z = 0\rangle = \frac{1}{\sqrt{2}}\{|+, -\rangle - |-, +\rangle\} \qquad (11.16)$$

The *Pauli principle* imposes to the *total wavefunction*, which is the product of an orbital state and of a spin state *to be antisymmetric* with respect to the permutation of the particles. There are four such wave functions complying with this principle, and which have the same energy $(\varepsilon_1 + \varepsilon_2)$. Namely, $\Phi_A \cdot |S = 1, S_z = 0$ or $\pm 1\rangle$, or $\Phi_S \cdot |S = 0, S_z = 0\rangle$. However, up to now, we have not taken into account the electrostatic repulsion term of the Hamiltonian $(1/4\pi\varepsilon0)e^2/r_{12}$. Assuming that this term is a "small" perturbation, its effect will be to lift the degeneracy between the triplet state with orbital component Φ_A and the singlet state with orbital component Φ_S. The energy difference between the triplet and the singlet is given by first-order perturbation theory (diagonalizing the corresponding 4×4 perturbation matrix):

$$\varepsilon_A - \varepsilon_S = -\frac{2}{4\pi\varepsilon_0} \int \phi_1^*(\vec{r}_1)\phi_2^*(\vec{r}_2)\frac{e^2}{r_{12}}\phi_1(\vec{r}_2)\phi_2(\vec{r}_1)d\vec{r}_1 \cdot d\vec{r}_2$$

$$(11.17)$$

which is a particular form of the exchange integral encountered in (10.29). It can be shown that this energy difference, if it favours the triplet state, by determining for it a lower energy, will favour the alignment of the neighbouring spins in the same direction, thus permitting the onset of a magnetization in the system. Besides, the exchange interaction energy involved, being of electrostatic nature, can have the right order of magnitude to explain a Curie temperature of 1000 K.

Hence, this *exchange interaction* between electrons of a solid *is a convincing microscopic origin of ferromagnetism.*

11.4 Superconductivity

11.4.1 Introduction

Besides the vanishing, mentioned in the introduction, of the solid's electrical resistance below T_c, (see Fig. 11.5) the signature of superconductivity includes another important phenomenon. When inserted in a region of space in which a magnetic field exists, one observes that the field is excluded from the volume of the superconducting sample (Fig. 11.6). The trajectories of the field lines are

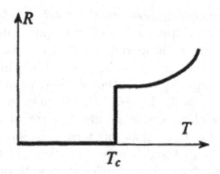

Fig. 11.5.   Complete vanishing of the resistivity of a superconductor below its transition temperature.

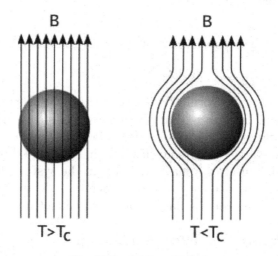

Fig. 11.6.   Meissner effect.

incurved in such a way as to surround the sample but do not penetrate it. This effect is termed the *Meissner effect*. To produce this effect, the sample generates, in reaction to the external field, intense permanent currents in a superficial layer, of thickness of the order of $\approx 1000\,\text{Å}$ which can circulate without loss, since the resistance to its passage is equal to zero. Recall that a substance has diamagnetic properties if an externally applied magnetic field induces in the sample, an internal field opposed to the externally applied field. From this standpoint, a superconductor is *perfectly diamagnetic*.

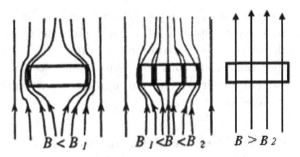

$$B < B_1 \qquad B_1 < B < B_2 \qquad B > B_2$$

Fig. 11.7.   Existence of two threshold magnetic fields in type II superconductor. The first one initiates a partial penetration in the sample through "vortices". The second suppresses the superconducting properties.

The two main signatures of superconductivity have important limitations. As already mentioned, the most important is the upper temperature limit T_c (the critical temperature) of the superconducting state. The superconducting solids known before 1987, were metals, and had critical temperatures which, at most, were of the order of $20\,\mathrm{K}$.[2] Several families of chemical substances, with specific crystal structures, have been discovered since, with critical temperatures between $30\,\mathrm{K}$ and $200\,\mathrm{K}$.

Another limitation concerns the maximum magnitude of the field which can be excluded from the volume of the superconducting sample. In this aspect, two types of superconductors can be distinguished. In so-called *type I* superconductors, the superconducting properties disappear above a certain value of external magnetic field. In *type II* superconductors, the offset occurs in two steps (Fig. 11.7). Above a first value of magnetic field B_1, which can be as low as a few hundred Gauss (10^{-2} Tesla), the magnetic field penetrates partially the sample, within volumes of cylindrical shape, of $\approx 1000\text{Å}$. These cylinders, parallel to the applied field, are termed *vortices* (each is a vortex). Each vortex is associated to a cylindrical surface of current circulating around the vortex. This current compensates, in the volume separating the vortices, the external field applied to the sample. It also determines within the volume of the vortex, a quantized value of the magnetic flux confined in the cylinder.

---

[2]It had been proved that an upper limit of T_c for this category of superconductors, was of the order of $30\,\mathrm{K}$.

Above a second limit B_2, generally much higher than B_1 (several Tesla, or even several tens of Tesla), the field penetrates the entire sample and destroys the superconducting state.

Other important limitations are found, such as the maximum value of the current which can pass through the superconductor without dissipative losses.

11.4.2   Microscopic mechanism, qualitative description

The microscopic mechanism of "conventional" superconductivity has been clarified between 1935 and 1950. Let us summarize here three important qualitative aspects of this mechanism. The first one, already mentioned in Chapter 1, is the quantum character of this mechanism. The superconducting state, in which permanent currents circulate in the sample, is a *stationary quantum state*, similar to the circulation without loss of the electron on its orbit in an atom. In a superconductor, this stationary state concerns the entire set of conduction electrons of the solid.

A second feature is the fact that, in contrast to the case considered for metals in Chapters 3–9, the electrons are not independent from each other. In a paradoxical way, considering their "natural" mutual electrostatic repulsion, they show an *effective attraction*. By this means, they form a set of particles bound to each other. The electric current in a superconductor corresponds to the motion of this entire set of bound particles in the solid. Another paradox is, as we have seen in Chapter 8, that in an ordinary metal, the resistive losses are mainly due to the interactions of the electrons with the oscillations of the ions/nuclei (or in an alternate formulation, to the collisions between electrons and phonons which limit the free path of the electron. However, in a superconductor, phonons do not have enough energy to perturb a set of N electrons bound to each other.

A third aspect is relative to the nature of the mechanism which underlies the effective attraction between electrons. This mechanism consists in an indirect interaction of two electrons by means of an interaction of each electron with the oscillations of the ions or nuclei. Its principle is the following (cf. Fig. 11.8). An electron, charged negatively, passing in the vicinity of of a positively charged ion will tend to displace it from its equilibrium position. This displacement induces a local excess of positive charge. As any perturbation in the solid,

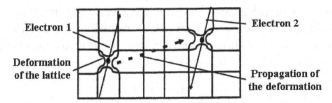

Fig. 11.8. Effective attraction between electrons through a "phonon exchange".

this excess of charge propagates, at the speed of sound (by means of an "acoustic" type of phonon), to reach another position in the solid. At this new location, the excess positive charge will attract another electron. The overall result of this mechanism is that an electron will have attracted another electron through a "mediator", namely the oscillation of the ions of the solid. This effective attraction can be larger than the "residual" electrostatic repulsion between electrons which has not been taken into account in the average crystal potential which determines the band structure.

In this process, the electron cedes or absorbs to the crystal lattice, the energy $\hbar\omega$ of an acoustic phonon which is smaller than a tenth of an electron-volt. The Pauli principle then requires, for this energy exchange to be possible, that the electron energy be close to the Fermi level, at a distance from this level of the order of $\hbar\omega$. Therefore, the electrons involved in this mechanism are necessarily conduction electrons.

A crucial point in the mechanism of superconductivity is the fact that the effective attraction thus described can lead to a bound state of the electrons. Indeed, when two particles attract each other they will form a bound state only if their attraction is sufficiently strong. It is relevant to show that the bound state in a superconductor is the result of the fact that the considered electrons have energies close to the Fermi level. The superconducting bound state is thus not only the result of a "mechanical" interaction. It puts also into play in an essential manner the Pauli principle.

11.4.3 Microscopic mechanism, quantitative description

Let us indicate schematically the essential steps of derivation of this property, as an illustration of the usefulness, in the study of a complex effect as superconductivity, of the various concepts introduced in this book. Consider two electrons of a free-electron gas, occupying states *above the Fermi level*, and attracting each other with a potential energy u. Let us show that they can form a bound state whose energy is situated at an interval ε *below the Fermi level*:

$$\varepsilon \approx \exp\left\{-\frac{2}{g(E_F) \cdot u}\right\} \tag{11.18}$$

This result will justify that an attraction, however weak, determines for the two bound electrons, a state which is *more stable* than the independent states of the free-electron gas. Since all the free-electrons of a metal are equivalent, this will prove that all these electrons are in a bound state.

In the absence of attraction between electrons, the state of the two considered electrons is represented by the product of two plane waves: $\psi_0 = \exp(i\vec{k}_1 \cdot \vec{r}_1) \cdot \exp(i\vec{k}_2 \cdot \vec{r}_2)$. For electrons with energies near the Fermi level, the total energy of this "pair" is $\approx 2E_F$.

The presence of a mutual attraction will only modify the *relative motion* of the two electrons. The center of gravity of the pair is still described by a plane wave. In order to describe the relative motion let us put:

$$\vec{r} = \vec{r}_2 - \vec{r}_1 \qquad \vec{k} = \frac{(\vec{k}_2 - \vec{k}_1)}{2} \tag{11.19}$$

The Schrödinger equation of the pair of electrons (mass $2\,m$) in the presence of an attracting potential $U(\vec{r})$ is

$$\frac{-\hbar^2}{m}\Delta\psi + U(\vec{r}) \cdot \psi(\vec{r}) = (2E_F + \varepsilon)\psi(\vec{r}) \tag{11.20}$$

The state representing the relative motion of the two particles can be developed as an infinite sum of plane waves defined by $\vec{k}$ vectors:

$$\psi(\vec{r}) = \int_{\vec{k}} g(\vec{k})\exp(-i\vec{k} \cdot \vec{r}) \cdot d\vec{k} \tag{11.21}$$

Substituting this form in the first term of the Schrödinger equation (11.20), we obtain:

$$\frac{-\hbar^2}{m}\Delta\psi = \frac{\hbar^2}{m}\int_{\vec{k}} k^2 g(\vec{k}) d\vec{k} \tag{11.22}$$

On the other hand let us denote:

$$u(\vec{k}_0 - \vec{k}) = \int_{\vec{r}} U(\vec{r}) \exp\left\{i(\vec{k}_0 - \vec{k}) \cdot \vec{r}\right\} \cdot d\vec{r} \qquad (11.23)$$

We can then write using Eqs. (11.20) and (11.23)

$$\int_{\vec{k}} u(\vec{k}_0 - \vec{k}) \cdot g(\vec{k}) \cdot d\vec{k} = \int_{\vec{r}} \exp(i\vec{k}_0 \cdot \vec{r}) d\vec{r}$$

$$\int_{\vec{k}} \left[2E_F + \varepsilon - \frac{\hbar^2}{m}k^2\right] g(\vec{k}) \exp(-i\vec{k} \cdot \vec{r}) d\vec{k} \qquad (11.24)$$

Since

$$\int_{\vec{r}} \exp(i\vec{k}_0 \cdot \vec{r}) d\vec{r} = \delta(\vec{k}_0) \qquad (11.25)$$

we obtain for Eq. (11.24) the form:

$$\int_{\vec{k}} u(\vec{k}_0 - \vec{k}) \cdot g(\vec{k}) \cdot d\vec{k} = \left[2E_F + \varepsilon - \frac{\hbar^2}{m}\right] \cdot k_0^2 \cdot g(\vec{k}_0)$$

$$(11.26)$$

As pointed out in the preceding paragraph, the effective attraction between electrons is mediated by the lattice oscillations which have a small energy as compared to the Fermi energy. If two electrons have too different energies, they will not interact. Hence, we assume that $u(\vec{k}_0 - \vec{k}) = (-u)$ is constant and *negative* in an energy interval δ above E_F, and *that this quantity vanishes outside this interval*. With this assumption, Eq. (11.26) takes the form:

$$-u \int_{\vec{k}} g(\vec{k}) d\vec{k} = \left\{2E_F + \varepsilon - \frac{\hbar^2}{m}k_0^2\right\} g(\vec{k}_0) = A(\vec{k}_0) \cdot g(\vec{k}_0)$$

$$(11.27)$$

The integral in the left member is a mere constant C. Thus, $g(\vec{k}_0) = -u \cdot C/A(\vec{k}_0)$, and likewise, for any value of $\vec{k}$, $g(\vec{k}) = -uC/A(\vec{k})$. We can write Eq. (11.27) in the form:

$$\frac{1}{u} = \int_{\vec{k}} \frac{g(\vec{k}) \cdot d\vec{k}}{(\hbar^2 \vec{k}^2/m) - 2E_F - \varepsilon} = \int_0^\delta \frac{g_1(E) dE}{2E - \varepsilon} \qquad (11.28)$$

The last member involves the passage of an integration on the $\vec{k}$ vector to an integration on the energy values. Since the density of states $g_1(E)$ is approximately constant near the Fermi level, $g_1(E) \approx g_1(E_F)$. Finally, by integrating the last term in Eq. (11.28):

$$\frac{g_1(E_F)}{2} \cdot Ln\left\{\frac{\varepsilon - 2\delta}{\varepsilon}\right\} = \frac{1}{u} \qquad (11.29)$$

which is equivalent to the expression (11.18) indicated above:

$$\varepsilon = E - 2E_F = -2\delta \exp\left\{-\frac{2}{g(E_F) \cdot u}\right\} \qquad (11.30)$$

The existence of a forbidden band of width ε between the bound states of the electron pair, and the free-electrons, determines for the quantum states of the electrons of a superconductor, the shape represented on Fig. 11.9. It allows to understand the "rigidity" of the superconductor state towards the external actions of electrical or magnetical natures (e.g. the perfect diamagnetism). The superconductor will respond to external actions, only if these actions give the electrons enough energy to overcome the width of the forbidden band ε, thus destroying the superconducting state.

It is the occurrence of this forbidden band between occupied states and empty states which explains the perfect conductivity of a superconductor. This is, paradoxically, the opposite of the situation of an insulator in which the insulating character has its origin in the forbidden band (the gap) between occupied states and empty states.

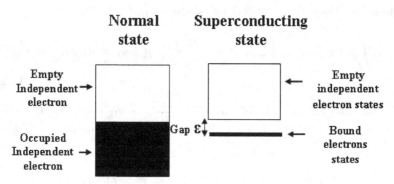

Fig. 11.9.   Schematic representation of the stabilization of the superconducting state by formation of bound states of electrons.

The distinction comes from the fact that here the "gap" separates *a bound state* from free-electron states, while in the case of insulators, the gap separates free-electron states.

Finally, let us point out that the interaction between electrons "mediated" by the oscillation of the lattice ions will be more efficient if the displacement of the ions is easier, i.e. if the solid is more deformable. Hence, superconductivity is likely to arise in "soft" materials as lead or tin, while copper, harder, does not become superconducting. Likewise, it can appear that the occurrence of superconductivity will be favoured by a stronger attraction between electrons and ions. This is only partly true. If the interaction is too strong the electrons will tend to stay in the vicinity of the ions, which cannot circulate in the solid, and thus the electron will not circulate either and will not contribute to the electric current (the solid will be insulating). This argument suggests that superconductivity requires a situation of "compromise" in which the attraction is neither too strong nor too weak. This compromise will determine the collective state of electrons which will persist at the highest temperature. In the framework discussed in this section, this optimal situation, the highest temperature of onset of superconductivity, has been quantitatively estimated in the range 25–30 K. The observed occurrence of substances in which the critical temperature is higher than 150 K shows that other mechanisms (not clarified at present) are required to account for their superconductivity.

Bibliography

Theorie du Magnétisme (In French), *A. Herpin*, Presses Universitaires de France (1968).

Ferromagnetism, *R.M. Bozorth*, John Wiley & Sons (1993).

Introduction to the Theory of Ferromagnetism, *A. Aharoni*, Clarendon Press (1996).

Superconductivity, *R.D. Parks*, CRC Press (1969).

Superfluidity and Superconductivity, *D.R. Tilley and J. Tilley*, CRC Press (1990).

Superconductivity, *C. Poole, H. Farach, R. Greswick, R. Prozorov*, Elsevier (2014).

Appendix A

Exercises

A.1   Examination Text No. 1 Hexagonal Boron Nitride

Although it contains two types of atoms, boron (B) and nitrogen (N), boron nitride possesses a simple enough crystal structure. It is therefore possible to calculated the energy dispersion curves $E(\overrightarrow{k}, n)$ of the electronic bands occupied by the valence electrons of the constituting atoms. This study will allow to compare this band structure to that of graphite which has a similar structure.

A.1.1   Direct and reciprocal lattices

Description of the structure

The structure is a succession of equally spaced parallel planes. In each plane the B and N atoms occupy the vertices of regular hexagons forming a "bee-hive" network. Each B atom is surrounded by three N atoms, and likewise each N atom is surrounded by three B atoms. The shortest B–N distance is $|\overrightarrow{\delta}| = 1.45$ Å. In two consecutive planes, the vertices of the hexagons are above each other, but their occupation is exchanged. Thus in the direction perpendicular to the planes, each B atom is comprised between two N atoms, and, conversely, each N atom is comprised between two B atoms. The interplanar distance is 3.30 Å. The frame of reference (x, y, z) is indicated on Fig. A.1.

(1)° (a) $(\overrightarrow{a}_1, \overrightarrow{a}_2)$ are the primitive translations which generate the crystal structure in a plane. Show that one can choose those translations, with $|\overrightarrow{a}_1| = |\overrightarrow{a}_2| = a$ and making an angle of 120°.

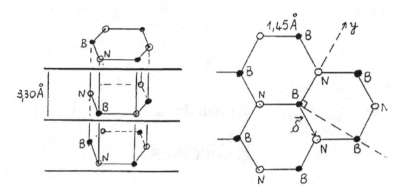

Fig. A.1.

Specifiy the modulus a. What is the type of two-dimensional Bravais lattice which is generated? How many nodes are closest to a given node? How many atoms are contained in the basis of this planar structure?

(b) Specify the direction and the modulus of the third primitive translation $\vec{a}_3$? Specify the nature of the three-dimensional Bravais lattice as well as the number of atoms in the primitive unit cell.

(2°) Determine the primitive translations $(\vec{a}_1^*, \vec{a}_2^*)$ of the reciprocal lattice corresponding to the structure of a plane. Represent the two first Brillouin zones in two dimensions. Denote Γ the origin of the reciprocal space, and P_i, Q_i, respectively, the vertices and the middle of the edges of the border of the first Brillouin zone. Specify as function of a the moduli of $\overrightarrow{\Gamma P_i}$ and of $\overrightarrow{\Gamma Q_i}$. Represent the orientations of these vectors with respect to the directions (1) of the B–N segments linking the neighbouring atoms; (2) of the primitive translations of the direct lattice; (3) of the x- and y-axes.

A.1.2 Electronic states in the tight binding approximation

In a first step of the study of the electronic structure of boron nitride, we assume that the energy bands are formed in the tight binding approximation, by the atomic wavefunctions of a single plane of the structure. More precisely, we consider a wavefunction of type p_z, centered on each of the boron and of the nitrogen atoms of a plane (Fig. A.2). Denote $p_B(\vec{r})$ and $p_N(\vec{r})$ respectively a p_z-type function

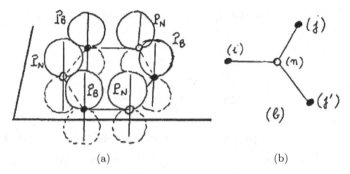

(a) (b)

Fig. A.2.

centered on a boron or nitrogen atom *situated at the origin*. The considered plane contains N unit cells. This two dimensional structure is made periodic by the Born–Von Karman method.

(3°) (a) Write the expressions of the two Bloch functions ϕ_B and ϕ_N constructed, in the tight binding approximation, from the p_B and p_N functions. The origin of coordinates can be chosen on a boron atom and the Bloch functions $\phi_B(\overrightarrow{k}, \overrightarrow{T_i}|\overrightarrow{r})$ and $\phi_N(\overrightarrow{k}, \overrightarrow{T_i}, \overrightarrow{\delta}|\overrightarrow{r})$ expressed as function of a $\overrightarrow{k}$ vector of the two-dimensional reciprocal space, of a vector $\overrightarrow{\delta}$ (cf. Fig. A.1) and of the translation vectors $\overrightarrow{T}_i$ which correspond to the unit cells in the plane.

(b) Assume that the wavefunctions p_B and p_n are normalized (i.e. $\|p_B\| = \|p_N\| = 1$) and that all the overlap integrals between different p_z functions are equal to zero (even between neighbouring atoms). Determine the values of the scalar products $\langle \phi_B | \phi_B \rangle$, $\langle \phi_N | \phi_N \rangle$, and $\langle \phi_B | \phi_N \rangle$.

(c) Show that any linear combination of ϕ_B and ϕ_N corresponding to a given $\overrightarrow{k}$ vector is a Bloch function.

(4°) Let $\widehat{H}$ be the Hamiltonian acting on an electron of the system. Denote:

$$H_{BB}(\overrightarrow{k}) = \langle \phi_B | \widehat{H} | \phi_B \rangle; H_{NN}(\overrightarrow{k}) = \langle \phi_N | \widehat{H} | \phi_N \rangle; H_{BN}(\overrightarrow{k})$$

$$= \langle \phi_B | \widehat{H} | \phi_N \rangle \tag{A.1}$$

Assume that the eigenfunctions of $\widehat{H}$ are of the form $(\phi_B + \lambda \cdot \phi_N)$. Write the corresponding eigenvalue equation. Deduce, by projecting on ϕ_B and ϕ_N the eigenvalues E, as function of H_{BB}, H_{NN}, and H_{BN}. How many energy bands does one obtain as a function of the p_z?

(5°) One has:

$$\widehat{H} = \frac{\widehat{p}^2}{2m} + \sum_n w_B^{(n)} + \sum_{n'} w_N^{(n')} \tag{A.2}$$

in which $w_B^{(n)}$ is the potential to which the electron is submitted in the *isolated* boron atom. Likewise, $w_N^{(n)}$ is the potential corresponding to the isolated nitrogen atom. We denote, by analogy, $p_B^{(i)}$ and $p_N^{(i)}$ the p_z functions centered respectively on the ith boron or nitrogen atom. We also assume for the sake of simplification:

$$\langle p_B^{(i)} | w_B^{(i)} | p_B^{(i)} \rangle \neq 0 \quad \langle p_N^{(i)} | w_N^{(i)} | p_N^{(i)} \rangle \neq 0 \tag{A.3}$$

We also assume that if the indices (i) and (n) correspond to neighbouring atoms:

$$\langle p_B^{(i)} | w_N^{(n)} | p_N^{(n)} \rangle = -\gamma' \tag{A.4}$$

$$\langle p_B^{(i)} | w_N^{(n)} | p_B^{(i)} \rangle = -\alpha_B \quad \langle p_N^{(i)} | w_B^{(n)} | p_N^{(i)} \rangle = -\alpha_N \tag{A.5}$$

Finally, if the atoms (i) and (j) are bound to the same atom (n) (cf. Fig. A.2(b)):

$$\langle p_B^{(i)} | w_N^{(n)} | p_B^{(j)} \rangle = -\gamma_B' \quad \langle p_N^{(i)} | w_B^{(n)} | p_N^{(j)} \rangle = -\gamma_N \tag{A.6}$$

Besides, all the matrix elements of the form $\langle p^{(i)} | w_B^{(n)} | p^{(j)} \rangle$ are equal to zero.

(a) Calculate H_{BB}, H_{NN}, H_{BN} as function of $\vec{k}$, α_B, α_N, γ_B, γ_N, γ' as well as function of the energies E_B^0 and E_N^0 of an electron in the isolated boron or nitrogen atom, respectively.

(b) Show that the possible energies of an electron in the considered bands are of the form:

$$2E = A \pm \{B^2 + 4\gamma'^2 |\Delta(\vec{k})|^2\}^{1/2} \tag{A.7}$$

with

$$A = \{E_B^0 + E_N^0 - 3\alpha_B - 3\alpha_N - (\gamma_B + \gamma_N)\sigma(\vec{k})\} \tag{A.8}$$

and

$$B = \{E_B^0 - E_N^0 + (3\alpha_N - 3\alpha_B) + (\gamma_N - \gamma_B)\sigma(\overrightarrow{k})\} \qquad (A.9)$$

Specify the form of $\sigma(\overrightarrow{k})$ and of $\Delta(\overrightarrow{k})$ and check that:

$$|\Delta(\overrightarrow{k})|^2 = 1 + 4\cos^2\left(\frac{k_x \cdot a}{2}\right) + 4\cos\left(\frac{k_x \cdot a}{2}\right) \cdot \cos\left(\frac{k_y \cdot a\sqrt{3}}{2}\right)$$
$$(A.10)$$

(6°) (a) Determine and represent graphically the variations of $|\Delta(\overrightarrow{k})|^2$ and of $|\Delta(\overrightarrow{k})|$ along the segments ΓP and ΓQ of the reciprocal space. At which points does $|\Delta(\overrightarrow{k})|^2$ reach its minimum value? Specify this value.

(b) Justify qualitatively that $|\gamma_B| \ll |\gamma'|$ and that $|\gamma_N| \ll |\gamma'|$. *In all the following questions, assume that* $\gamma_B = \gamma_N = 0$. Represent then qualitatively the variations of the energies $E(\overrightarrow{k})$ of the bands originating from p_z, on the closed path $\Gamma P_i Q_i \Gamma$. (P_i and Q_i are respectively the vortex and the middle of adjacent edges.) What is minimum energy interval which separates these bands. At which point of reciprocal space this minimum occurs?

A.1.3 Occupation of the bands of boron nitride

The electronic configuration of the boron atom B is $(1s)^2(2s)^2(2p)^1$ and that of nitrogen N is $(1s)^2(2s)^2(2p)^3$. The less bounded states are (2s) and (2p).

The structure of the atomic planes of boron nitride suggests that, in these planes, the atoms are linked by (sp^2) orbitals, linear combinations of the six orbitals $(2s), (2p_x), (2p_y)$ of boron and of nitrogen pointing towards the three neighbouring atoms. These orbitals form a "bonding" band (sp^2) with a very low negative energy, and an "antibonding" band (sp^2) separated from the other, and having a higher energy.

(7°) (a) How many electrons per unit cell occupy the set of bands formed by the orbitals $(2s, 2p_x, 2p_y, 2p_z)$. How many states, taking the spin into account, are there empty states in the bonding band (sp^2)? How many empty states are there in the bands formed by the p_z orbitals?

(b) What argument, based on consideration of the crystal structure, can be invoked to justify the fact that the p_z bands determined in the preceding questions, have energies situated between the energies of the bonding and antibonding bands (sp^2)?

(c) Based on (b), indicate if boron nitride is an insulator or a conductor. Which parameters, reflecting the different properties of boron and of nitrogen explain this property?

A.1.4 Comparison with graphite

(8°) We consider now graphite whose crystal structure is obtained from that of boron nitride by replacing the B and N atoms by carbon atoms C. The electronic configuration of carbon is $(1s)^2(2s)^2(2p)^2$, explain why one can determine, on the basis of the results obtained for boron nitride, if graphite is insulating or conducting. Represent the variations of the energies $E(\overrightarrow{k})$ of the bands based on the p_z functions, on the closed circuit $\Gamma P_i Q_i \Gamma$.

(9°) We wish to deduce the shape of the density of states $g(E)$ relative to the bands generated by p_z, from the expression of $E(\overrightarrow{k})$ determined for these bands in questions (5°) and (6°).

(a) How can the following equation:

$$dw^* = dE \int_{\text{surface } U=E} \frac{dS(\overrightarrow{k})}{|\overrightarrow{\nabla}_k E(\overrightarrow{k})|} \qquad g(E) \cdot dE = \frac{\Omega}{4\pi^3} dw^* \quad \text{(A.11)}$$

which is Eq. (5.34) in Chapter 5, can be adapted to calculate the density of states of a two-dimensional crystal.

(b) What is the shape of the curves $E = $ constant for energies close to the energy values near the point Γ. Show that in the neighbourhood of Γ, the density of states is independent of E.

(c) Consider the curve $E = E_Q$ where E_Q is the energy for one of the Q_i points defined in question (6°) (b). Show that this curve is formed of portions of straight lines. Based on the symmetry of the curves $E = $ constant, show that the curve $E = E_Q$ is a hexagon.

(d) Show that on one segment of the curve $E = E_Q$ that the density of states goes to infinity for $E = E_Q$.

(e) Represent graphically the variations of $g(E)$ for the bands p_z.

(10°) We now take into account, in the calculation of the p_z bands, all the planes of the structure.

(a) How many Bloch functions, associated to the p_z orbitals, must be considered for each $\overrightarrow{k}$ vector of the reciprocal space? How many energy bands will then be generated by the p_z orbitals?

(b) Considering the crystal structure, discuss the qualitative consequences of taking into account the interaction between planes, to determine the insulating or conducting character of boron nitride and of graphite, as well as on the characteristics of the density of states.

A.2 Examination Text No. 2 Metallic Binary Alloys

The experimental studies of metallic binary alloys A–B such as Cu–Zn, Cu–Al, etc. show that these alloys, in which A and B have different valencies, have different structures when the proportions of the two components (e.g. Cu and Zn) are modified, or if the temperature is varied. The present problem is to examine two types of questions pertaining to these alloys. In Section A.2.1, one will consider the changes of the X-ray diffraction spectrum (cf. Chapter 4) of the alloy which are observed when the alloy changes its structure (cf. Chapter 2) for a given alloy composition. In Section A.2.2, which is independent of Section A.2.1, one will study the relationship between the electronic band structure of the alloy and the stability of each of its structures (cf. Chapter 5).

A.2.1 Diffraction

Brass, whose composition is CuZn (50% of Cu atoms and 50% of Zn atoms), can be found with two possible crystal structures. The first one denoted β-brass is stable above $T_c = 470°$C. It has a cubic centered structure (CC) whose conventional cubic unit cell has the edge $a \approx 3$ Å. At each node of its Bravais lattice there is a single atoms which is either copper (Cu) or zinc (Zn). For the sake of simplicity, one will consider that all the nodes are occupied by identical "average" atoms (50% Cu, and 50% Zn).

The second structure, denoted β'-brass is stable below T_c. In this structure, *the vertices and the centers* of the conventional cubic unit cell are occupied by two different "atoms" denoted S and C. The atom S has the probability $(1 - p) \neq 1/2$, to be a copper (Cu) atom and the probability p to be a zinc (Zn) atom. Conversely, the C atoms has respectively the probabilities p and $(1-p)$ to be a copper or zinc atom. As emphasized, in the β-brass structure S and C are identical $(p = 1/2)$.

(1°) The X-ray diffraction spectrum of a crystal expresses the variations of the diffracted intensity $I(\vec{K})$, as function of $\vec{K}$, in the reciprocal space. Explain briefly why this intensity is non-zero *only* if $\vec{K}$ defines the nodes of a lattice $(h\vec{a}_1^* + k\vec{a}_2^* + l\vec{a}_3^*)$. Specify for the structure of β-brass the type of lattice, the dimensions of the conventional unit cell, and the components of the $\vec{a}_i^*$ vectors.

(2°) The amplitude of the wave diffracted by an atom is proportional to the atomic scattering factor $f(|\vec{K}|)$ of this atom. If an atom which has the probability $(1 - p)$ to be a copper atom and a probability p to be a zinc atom, we assume that its scattering factor is $f = (1 - p)f_{Cu} + pf_{Zn}$ in which f_{Cu} and f_{Zn} are respectively the scattering factors of copper and zinc. Show that for the structure of β-brass, using the table hereunder, and interpolating, the ratio of the two intensities:

$$r = \frac{I(h = 1, k = 0, l = 0)}{I(h = 2, k = 2, l = 2)} \qquad \text{(A.12)}$$

in which (h, k, l) are defined in question (1°).

$|\vec{K}|$ (Å^{-1})	0	1	2	4	6	8	10	12	14	16
f_{Cu}	29	28	25	19.2	14.2	10.9	8.7	7.4	6.6	6
f_{Zn}	30	29.1	25.8	20	15.1	11.5	9.1	7.8	6.8	6.1

(3°) What is the type of the Bravais lattice of the β' structure? How many atoms does the elementary unit cell contain? Specify the type and the elementary unit cell of the reciprocal lattice $(h'\vec{b}_1 + k'\vec{b}_2 + l'\vec{b}_3)$ of this structure.

(4°) Find, within the reciprocal lattice of the β'-structure, the location of the X-ray diffraction $\vec{K}$ vectors of the β-brass structure. Specify the relationship between the h', k', l' defining these vectors. Show that additional directions of diffraction distinguish the β'-structure from the β-brass structure. Represent on a drawing of the $\vec{b}_i^*$ unit cell the points corresponding to the diffractions common to the two structures, and those which only correspond to the β' structure.

(5°) (a) The atomic scattering factors of the S and C atoms are defined as in question (2°). Determine the expression of the structure factor $F(\overrightarrow{K})$ of the β' structure as a function of the probability p, and of the atomic scattering factors f_{Cu} and f_{Zn} and of the components (h', k', l').

(b) Show that the intensities of the additional diffractions which are relative to the β' structure are proportional to $(p - \frac{1}{2})^2$.

(6°) What can be said of the diffractions common to the two structures β, and β'? How can a diffraction experiment allow to determine the probability of occupation p of a node of the lattice by a copper or a zinc atom?

A.2.2 Electronic energy and stability of the alloys

In this part, the stability of an alloy of composition Cu–B is studied as a function of its composition (B being either zinc or aluminum). The composition of the alloy will be denoted $Cu_{1-x}B_x$, in which x is the concentration of the B atoms $(0 < x < 1)$. As in part one, it will be considered that the alloy is a perfect monoatomic crystal whose nodes of the Bravais lattice are occupied by "average" identical atoms, each of these atoms having a probability $(1 - x)$ to be a copper atom and a probability x to be a B atom.

The valence of copper is one. This means that each copper atom contributes by a single electron to the number of electrons of the conduction band. The valence of zinc is 2, and the valence of aluminum is 3. Two different crystal structure will be compared for the alloy. The first one is a face centered cubic structure (FCC). The second is a cubic centered (CC) structure.

(1°) Let a_f and a_c the respective edges of the conventional cubic unit cells of the two structures. Assume that the average volume of a given atom in the two structures is the same. What is then the value of the ratio (a_f/a_c). In the following questions, it will be assumed that this ratio has the preceding value and the value of a_c will be taken as $a_c \approx 3$ Å.

(2°) Express as function of x the average number $\nu(x)$ of conduction electrons per atom in the alloy $Cu_{1-x}B_x$ for B = Zn, and B = Al. In what interval of values does this number vary?

(3°) Assume that the conduction electrons are independent electrons submitted to a uniform potential. The temperature of the alloy is $T = 0\,\mathrm{K}$.

(a) Determine the radius k_F of the Fermi sphere for the electrons of the FCC structure as a function of x and of $(2\pi/a_j)$ for $\mathrm{B = Zn}$ and for $\mathrm{B = Al}$.

(b) What can be said of the value of k_F in the case of the cubic centered alloy.

(c) Represent schematically the variations of the density of states $g(E)$ and the location of the Fermi level, for a given value of x for the two structures FCC and CC.

(4°) (a) Recall the type and the conventional unit cell of the reciprocal lattices of the two structures FCC and CC. What is, for each structure, the shortest distance between the origin and a node of the reciprocal lattice.

(b) Show that, for $x = 0$, the Fermi sphere is entirely contained inside the first Brillouin zone of the structures FCC and CC.

(5°) When the number ν of electrons per atom increases, the radius of the Fermi sphere increases. Determine the "critical" values of ν which correspond to a Fermi sphere tangent to the surface of the first Brillouin zone closest to the origin. Denote $\nu(FCC)$ and $\nu(CC)$ the critical values relative to the two structures. Deduce the corresponding values of x for $\mathrm{B = Zn}$ and $\mathrm{B = Al}$. They will be denoted $x(FCC, Zn), x(FCC, Al), x(CC, Zn), x(CC, Al)$.

(6°) The electrons are now submitted to a periodic potential $V(\overrightarrow{r})$.

(a) Represent schematically the shape of the curve $E(k)$ along the direction of the vector (k_0) in reciprocal space of the surface of the first Brillouin zone closest to the origin. How does this shape differ from that of free electrons in a uniform potential.

(b) On the basis of the curve $E(k)$ explain the shape of the density of states represented (Fig. A.3).

(7°) (a) Assume that the Fermi surface can still be considered as spherical. What is the radius k_F which corresponds to the point G of the density of states.

(b) The value of the matrix element $\langle k_0|V(\overrightarrow{r})|k_0\rangle = 0.5\,\mathrm{eV}$, for the two considered structures (FCC and CC). What is approximately the value of the energy difference $(E_G - E_M)$.

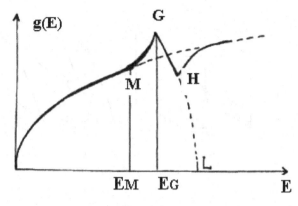

Fig. A.3.

(8°) (a) Represent schematically, on the same chart, the variations of the density of states $g_{FFC}(E)$ and $g_{CC}(E)$ and place the points M and G relative to the two structures. Discuss qualitatively, for each structure, as function of the value of x, the compared locations of the Fermi level corresponding to the free-electron in a uniform potential and of the quasi-free one corresponding to the $V(\overrightarrow{r})$ potential.

(b) Compare, as function of the value of x the locations of the Fermi level in the FCC and CC structures.

(9°) Assume that for $T = 0\,\mathrm{K}$ the stable structure of the alloy is that corresponding to the minimum of the total electronic energy. Which of the structures is the stable one in the following cases: $x(Zn) = 0.2$; $x(Al) = 0.17$; $x(Al) = 0.24$.

(10°) The actual periodic potential is taken into account and the Fermi surface is no longer assimilated to a sphere. Indicate towards which values of x the stability limits obtained would be displaced.

A.3 Examination Text No. 3 Properties of Bismuth

Bismuth is a metal with unusual electronic properties: Very small specific heat, very high Hall coefficient, etc. The subject of this examination is to study, in the framework of the quasi-free electrons in a periodic potential, the band structure which determines these unusual properties.

Indications: (1°) *If the questions marked by an asterisk (∗) are not solved, the student can nevertheless proceed to solve the other*

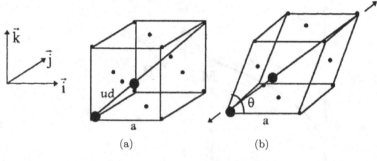

Fig. A.4.

questions. (2°) *The only lengthy calculations are in questions* (4°) (b), (4°) (d) *and* (7°) (a). (3°) *It is recalled that* $\hbar^2/2m = 3.815\,\text{eV}$. A^2. *For* $a = 6.676\,\text{Å}$, *one has* $(2\pi/a) = 0.941\,\text{Å}^{-1}$, *and* $(2\pi/a)^2 = 0.886\,\text{Å}^{-2}$.

A.3.1   Part I: Crystal structure, Bravais and reciprocal lattices

The Bravais lattice of bismuth can be considered, as a first approximation, as a face centered cubic lattice (FCC) whose conventional unit cell (Fig. A.4) has the parameter $a = 6.676\,\text{Å}$. The three orthogonal unit vectors $\vec{i}$, $\vec{j}$, $\vec{k}$ parallel to the edges of this cell will be the reference frame in all the problem. The atomic basis (referred to the *elementary unit cell*) contains two atoms situated respectively at the origin and on the diagonal passing through the origin, at a distance $u \cdot d$, d being the length of the diagonal of the conventional cubic cell, and $0 < u < 0.5$ a parameter. $u = 0.5$ means that the second atom is at the center of the conventional cell. We denote $\overrightarrow{K}$ the reciprocal space vectors.

(1°) (a) Recall, as function of a, and $\vec{i}$, $\vec{j}$, $\vec{k}$, the expression of the vectors $(\vec{a}^c_1, \vec{a}^c_2, \vec{a}^c_3)$ defining the elementary FCC unit cell. What is the volume ω_c of this cell?

(b) What is the expression of the vectors $\vec{a}^{c,*}_j$ which define the elementary unit cell of the reciprocal lattice? What is the type of this lattice? What is the edge of its conventional cubic reciprocal cell? Express the volume ω_c^* of the elementary reciprocal cell as function of ω_c.

(*c) Show that for $u = 0.5$, the structure can be described as a simple cubic structure with a single atom in the basis. Specify the dimension of the cell. What is the type of the reciprocal lattice and the volume of its unit cell.

(2°) The actual structure of bismuth differs from the approximate structure in question (1°) by a lengthening of the unit cell along the (111) diagonal. This lengthening induces an angle change ($\pi/2 \to \theta$) between the edges of the cube, but does not change their length a, nor the value of the u parameter (cf. Fig. A.4(b)). It is indicated that the elementary unit cell of the structure is then defined by the vectors:

$$\vec{d}_1 = a_0(1, 1, \varepsilon) \quad \vec{d}_2 = a_0(\varepsilon, 1, 1) \quad \vec{d}_1 = a_0(1, \varepsilon, 1) \qquad (A.13)$$

with $a_0 = |\vec{d}_i^{\,c}|/\sqrt{2}$ and $\varepsilon = 0.04$ ($\varepsilon \langle\langle 1\rangle$).

(a) Express as function of a_0 and ε, to first-order in ε, the volumes ω and ω^* of the elementary direct and reciprocal cells of the structure. Evaluate their numerical values in Å^3 and in Å^{-3}.

(b) Show that one of the vectors of the reciprocal lattice is, in the reference frame $\vec{i}, \vec{j}, \vec{k}$,

$$\vec{d}_1^* = \alpha(\varepsilon) \cdot (1, 1, -1 - \varepsilon) \qquad (A.14)$$

in which $\alpha(\varepsilon)$ is a function of $2\pi/a$ and of ε which will be determined by a first-order expansion in ε. Indicate, by analogy, the expressions of $\vec{d}_2^*$ and of $\vec{d}_3^*$.

A.3.2 Part II: Band structure in the free-electrons approximation

Each of the two atoms of the basis has five valence electrons which are likely to be part of the metal conduction electrons. In this section the conduction electrons are considered as free-electrons in a uniform potential.

(3°) Show that if the crystal lattice is of the FCC type, the radius of the Fermi sphere is $k_F = 1.684 \cdot (2\pi/a)$. Calculate, to first-order in ε, the value of k_F relative to the deformed structure ($\varepsilon \neq 0$) as well as the volume Ω_F^* of the Fermi sphere. Compare Ω_F^* to the volume ω^* of the reciprocal elementary unit cell. Explain the result thus obtained.

(4°) We consider that the crystal has the FCC lattice, and that its reciprocal lattice is defined by the $\overrightarrow{a}_i^{*,c}$ vectors.

(a) Let $\overrightarrow{G}$ be a vector of the reciprocal lattice. Recall the definition of a Brillouin plane defined by $\overrightarrow{G}$. What is its distance to the origin of the reciprocal lattice?

(b) List the vectors $\overrightarrow{G}_i$ of the reciprocal lattice associated to Brillouin planes which intersect the Fermi sphere. Give their expressions as function of the set $(\overrightarrow{i}, \overrightarrow{j}, \overrightarrow{k})$. Denote the $\overrightarrow{G}_i = (2\pi/a)(n_1, n_2, n_3)$. Explain the importance of determining these planes to study the quantum spectrum of a free-electron.

(c) How many different distances are found between the origin Γ of the reciprocal lattice and these planes? Each distance defines a class of Brillouin planes. Indicate for each class the number of Brillouin planes thus determined.

(d) Note, on the basis of the results of (b) and (c), that the Brollouin plane defined by $\overrightarrow{G}_1^m = (2\pi/a)(1, 1, 3)$ belongs to the class furthest from Γ and which intersect the Fermi sphere. Denote $\overrightarrow{G}_i^m$ the 24 vectors of the reciprocal lattice defining this class of Brillouin planes. Express the $\overrightarrow{G}_i^m$ as function of the $\overrightarrow{a}_i^{*,c}$.

(5°) Keeping the assumption that the crystal is strictly of the FCC type, the shape of the first Brillouin zone is recalled hereunder. Referred to the frame $(\overrightarrow{i}, \overrightarrow{j}, \overrightarrow{k})$, the coordinates of points L (center of a hexagon) and X (center of a square) are (Fig. A.5):

$$L = (2\pi/a)(1/2, 1/2, 1/2) \quad \text{and} \quad X = (2\pi/a)(0, 1, 0).$$

(a) How many points similar to L are at the center of hexagons? Which one are equivalent to L, i.e. are related to L by a reciprocal lattice vector? Denote T_i the *non-equivalent ones*.

(b) Express $\overrightarrow{\Gamma L}$, $\overrightarrow{\Gamma X}$, and one of the $\overrightarrow{\Gamma T}_i$, denoted more simply $\overrightarrow{\Gamma T}$ as function of the $\overrightarrow{a}_i^{*,c}$.

(c) Show that each of the vectors $\overrightarrow{\Gamma L}$ and $\overrightarrow{\Gamma T}$ is equivalent to six vectors $\overrightarrow{G}_i^m/2$ (defined in question (4°) (d) which will be listed). Admit, without proof, that each of the vectors $\overrightarrow{G}_i/2 \neq \overrightarrow{G}_i^m/2$ defined in question (4°) (b) are equivalent to $\overrightarrow{\Gamma X}$.

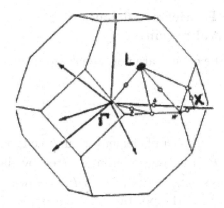

Fig. A.5.

(6°) (a) Recall the definition of the representation of the quantum spectrum $E(\overrightarrow{K}, n)$ of a solid in *a reduced zone scheme*.

(b) On the basis of questions (5°) (a) and (5°) (c), show that in a reduced zone scheme, the two lowest energy levels $E_1(L)$ and $E_2(L)$ of the free electron, at the point L, have respectively a two-fold and a six-fold degeneracy. Calculate, first as a function of $(2\pi/a)$ and then in electron-volts, the values of $E_1(L)$ and $E_2(L)$. Explain why, in the case of a strict FCC structure, the quantum spectrum is the same at the points T_i.

(7°) Consider now the actual structure deformed FCC structure ($\varepsilon \neq 0$). Admit that the vectorial relations in questions (4°) (d) and (5°) (b) are valid at the condition of substituting the vectors $\overrightarrow{a}_i^{*}$ to the vectors $\overrightarrow{a}_i^{c,*}$.

(a) Show that the moduli of the vectors $\overrightarrow{G}_i^m/2$, $\overrightarrow{\Gamma L}$, and $\overrightarrow{\Gamma T}$ are no longer identical. There are three different moduli which will be calculated to first-order in ε. Also calculate the moduli of $\overrightarrow{\Gamma L}$ and $\overrightarrow{\Gamma T}$.

(b) To first-order in ε, and in eV, calculate the energy shift of $E_1(L)$, and $E_2(L)$. Show that their degeneracies are not changed.

(c) Calculate the shift of $E_1(T)$. Show that $E_2(T)$ is decomposed into two levels $E_2'(T)\langle E_2''(T)$. Calculate their shifts in eV. Specify their respective degeneracies. Represent the relative positions of the levels $E_1(L), E_2(L), E_1(T), E_2(T)$.

A.3.3 Part 3: Fourier coefficients of the crystal potential

The electrons of the crystal are now *Bloch electrons* submitted to the potential:

$$V(\overrightarrow{r}) = \frac{1}{N_M} \sum_{i,j} \mathbf{v}(\overrightarrow{r} - \overrightarrow{R}_{i,j}) \qquad (A.15)$$

in which N_M is the number of elementary unit cells of the crystal, and in which $\mathbf{v}(\overrightarrow{r} - \overrightarrow{R}_{ij})$ is the potential exerted by the bismuth atom situated at positions $\overrightarrow{R}_{ij} = \overrightarrow{T}_i + \overrightarrow{r}_j$. ($\overrightarrow{T}_i$ is a primitive translation of the crystal, and $\overrightarrow{r}_j$ the position of the atom j in the unit cell at the origin.) The Fourier transform of $\mathbf{v}(\overrightarrow{r})$ is $\mathbf{v}(\overrightarrow{K})$:

$$\mathbf{v}(\overrightarrow{K}) = A \int \mathbf{v}(\overrightarrow{r}) \exp(-i\overrightarrow{K} \cdot \overrightarrow{r}) \cdot d\overrightarrow{r} \qquad (A.16)$$

A is a normalization factor.

(8°) (a) Show that the Fourier coefficients V_G of the potential, which determine the quantum spectrum of the electron in the approximation of the quasi-free electron can be written in the form:

$$V_G = \mathbf{v}(|\overrightarrow{G}|) \cdot \mathbf{F}(\overrightarrow{G}) \qquad (A.17)$$

in which $F(\overrightarrow{G})$ is the structure factor corresponding to a scattering factor $f(|\overrightarrow{K}|) = 1$ which appears in the expression of the intensity of diffracted X-rays by a crystal.

(b) Let $\overrightarrow{G} = n_1 \overrightarrow{a}_1^* + n_2 \overrightarrow{a}_2^* + n_3 \overrightarrow{a}_3^*$. Show that, as function of the u parameter defined in Appendix A.3.1, the modulus of the structure factor has the form:

$$|F(\overrightarrow{G})| = 2|\cos\{\pi u(n_1 + n_2 + n_3)\}| \qquad (A.18)$$

(9°) (a) Consider the vectors $\overrightarrow{G} = 2\overrightarrow{\Gamma L}$, and $\overrightarrow{G} = 2\overrightarrow{\Gamma T}$ as well as the vectors $\overrightarrow{G}_i^m$ defined in question (5°) (c). Indicate in a table, for these different vectors (or groups of vectors) the value of $(n_1 + n_2 + n_3)$, the expression of $|F(\overrightarrow{G})|$, as a function of u, and the numerical value of $|F(\overrightarrow{G})|$ for $u = 0.468$, as well as the value of V_G expressed in eV. Figure A.6 reproduces the values of $\mathbf{v}(|\overrightarrow{K}|)$ as function of $\kappa = |\overrightarrow{K}|^2 \cdot (a^2/4\pi^2)$.

(*b) Explain why the $|F(\overrightarrow{G})|$ thus calculated have zero value when $u = 0.5$.

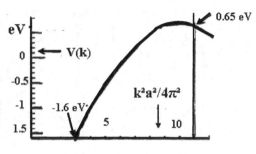

Fig. A.6.

A.3.4 Part 4: Effect of the potential on the properties of bismuth

(10°) We consider again the standard FCC lattice for $u = 0.468$. We examine the lifting of the degeneracy of the levels $E_1(L), E_1(T), E_2(L), E_2(T)$ considered in question (6°) (b) by the crystal potential $V(\overrightarrow{r})$.

(a) Determine on the basis of the results of question (9°) the degeneracy lifting (expressed in eV) of level $E_1(L)$. Denote $E_1^-(L)$ and $E_1^+(L) > E_1^-(L)$ the two resulting levels. Represent, along the segment ΓL, using the results of (6°) (b) the shape of the energy band $E(k)$ which ends at $E_1^+(L)$. Same question for $E_1(T)$ and the corresponding bands on the segment ΓT.

(b) What are the dimensions of the matrices which must be diagonalized to determine the degeneracy lifting of the levels $E_2(L)$ and $E_2(T)$.

(c) We assume, to simplify, that $E_2(L)$ separate into two levels $E_2^-(L)$ and $E_2^+(L)$ having the same degeneracy. Their separation is determined by the Fourier coefficient relative to the $\overrightarrow{G}_i^m$ associated to L. (cf. question (5°) (c)). Determine their separation as well as their position in the spectrum (in eV). What is, on the segment ΓL in the vicinity of L the possible shape, deduced by analogy with question (10°) (a), of the bands $E(\overrightarrow{K})$ which end at the levels $E_2^-(L)$ and $E_2^+(L)$.

(d) Determine, similarly, the degeneracy lifting of $E_2(T)$ and the shape of $E(\overrightarrow{K})$ near T. In this case assume that $E_2(T)$ is the superposition of $E_2'(T)$ and of $E_2''(T)$ and study separately into two levels of each of the former levels.

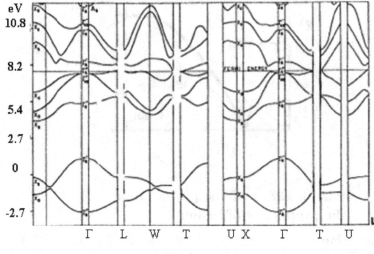

Fig. A.7.

(11°) *We assume that the physical properties of bismuth are entirely determined by the band structure near the points L and T.*

(a) The lattice is undeformed and $u = 0.5$. Explain why bismuth has then the properties of an ordinary metal such as sodium.

(b) The lattice is undeformed and $u = 0.468$. On the basis of the results of question (10°), show that the electronic spectrum (represented entirely by the vicinity of L and T) comprises a "global" forbidden band. Specify its width in eV. How many non-degenerate energy bands are situated below this forbidden band, near L and T. A degenerate band will be considered as a superposition of non degenerate bands. How many electrons can occupy a non-degenerate band? Deduce from this result if bismuth is a conductor or an insulator.

(c) The deformation of the lattice ($\varepsilon \neq 0$) *and* the real position of the atoms ($u = 0.468$) are taken into account. We assume that the degeneracy liftings are the same as those calculated in question (10°). Show, using the results of questions (7°) (b) and (7°) (c) that the positions of the bands $E(\overrightarrow{K})$ represented in questions (10°) (b) and (10°) (c) are modified. Explain why bismuth is a conductor (it will be necessary to use the values of energy precisely determined).

(12°) (a) Indicate qualitatively the position of the Fermi level relatively to the bands represented near L and T. How many bands are totally or partially occupied near L? Near T?

(b) Why can one say that in bismuth there is an "electron band" and a "hole band"? Near which points of the Brillouin zone are respectively situated these bands?

(c) Why does bismuth contain very few carriers ($10^{23}/m^3$ instead of $10^{29}/m^3$ in copper)? Explain why the Hall effect is specially easy to detect in bismuth.

(d) A more detailed calculation provides the results represented on Fig. A.7. How these results, relative to $E(\overrightarrow{K})$ along ΓL and ΓT, compare to those obtained here (separation of levels, position of the Fermi level)?

Appendix B

Solutions of Exercises

B.1 Examination Text No. 1 Hexagonal Boron Nitride

(1°) (a) We have: $\overrightarrow{a}_1 = a.\overrightarrow{i}$ and $\overrightarrow{a}_2 = (-a/2)\overrightarrow{i} + (a\sqrt{3}/2)\overrightarrow{j}$. With $a = 1.45\sqrt{3} = 2.51$Å. The Bravais lattice is hexagonal (triangular would also be an acceptable answer) with six nodes surrounding a given node. The surface of the unit cell is $s = a^2\sqrt{3}/2$. There are two atoms in the basis (B and N) (Fig. B.1).

(b) $\overrightarrow{a}_3 = c\overrightarrow{k}$ with $c = 6.60$Å. The Bravais lattice is of the hexagonal type with 4 atoms in the unit cell.

(2°) $\overrightarrow{a}_i^*$ is defined by the relations $\overrightarrow{a}_i^*.\overrightarrow{a}_j = 2\pi\delta_{ij}$, thus:

$$\overrightarrow{a}_1^* = \frac{2\pi}{s}\overrightarrow{a}_2 \times \overrightarrow{k} = \frac{4\pi}{a\sqrt{3}}\left(\sqrt{\frac{3}{2}}\overrightarrow{i} + \frac{\overrightarrow{j}}{2}\right)$$

$$\overrightarrow{a}_2^* = \frac{2\pi}{s}\overrightarrow{k} \times \overrightarrow{a}_1 = \frac{4\pi}{a\sqrt{3}}\overrightarrow{j} \tag{B.1}$$

The Brillouin zones have the shape and orientation represented hereunder (Fig. B.2), with $|\Gamma P_i| = (4\pi/3a)$ and $|\Gamma Q_i| = (2\pi/a\sqrt{3})$.

(3°)(a)

$$\phi_B = \frac{1}{\sqrt{N}}\sum_i \exp(i\overrightarrow{k}.\overrightarrow{T}_i).p_B(\overrightarrow{r} - \overrightarrow{T}_i) \tag{B.2}$$

$$\phi_N = \frac{1}{\sqrt{N}}\sum_i \exp(i\overrightarrow{k}.\overrightarrow{T}_i).p_N(\overrightarrow{r} - \overrightarrow{T}_i) \tag{B.3}$$

Physics of Electrons in Solids

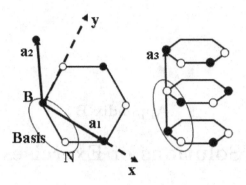

Fig. B.1.

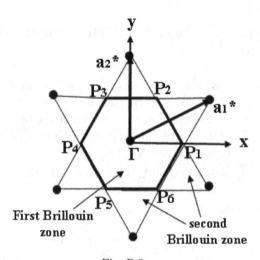

Fig. B.2.

(The normalization was not required)

(b) $\langle \phi_B | \phi_B \rangle = 1$, $M \langle \phi_N | \phi_N \rangle = 1$, $\langle \phi_B | \phi_N \rangle = 0$

(c) $\phi_B + \lambda \phi_N = \exp(i \vec{k} . \vec{r}) . \{ u_{B, \vec{k}}(\vec{r}) + \lambda u_{N, \vec{k}}(\vec{r}) \}$. If $u_{B, \vec{k}}(\vec{r})$ and $u_{N, \vec{k}}(\vec{r})$ are periodic and of period $\vec{T}_i$, the function between $\{ \ \}$ is also periodic.

(4°) The projections on ϕ_B and ϕ_N of the eigenvalue equation of $\widehat{H}$ are: $H_{BB} + \lambda H_{BN} = E$, and $H_{BN} + \lambda H_{NN} = \lambda E$. Eliminating λ between the two equations leads to:

$$E = \frac{1}{2}(H_{NN} + H_{BB}) \pm \frac{1}{2}\sqrt{(H_{NN} - H_{BB})2 + 4|H_{BN}|^2} \qquad (B.4)$$

For each value of $\vec{k}$ there are therefore two values of the energy: the p_z states generate two energy bands.

(5°)(a) We have:

$$H_{BB} = \frac{1}{N} \sum_{i,j} \exp\{i\vec{k}(\vec{T}_j - \vec{T}_i)\}\langle p_B^{(i)} | \frac{p^2}{2m} + w_B^{(i)}$$

$$+ \sum_{n \neq i} w_B^{(n)} + \sum_{n'} w_N^{(n')} | p_B^{(i)} \rangle \tag{B.5}$$

The term $(p^2/2m + w_B^{(i)})$ determines an energy E_B^0. The first following sum determines a contribution equal to zero. The last sum has two contributions. That corresponding to $i = j$ is equal to $-3\alpha_B$, since there are three (n) neighbours for each value of (i), and that the exponential prefactor is equal to unity for $i = j$. The second contribution which corresponds to i and j second neighbours is the following, as there are six second neighbours and that the exponential prefactor contains one of the six translations joining a node of the lattice to its neighbours:

$$-\gamma_B.\sigma(\vec{k}) = -\gamma_B \sum_{\mu} \exp(i\vec{k}.\vec{T}_\mu) \tag{B.6}$$

with:

$$\vec{T}_\mu = \pm\vec{d}_1; \pm\vec{d}_2; \pm(\vec{d}_1 + \vec{d}_2) \tag{B.7}$$

Thus:

$$\sigma(\vec{k}) = 2\left\{ 2\cos^2\left(\frac{k_x.a}{2}\right) 2\cos\left(\frac{k_x.a}{2}\right).\cos\left(\frac{k_y.a\sqrt{3}}{2}\right) - 1 \right\} \tag{B.8}$$

and finally:

$$H_{BB} = E_B^0 - 3\alpha_N - \gamma_B.\sigma(\vec{k}) \tag{B.9}$$

Likewise:

$$H_{NN} = E_N^0 - 3\alpha_N - \gamma_N.\sigma(\vec{k}) \tag{B.10}$$

On the other hand, in the calculation of H_{BN} the first and the last term of the Hamiltonian have zero contributions. The non-zero contribution is of the form:

$$H_{BN} = -\gamma'.\Delta(\vec{k}) = -\gamma' \sum \exp(i\,\vec{k}.\vec{T}_\mu) \qquad \text{(B.11)}$$

in which the μ index corresponds to the unit cell containing one of the three nitrogen atoms closest to the boron atom situated at the origin. We have $\vec{T}_\mu = 0; -\vec{a}_1; -(\vec{a}_1 + \vec{a}_2)$. Hence:

$$\Delta(\vec{k}) = 1 + \exp(-i\,\vec{k}.\vec{a}_1) + \exp[-i\,\vec{k}\,(\vec{a}_1 + \vec{a}_2)]$$
$$= 1 + \exp(-ik_x.a) + \exp\left[-i\left(\frac{k_x.a}{2} + \frac{k_y.a\sqrt{3}}{2}\right)\right] \qquad \text{(B.12)}$$

(b) Reporting the expressions of H_{BB}, H_{NN}, H_{BN} in the equation in question (4°) provides the result expected in question (5°)(b). As well, using the expression of $\Delta(\vec{k})$ obtained hereabove determines the form expected for $|\Delta(\vec{k})|^2$ (Fig. B.3).

(6°)(a) For symmetry reasons, the variations of Δ along all the segments ΓP_i is the same. Examine this variation along ΓP_1 parallel to x. We have $k_y = 0$, and $0\langle k_x\langle 4\pi/3a$. Thus:

$$\Delta^2 = \left(1 + 2\cos\frac{k_x a}{2}\right)^2; \quad |\Delta| = \left|1 + 2\cos\frac{k_x a}{2}\right| \qquad \text{(B.13)}$$

Their variations are reproduced on the figures (Fig. B.3). Note that Δ^2 vanishes in P with a slope equal to zero, while $|\Delta|$ vanishes with a finite slope.

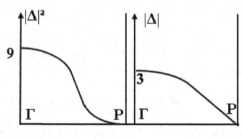

Fig. B.3.

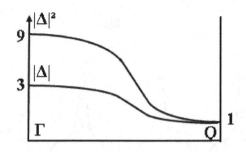

Fig. B.4.

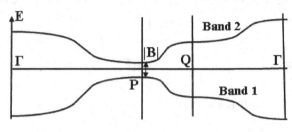

Fig. B.5.

Along ΓQ_i the variations of these two quantities are determined by:

$$\Delta^2 = \left[1 + 8\cos^2\left(\frac{ka\sqrt{3}}{4}\right)\right] ; \quad 0 < k < \frac{2\pi}{a\sqrt{3}} \qquad (B.14)$$

The corresponding variations of Δ^2 and $|\Delta|$ are reproduced on the figures (Figs. B.4 and B.5).

(b) γ' involves overlaps between nearest neighbours while γ_B and γ_N involve overlaps between second neighbours. With the assumption made, the energy depends on $\overrightarrow{k}$ only through the variations of Δ^2. As a consequence, the two bands determined by the expression of th energy, the variations sketched hereunder. The minimum interval between the two bands occurs at the points P_i. The magnitude of this interval is:

$$|B| = |E_B^0 + 3\alpha_N - 3\alpha_B| \qquad (B.15)$$

(7°)(a) There are $(3 + 5) = 8$ valence electrons in the elementary unit cell. Each elementary band generated by a non-degenerate atomic orbital contains $2N$ states (spin states included). The bonding

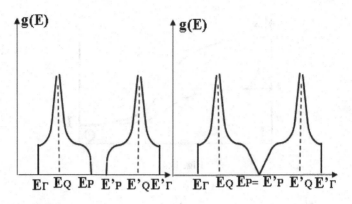

Fig. B.6.

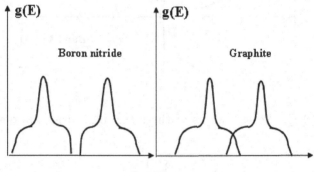

Fig. B.7.

band $(\text{sp})^2$ formed by three atomic orbitals thus contains $6N$ states. The two p_z bands are formed by $2p_z$ orbitals. Each contains $2N$ states.

(b) The distances between atoms in the planes are shorter than the distances perpendicular to the planes. Besides, the combinations of orbitals (s, p_x, p_y) being oriented in the plane, their overlap are more important. The energy shifts induced by these overlaps will be larger (downwards for the bonding orbitals, upwards for the anti-bonding ones). In contrast, the p_z orbitals have a small "side" overlap and no overlap in the vertical direction. Their average energy will therefore remain close to the energies of the atomic levels.

(c) At $T = 0$ K, the $8N$ valence electrons occupy the lowest energy levels. They will first occupy the (sp^2) bonding states which therefore form a "filled" band occupied by $6N$ electrons. There remains $2N$

electrons which will fill exactly the p_z band of lowest energy. As seen in question (5°), this band is separated from the other p_z band by a forbidden band of width $|B|$. The last occupied state is therefore the last available state of this band which is separated from the next band by a gap. *The considered solid is therefore an insulator.*

It is the fact that $|B| \neq 0$ which determines the insulating character. B is the result of the differences between the energies E^0 and α relative to the atoms B and N. The value of $\varepsilon = (E_B^0 - E_N^0) \neq 0$ because the energy of en electron in the atom N is different from the energy of an electron in the atom B. The difference $(\alpha_N - \alpha_B)$ is much smaller and cannot compensate ε.

(8°) In graphite, the crystal structure, the atomic orbitals, and the number of electrons are the same as in boron nitride. All the calculations remain valid. The only difference resides in the fact that the two atoms are identical. Thus $B = 0$. The two p_z bands are in contact at the P point of the Brillouin zone. The last occupied states is thus adjacent to the continuum of states of the second p_z band. Graphite is therefore a conductor (cf. Fig. B.6).

(9°)(a) In two dimensions the formula providing the density of states must be replaced by:

$$g(E) = \frac{S}{2\pi^2} \int_{l(E=cst)} \frac{dl(\vec{k})}{|\nabla_{\vec{k}} E|} \qquad (B.16)$$

In which S (surface of the crystal) is $(Na^2\sqrt{3}/2)$.

(b) In the two-dimensional reciprocal space the lines of equal energy correspond to $|\Delta(\vec{k})| = cst$. When the energy is close to that of the Γ point, the values of $\vec{k}$ are close to zero. In this neighbourhood the expansion of $|\Delta(\vec{k})|$ is therefore:

$$|\Delta|^2 = 9 - \frac{3}{2}a^2(k_x^2 + k_y^2) \qquad (B.17)$$

The equal energy curves are therefore circles. In the expression of $g(E)$, the integration element is then $dl(\vec{k}) = |\vec{k}| d\theta$ and the energy gradient is $|\nabla_{\vec{k}} E| = b.|\vec{k}|$, thus:

$$g(E) = \frac{S}{2b\pi^2} \int d\theta = \frac{S}{b\pi} = \text{Constant} \qquad (B.18)$$

The density of states is independent of E in the vicinity of $E(\Gamma)$. More precisely, we have $S = (Na^2\sqrt{3}/2)$ and $b = 108a^2/\sqrt{B^2 + 36}$.

(c) For $E = E^0$, we have $|\Delta|^2 = 1$, which leads to the equation:

$$\cos\left(\frac{k_x a}{2}\right)\left\{\cos\frac{k_x a}{2} + \cos\frac{k_y a\sqrt{3}}{2}\right\} = 0 \qquad (B.19)$$

The vanishing of $\cos(\frac{k_x a}{2})$ yields $k_x = \pi/a$. This is the segment perpendicular to x and joining to consecutive Q_i points. For symmetry reasons, all the similar segments between consecutive Q_i points correspond to the same energy. They form an hexagon included in the first Brillouin zone.

(d) To show that $g(E)$ goes to infinity it is sufficient to show that the contribution to $g(E)$ relative to the vicinity of Q_i has this characteristics. We have for instance:

$$k_x = \frac{\pi}{a}; k_y = \frac{\pi}{a\sqrt{3}} - \kappa \qquad (B.20)$$

with

$$dl = d\kappa \quad and \quad |\nabla E| = |\frac{\partial E}{\partial k_x}| \propto \cos\frac{k_y a\sqrt{3}}{2} = C.\sin\kappa\frac{a\sqrt{3}}{2} \quad (B.21)$$

Thus finally:

$$g(E) \propto \int_0^{\kappa_0} \frac{d\kappa}{\sin\kappa\frac{a\sqrt{3}}{2}} \propto \int_1^{u_0} \frac{du}{1 - u^2} \propto \{\frac{Ln(1+u)}{Ln(1-u)}\}_1^{u_0} \qquad (B.22)$$

which has a logarithmic divergence.

(e) The variations of $g(E)$ are determined, on the one hand, by the constance of $g(E)$ near $E(\Gamma)$, the divergence for $E(Q_i)$ and the vanishing for $E = E(P_i)$. The resulting curve for the two considered bands is reproduced (Fig. B.7) for BN and for graphite. It can be shown that in the latter case the vanishing at P_i occurs with a finite slope.

($10°$)(a) When the three-dimensional structure is taken into account there are four atoms in each unit cell. One must then consider 4 Bloch functions: $\phi_B^{(1)}$, $\phi_B^{(2)}$, $\phi_N^{(1)}$, $\phi_N^{(2)}$. The resulting secular equation resulting, as in question ($4°$), of the projection of the eigenvalue equation. It will then determine four bands.

(b) If there was no overlap between wavefunctions of neighbouring planes, the four bands obtained would form two degenerate bands (since each plane will determine two identical bands). The overlap is small, since the vertical interatomic distance is twice as large as in the planes. The situation will therefore differ little from that of a single plane. Each band determined for a single plane will be slightly shifted with respect to that of the adjacent plane. There will be a larger spread of the bands. In the case of boron nitride this effect will reduce the gap between bands. This effect being small the solid will remain an insulator. In the case of graphite, there will be an overlap between the two "plane" bands instead of a mere contact between bands. The conducting character of graphite is then increased. On the other hand, the divergence of the density of states at E_Q is suppressed: the integration performed above on a line must be performed on a surface. This has the effect of adding to the "singular" direction Q_i a large number of other directions, in which the function to integrate remains finite. There will only be a "peak" in the density of states.

B.2 Examination Text No. 2 Metallic Binary Alloys

B.2.1 Part 1

($1°$) $I(\overrightarrow{K})$ is proportional to (Chapter 4)

$$I \propto \prod_{j=1}^{3} \frac{\sin^2(N_j \overrightarrow{K}.\overrightarrow{a_j} / 2)}{N^2 \sin^2(\overrightarrow{K}.\overrightarrow{a_j} / 2)} \qquad (B.23)$$

which for large values of N is only non-zero if $\overrightarrow{K}$ is a vector of the reciprocal lattice of the structure of β-brass. Its lattice being of the CC type, the reciprocal lattice is of the FCC type. The conventional unit cell of the direct lattice is a cube of edge a. If we denote $(\overrightarrow{i}, \overrightarrow{j}, \overrightarrow{k})$ the three unit vectors parallel to the edges of the conventional unit cell, the *elementary* unit cell is

$$\frac{a}{2}(\overrightarrow{i} + \overrightarrow{j} - \overrightarrow{k}); \quad \frac{a}{2}(-\overrightarrow{i} + \overrightarrow{j} + \overrightarrow{k}); \quad \frac{a}{2}(\overrightarrow{i} - \overrightarrow{j} + \overrightarrow{k}) \qquad (B.24)$$

The conventional unit cell of the reciprocal lattice is a cube of edge $(4\pi/a)$. The corresponding elementary unit cell is for this FCC lattice:

$$\vec{d}_1^* = \frac{2\pi}{a}(\vec{i} + \vec{j}); \quad \vec{d}_2^* = \frac{2\pi}{a}(\vec{j} + \vec{k}); \quad \vec{d}_3^* = \frac{2\pi}{a}(\vec{i} + \vec{k})$$
(B.25)

(2°) For a monoatomic crystal, the intensities are proportional to the square of the atomic scattering factor (cf. Chapter 4). This factor depends on the modulus $|\vec{K}|$. The two considered diffractions correspond respectively to the vectors $\vec{K}_1 = \vec{d}_1^*$ and $\vec{K}_2 = 2(\vec{d}_1^* + \vec{d}_2^* + \vec{d}_3^*)$ whose moduli are respectively, for $a = 3\text{Å}$, $|\vec{K}_1| = 2.96\text{Å}^{-1}$ and $|\vec{K}_2| = 14.5\text{Å}^{-1}$. For these values, the table determines the values $f_1 = (f_{Cu} + f_{Zn})/2 = 22.6$ and $f_2 = 6.5$. Finally:

$$r = \frac{f_1^2}{f_2^2} \approx 12$$
(B.26)

(3°) The β' structure has a simple cubic Bravais lattice of edge a, with two atoms per unit cell (S and C). The reciprocal lattice is also simple cubic with edge $(2\pi/a)$. Its unit cell is defined by

$$\vec{b}_1^* = \frac{2\pi}{a}\vec{i}^*; \ \vec{b}_2^* = \frac{2\pi}{a}\vec{j}^*; \ \vec{b}_3^* = \frac{2\pi}{a}\vec{k}^*$$
(B.27)

(4°) The diffraction reflections of the β structure are those of the reciprocal lattice of the centered cubic structure. Thus:

$$\vec{K} = \frac{2\pi}{a}[h(\vec{i} + \vec{j}) + k(\vec{j} + \vec{k}) + l(\vec{i} + \vec{k})]$$
(B.28)

or

$$\vec{K} = (h + l)\vec{b}_1^* + (h + k)\vec{b}_2^* + (k + l)\vec{b}_3^*$$
(B.29)

Hence, $(h' + k' + l') = 2m$. While for the β' structure all the diffractions nodes $\vec{b}_i^*$ of the reciprocal lattice are relevant. There are additional reflections corresponding to $(h' + k' + l') = 2m + 1$ (Fig. B.8).

(5°) The structure factor is determined by $F(\vec{K}) = \sum_j f_j \exp(-i\vec{K}.\vec{r}_j)$. The sum is over the atoms of the basis. There are two such atoms (S and C) which can be placed at the origin and at the position $\vec{r} = (a/2)(\vec{i} + \vec{j} + \vec{k})$. On the other hand, the

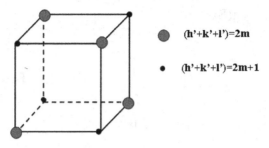

(h'+k'+l')=2m

(h'+k'+l')=2m+1

Fig. B.8.

scattering factors of these atoms are, according to the assumption of question $(2°)$, $\{(1-p)f_{Cu}+pf_{Zn}]$ and $[pf_{Cu}+(1-p)f_{Zn}]$. Thus we obtain:

$$F(\vec{K}) = f_{Cu}[1-p\{1-(-1)^{h'+k'+l'}\}] + f_{Zn}[p+(1-p)(-1)^{h'+k'+l'}] \tag{B.30}$$

The additional diffractions correspond to $(h'+k'+l')$ odd. The intensity being proportional to the square of the structure factor, finally:

$$I = |F^2| = 4\left(\frac{1}{2}-p\right)^2 (f_{Cu}-f_{Zn})^2 \tag{B.31}$$

f_{Cu} having a value close to f_{Zn}, this intensity is weak.

(6°) The formula providing the structure factor for $(h'+k'+l')$ even shows that the intensities of the diffractions common to the two structures β and β' do not depend of the p probability. They are identical for the two structures. To determine p one can measure the intensity of a diffraction specific of the β' structure and of a diffraction common to the two structures. Their ratio only depends of the unknown parameter p.

B.2.2 Part 2

(1°) For a monoatomic crystal the average volume of an atom is equal to that of the elementary unit cell. This volume is $(a_{fcc})^3/4$ for th FCC structure and $(a_{CC})^3/2$ for the cubic centered one. The quality of these two values yields $(a_{fcc}/a_{CC}) = 2^{1/3} = 1.26$. Thus, for $a_{CC} = 3$ Å, $a_{fcc} = 3.78$ Å.

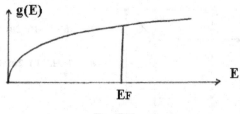

Fig. B.9.

(2°) Each atom has, in average, $(1 - x)$ electrons from the Cu atom and $2x$ electrons from the Zn atom (or $3x$ electrons from the Al atom). For B = Zn one has therefore $\nu(x) = (1 + x)$ for B=Zn and $\nu(x) = (1 + 2x)$ for B = Al. Hence, $1 < \nu < 2$ in the first case and $1 < \nu < 3$ in the second case.

(3°) (a) The number of states in the Fermi sphere is equal to half the number of electrons (since one orbital state can be occupied by two electrons with opposite spins). Thus,

$$\frac{(4\pi/3)k_F^3}{8\pi^3/N\omega} = \frac{N(1+y)}{2} \tag{B.32}$$

where ω is the volume of the elementary unit cell and $y = x(\text{resp.}2x)$ if the B atom is Zn or Al. In the case of the FCC structure, one has $\omega = (a_{fcc}^3/4)$. Hence:

$$k_F = \frac{2\pi}{a_{fcc}} \cdot \left[\frac{3(1+y)}{2\pi}\right]^{1/3} \tag{B.33}$$

(b) k_F has the same value for the CC structure since the elementary unit cell has the same volume (cf. question (1°)).

(c) The Fermi level is the same for the two structures. The density of state $g(E)$ is the parabolic curve (Fig. B.9) representing the function:

$$g(E) = \frac{4\pi N\omega}{h^3}(2m)^{3/2}\sqrt{E} \tag{B.34}$$

(4°)(a) The reciprocal lattices of the FCC and CC structures are respectively of the CC and FCC types with conventional cubic cells equal respectively to $(4\pi/a_{fcc})$ and $(4\pi/a_{CC})$. In the first case, the shortest distance is between the center of the conventional cell and a corner. In the second case, it is between a corner and the center of a face. Thus one has

$$d_{FCC} = \frac{4\pi}{a_{fcc}} \cdot \frac{\sqrt{3}}{2} \qquad \text{(B.35)}$$

and

$$d_{CC} = \frac{4\pi}{a_{CC}} \cdot \frac{\sqrt{2}}{2} \qquad \text{(B.36)}$$

(b) The distance between the origin of the reciprocal lattice and the closest plane of the first Brillouin zone is equal to half the distances calculated in (4°)(a) for each structure. For the Fermi sphere to be included in the first Brillouin zone, one must then have $k_F < d/2$. For $y = 0$, this condition yields:

$$k_F = 0.781 \frac{2\pi}{a_{fcc}}; \quad \frac{d_{FCC}}{2} = 0.866 \frac{2\pi}{a_{fcc}}; \quad \frac{d_{CC}}{2} = 0.89 \frac{2\pi}{a_{fcc}} \qquad \text{(B.37)}$$

The inclusion condition is thus verified in the two cases.

(5°) The critical values of ν are determined by the condition $k_F(\nu) = (d/2)$. One finds $\nu(FCC) = 1.363$ and $\nu(CC) = 1.48$. One deduces from the relations established in question (2°) that:

$$x(FCC, Zn) = 0.36; \quad x(FCC, Al) = 0.18 \qquad \text{(B.38)}$$
$$x(CC, Zn) = 0.48; \quad x(CC, Al) = 0.24 \qquad \text{(B.39)}$$

(6°) The curve keeps the form it has for free electrons in a uniform potential except for k close to k_0 where it has a horizontal tangent. With $E(k_0) = E_0 - |V_G|$ with $E_0 = \hbar^2 k_0^2 / 2m$ and $V_G = \langle k_0 | V(\vec{r}) | k_0 \rangle$. The "gap" in k_0 is equal to $2|V_G|$. Figure B.10 represents $E(k)$ in an extended zone scheme (cf. Chapter 3).

(b) The curve (Fig. B.10) has four parts.

(i) Between the origin and M the energy is lower than $E_{M'}$. The corresponding states are almost equal to plane waves for which the density of states has the parabolic form of free electrons.

(ii) Between M and G the curve $E(k)$ is deformed in the k_0 direction because there are a larger number of states per energy interval. In the other directions $E(k)$ is close to the behaviour of free-electrons. As a consequence, the density of states increases.

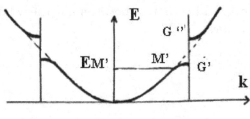

Fig. B.10.

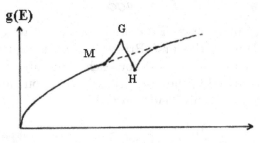

Fig. B.11.

(iii) Between G and H, the energy is within the gap in the k_0 direction. This direction and the neighbouring ones do not contribute to the density of states, thus $g(E)$ decreases.

(iv) Above H, the energy is larger than G. There are again states in the k_0 direction and the neighbouring directions and $g(E)$ increases again (we have assumed the magnitude of the gap small enough) (Fig. B.11).

(7°) (a) If the Fermi surface is assumed to be spherical, the G point corresponds to the situation in which the Fermi surface is in contact with the surface of the Brillouin zone in k_0, whose value has been specified in question (5°).

(b) The energy E_G is (for a spherical Fermi surface) $E_G = \hbar^2 k_F^2 / 2m$. The M point corresponds to the M' point of the $E(k)$ curve in the direction k_0: $E_M = \hbar^2 k_M^2 / 2m$. It is therefore of interest to determine the M' point. It is the point where the curve $E(k)$ differs significantly from the free electron curve. We can define it by the condition $(E_{\text{free}} - E_{\text{qu-free}}) = |V_G|/n$ in which $n > 2$.

One cannot calculate the position of M' from the expansion near $k_0 = k_F$ as performed in Chapter 5 since this expansion assumes that $|E(k + G) - E(k)|$ is small with respect to $|V_G|$. Actually in M', the two terms are comparable. One must then use the general formula based on the diagonalization of the Hamiltonian in the set of states $(k, k + G)$. This formula is

$$E = \frac{\varepsilon(k + G) + \varepsilon(k)}{2} - \sqrt{\frac{[\varepsilon(k + G) - \varepsilon(k)]^2 + 4V_G^2}{2}} \qquad \text{(B.40)}$$

We have $k_0 = (-G/2)$, and $\varepsilon(k) = \hbar^2 k^2/2m$. Let us put $k = k_F - \Delta k$. The position of M' is determined by $E = |V_G|/n$. We then find:

$$\frac{\Delta k}{k_F} = \frac{n|V_G|(1 - \frac{1}{n^2})}{2E(k_F)} \qquad \text{(B.41)}$$

For $n = 2$ we obtain:

$$\frac{\Delta k}{k_F} = \frac{3|V_G|}{4E(k_F)} \qquad \text{(B.42)}$$

$E(M)$ is the energy of the free electron corresponding to the vector $(k_F - \Delta k)$. The numerical evaluation (using the values in question $(1°)$) yields:

$$\Delta k = 0.0468 k_F; \quad (E_G - E_M) - 0.09 E_G \qquad \text{(B.43)}$$

Thus, with $E_G(FCC) \approx 8$ eV, and $E_G(CC) \approx 8.45$ eV, $(E_G - E_M) \approx 0.7$ eV.

$(8°)$ (a) With the values found for the positions of points M and G for the two structures, the density of states have the shape represented hereunder (dashed line for the CC structure, and dotted line for free electrons) (Fig. B.12).

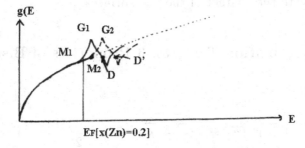

Fig. B.12.

When x is chosen such as k_F is below points M_1 or M_2 depending of the structure, the density of states is the same as for free-electrons and the Fermi level is also close. When x is such as to make k_F between M_i and G_i ($i = 1, 2$) the density of states is larger than for free electrons, and, for a given total number of electrons, the last occupied state will have lower energy than for free electrons: the Fermi level will be lower. Above G_i, the Fermi level will increase progressively towards the level of free-electrons.

(b) Let D be a point such as the integral of the density of states between the origin and D is the same for the two structures. For a value of x corresponding to k_F situated between M_i and D, the Fermi level of the FCC structure will be lower than that of the CC structure. Above D and up to D' the Fermi level of the CC structure will be lower than that of the FCC structure.

(9°) The total energy of the electrons varies in the same manner as the Fermi energy. The most stable structure will therefore be that corresponding to the lowest Fermi level. For $x(Zn) = 0.2$ one is slightly above M_1. The stable structure is FCC. The result is the same for $x(Al) = 0.17$ (one is between points M_1 and G_1). In contrast it is the CC structure which is stable for $x(Zn) = 0.45$ and $x(Al) = 0.24$ which correspond on the curve to points close to G_2.

(10°) If we consider a Fermi surface corresponding to quasi-free electrons in a periodic potential, the Fermi sphere is deformed in the direction of k_0 and equivalent directions. There will be outward extensions of the surface in these directions. The surface will "touch" the surface of the Brillouin zone for values of k_F smaller than those calculated previously. The energy E_G is displaced to the left. The density of states $g(E)$ decreases for a lower value of the energy. The calculation becomes more difficult because the G point does no longer correspond to the contact of the two surfaces.

B.3   Examination Text No. 3 Properties of Bismuth

(1°) (a)

$$\vec{d}_1^{\,c} = \frac{a}{2}(\vec{i} + \vec{j}) \quad \vec{d}_2^{\,c} = \frac{a}{2}(\vec{j} + \vec{k}) \quad \vec{d}_3^{\,c} = \frac{a}{2}(\vec{i} + \vec{k}) \quad \omega_c = \frac{a^3}{4}$$

(B.44)

(b)

$$\vec{a}_1^{c,*} = \frac{2\pi}{a}(\vec{i} + \vec{j} - \vec{k}) \quad \vec{a}_2^{c,*} = \frac{2\pi}{a}(-\vec{i} + \vec{j} + \vec{k})$$

$$\vec{a}_3^{c,*} = \frac{2\pi}{a}(\vec{i} - \vec{j} + \vec{k}) \tag{B.45}$$

The reciprocal lattice is cubic centered with a conventional cell of edge $(4\pi/a)$. The volume of the elementary reciprocal space is $\omega_c^* = (8\pi^3/\omega_c)$.

(c) The atoms are at the vortices of a simple cubic lattice of edge $(a/2)$. The atoms generated from the first atom occupy half the vortices and the ones generated from the second atom occupy the other half. The reciprocal lattice is also simple cubic with a dimension of cell $(4\pi/a)$.

(2°) (a) We have $a_0 = |\vec{a}_i^c|/\sqrt{2} = a\sqrt{2}/\sqrt{2} = a/2$. The direct unit cell has the volume $\omega = (\vec{a}_1, \vec{a}_2, \vec{a}_3)$ thus equal to $2a_0^3(1 - \frac{3\varepsilon}{2}) = \omega_c(1 - \frac{3\varepsilon}{2}) = 69.92 \text{ Å}^3$. The reciprocal cell is $\omega^* = (8\pi^3)/\omega = 3.55 \text{Å}^{-3}$.

(b) The definition of primitive translations yields:

$$\vec{a}_1^* = \frac{2\pi}{\omega}(\vec{a}_2 \wedge \vec{a}_3) = \frac{2\pi}{a}\left(1 + \frac{\varepsilon}{2}\right)(1,1;-1-\varepsilon) \tag{B.46}$$

The other vectors are deduced by circular permutation.

$$\vec{a}_2^* = \frac{2\pi}{a}\left(1 + \frac{\varepsilon}{2}\right)(-1-\varepsilon,1;1) \quad \vec{a}_3^* = \frac{2\pi}{a}\left(1 + \frac{\varepsilon}{2}\right)(1,-1-\varepsilon,1) \tag{B.47}$$

(3°) The radius of the Fermi sphere is determined by $k_F = (3\pi^2 n/\omega)^{1/3}$ in which n is the number of electrons in the elementary unit cell, and ω the volume of this cell. There are two atoms each having five valence electrons, thus $n = 10$. In the case of the FCC structure, the volume of the cell is ω_c while for the deformed structure the volume is ω. One has therefore:

$$k_F = \frac{2\pi}{a} \cdot \left(\frac{120}{8\pi}\right)^{1/3} = 1.684\left(\frac{2\pi}{a}\right) \tag{B.48}$$

and

$$k_F = \frac{2\pi}{a} \cdot \left(\frac{120}{8\pi}\right)^{1/3}\left(1 + \frac{\varepsilon}{2}\right) = 1.718\left(\frac{2\pi}{a}\right) \tag{B.49}$$

On the other hand the volume of the Fermi sphere is $(4\pi k_F^3)/3$, thus:

$$\frac{8\pi^3}{a^3}\left(\frac{4\pi}{3}\cdot\frac{120}{8\pi}\right)\left(1+\frac{3\varepsilon}{2}\right) = 5.\frac{8\pi^3}{a^3/4}\cdot\left(1+\frac{3\varepsilon}{2}\right) = 5\omega^* \qquad (B.50)$$

Each reciprocal cell contains N_M states which can be occupied by $2N_M$ electrons. The Fermi sphere contains all the states occupied by the $10N_M$ electrons of bismuth. The density of states being uniform in the reciprocal space, the volume of the Fermi sphere is thus equal to that of five elementary unit cells of the reciprocal lattice.

(4°) (a) A Brillouin plane is the mediator plane of the vector $\vec{G}$ whose origin coincides with the origin Γ of the reciprocal lattice. This plane is defined by $(-2\vec{k}.\vec{G} + \vec{G}^2 = 0$. The distance of the plane to the origin is $|\vec{G}/2|$.

(b) The considered vectors are such as the distance of the corresponding Brillouin planes to the origin Γ is smaller than the radius of the Fermi sphere. Thus: $\vec{G}^2 < 4k_F^2$. For the FCC lattice this condition is

$$(n_1^2 + n_2^2 + n_3^2) < 11.34 \qquad (B.51)$$

On the other hand, these vectors must be linear combinations with integer coefficients of the $\vec{d}\, i^{c,*}$. The required vectors correspond to the following n_i values:

$$(\pm1,\pm1,\pm1); (\pm2,0,0), \quad (0,\pm2,0), (0,0,\pm2) \qquad (B.52)$$
$$(\pm2,\pm2,0), (\pm2,0,\pm2), (0,\pm2,\pm2), \qquad (B.53)$$
$$(\pm3,\pm1,\pm1), (\pm1,\pm3,\pm1), (\pm1,\pm1,\pm3) \qquad (B.54)$$

In the quasi-free electrons in a periodic potential, the Brillouin planes are associated to discontinuities of the energy curves. These discontinuities determine partial or global forbidden bands in the electronic spectrum. It is near these planes that the spectrum differs from that of the free-electron.

(c) The four preceding groups of $\vec{k}$ vectors correspond to four distances between a Brillouin plane and Γ. These distances are $(2\pi/a)^2(n_1^2 + n_2^2 + n_3^2)$ thus respectively $(n_1^2 + n_2^2 + n_3^2) = 3, 4, 8, 11$. These distances correspond respectively to $8, 6, 12,$ and 24 equivalent vectors.

(d) The expression of the 24 vectors is, denoting to simplify, $(u, v, w) = (\overrightarrow{a}_1^{c,*}, \overrightarrow{a}_2^{c,*}, \overrightarrow{a}_3^{c,*})$, and $G(i) = \overrightarrow{G}_i^m$.

$$G(1), G(2) = \pm(u + 2v + 2w); G(3),$$

$$G(4) = \pm(2u + v + 2w) \tag{B.55}$$

$$G(5), G(6) = \pm(2u + 2v + w), G(7),$$

$$G(8) = \pm(v + 2w) \tag{B.56}$$

$$G(9), G(10) = \pm(2u + w), G(11), G(12) = \pm(u + 2v) \tag{B.57}$$

$$G(13), G(14) = \pm(2v + w), G(15), G(16) = \pm(u + 2w) \tag{B.58}$$

$$G(17), G(18) = \pm(2u + v), G(19), G(20) = \pm(u - v - w) \tag{B.59}$$

$$G(21), G(22) = \pm(-u + v - w), G(23),$$

$$G(24) = \pm(-u - v + w) \tag{B.60}$$

($5°$)(a) There are eight points corresponding to the eight diagonals of the cube. The L point $(2\pi/a)(1/2, 1/2, 1/2)$ is at one-fourth of the diagonal of the conventional cell of the reciprocal lattice. This point being at the center of a "face" of the surface of the Brillouin zone, is only equivalent to the L' point defined by the opposite k-vector $(2\pi/a)(-1/2, -1/2, -1/2)$. The six other points T_i are pairs of equivalent vectors.

(b) One can take $\overrightarrow{\Gamma T} = \overrightarrow{a}_1^{c,*}/2$. Then $\overrightarrow{\Gamma L} = (\overrightarrow{a}_1^{c,*} + \overrightarrow{a}_2^{c,*})/2$ and $\overrightarrow{\Gamma X} = (\overrightarrow{a}_1^{c,*} + \overrightarrow{a}_1^{c,*} + \overrightarrow{a}_1^{c,*})/2$.

(c) One can check easily that the difference $(\overrightarrow{\Gamma L} - \overrightarrow{G}(i)/2)$ is a vector of the reciprocal lattice for each of the following six $G(i)$: $G(19)/2, G(20)/2, \ldots, G(24)/2$. Likewise, $\overrightarrow{\Gamma T}$ is equivalent to the six vectors $(1)/2, G(2)/2, G(11)/2, G(12)/2, G(15)/2, G(16)/2$.

($6°$) In the representation in "reduced zone" all the states are represented in the first Brillouin zone, by representing $E(\overrightarrow{K} + \overrightarrow{G})$ as $E(\overrightarrow{K})$, with $\overrightarrow{K}$ belonging to the first Brillouin zone.

The lowest energy levels at L are those: (1) corresponding to L or the equivalent $\overrightarrow{K}$ vectors. (2) Which correspond to the smallest K vectors since the energy is the kinetic energy $\hbar^2 K^2/2m$.

This defines the vectors $\overrightarrow{\Gamma L}$ and the equivalent vectors found in question ($5°$) (c). Hence there are two energy levels. The lowest corresponds to $|\overrightarrow{\Gamma L}|$ and the other to the modulus of the 6 vectors $\overrightarrow{G}_i^m/2$.

Thus:

$$E_1(L) = \frac{\hbar^2}{2m}(|\overrightarrow{\Gamma L}|^2 = \frac{\hbar^2}{2m}\frac{4\pi^2}{a^2}\cdot\frac{3}{4} = 2.535 \text{ eV} \qquad (B.61)$$

$$E_2(L) = \left(\frac{\overrightarrow{G_i^m}}{2}\right)^2 = \frac{\hbar^2}{2m}\frac{4\pi^2}{a^2}\cdot\frac{11}{4} = 9.295 \text{ eV} \qquad (B.62)$$

The energy levels are the same at the Ti points since the moduli of the different vectors are the same, due to the cubic symmetry which relates the different diagonals as in L.

(7°) (a) The expressions of the $\overrightarrow{\Gamma L}$ and the $\overrightarrow{G_i^m}$ as function of the basis vectors of the deformed reciprocal cell are the same as above by substituting the $\overrightarrow{a}_i^*$ to the $\overrightarrow{a}_i^{c,*}$. We have:

$$\overrightarrow{\Gamma L} = \frac{2\pi}{a}\left(1 + \frac{\varepsilon}{2}\right)(1 - \varepsilon, 1 - \varepsilon, 1 - \varepsilon) \quad \text{with} \quad |\overrightarrow{\Gamma L}|^2 = \frac{3\pi^2}{a^2}(1 - \varepsilon)$$
$$(B.63)$$

The $\overrightarrow{G_i^m}/2$ vectors equivalent to $\overrightarrow{\Gamma L}$ all have the modulus $(11\pi^2/a^2)(1 + 21\varepsilon/11)$. Likewise:

$$|\overrightarrow{\Gamma T}|^2 = \frac{3\pi^2}{a^2}\left(1 + \frac{5\varepsilon}{3}\right) \qquad (B.64)$$

There are two groups of equivalent $\overrightarrow{G_i^m}$ vectors with different moduli:

$$\left|\frac{\overrightarrow{G_i^m}}{2}\right|^2 = \frac{11\pi^2}{a^2}\left(1 - \frac{3\varepsilon}{11}\right) \qquad (B.65)$$

$$\left|\frac{\overrightarrow{G_i^m}}{2}\right|^2 = \frac{11\pi^2}{a^2}\left(1 + \frac{13\varepsilon}{11}\right) \qquad (B.66)$$

(b) The six vectors $\overrightarrow{G_i^m}/2$ equivalent to $\overrightarrow{\Gamma L}$ having the same modulus, The $E_2(L)$ level keeps the same degeneracy equal to 6. As $|\Gamma L| = |\Gamma L'|$ the level $E_1(L)$ remains doubly degenerate. The calculated energy shifts are:

$$\Delta E_1(L) = -\varepsilon.E_1(L) = -0.101 \text{ eV};$$

$$\Delta E_2(L) = \frac{21\varepsilon}{11}E_2(L) = 0.710 \text{ eV} \qquad (B.67)$$

(c) The separation of the $\overrightarrow{G_i^m}/2$ vectors into two groups with distinct moduli leads to a separation of the $E_2(T)$ level into two levels, one E' being doubly degenerate and the other E'' four-fold degenerate. The two-fold degeneracy of $E_1(T)$ is preserved. The shifts in energy are as follows:

$$\Delta E_1(T) = \frac{5\varepsilon}{3} E_1(T) = 0.169 \text{ eV}; \quad \Delta E'_2(T) = -0.101 \text{ eV};$$

$$\Delta E''_2(T) = +0.439 \text{ eV} \tag{B.68}$$

(8°)(a) The Fourier transform of $V(r)$ is:

$$V_G = A \int V(\vec{r}) \exp(-i\vec{G}.\vec{r}).d\vec{r}$$

$$= \frac{1}{N_M} \sum_{ij} A \int \mathbf{v}(\vec{r} - \vec{R}_{ij}) \exp(-i\vec{G}.\vec{r}).d\vec{r} \tag{B.69}$$

Thus

$$V_G = \frac{1}{N_M} \sum_{ij} \exp(-i\vec{G}.\vec{R}_{ij}) A \int \mathbf{v}(\vec{r}) \exp(-i\vec{G}.\vec{r}'.d\vec{r}' \tag{B.70}$$

and

$$V_G = \frac{1}{N_M} \left[\sum_i \exp(-i\vec{G}.\vec{T}_i) \right] \left[\sum_{j=1}^{2} \exp(-i\vec{G}.\vec{r}_j) \right] \mathbf{v}(|\vec{G}| \tag{B.71}$$

in which $\vec{G}$ is a reciprocal lattice vector. The exponential functions of the $\vec{T}_i$ are equal to unity. The sum of the factors is therefore equal to the number of unit cells N_M. Finally:

$$V_G = \mathbf{v}.|\vec{G}|. \sum_{j=1}^{2} \exp(-i\vec{G}.\vec{r}_j) \tag{B.72}$$

in which one recognizes the structure factor:

$$F(\vec{G}) = \sum_{j=1}^{2} \exp(-i\vec{G}.\vec{r}_j) \tag{B.73}$$

(b) For the two atoms in the structure basis, the $\vec{r}_j$ are equal to 0 and to $u(\vec{d}_1 + \vec{d}_2 + \vec{d}_3)$: The calculation then yields:

$$|F(\vec{G})| = |1 + \exp[-2\pi i u(n_1 + n_2 + n_3)]|$$

$$= 2|\cos \pi u(n_1 + n_2 + n_3)| \tag{B.74}$$

(9°) (a) The moduli correspond to equal values of κ, thus respectively 3 or 11. The corresponding energies are respectively -1.6eV and $+0.65$eV. Therefore, the following table.

$\vec{G}$	$n_1 + n_2 + n_3$	$|F(\vec{G})|(u)$	$|F(\vec{G})|(0.468)$	$|\mathbf{v}(|\vec{G}|)|$	$|V_G|$
$2\vec{\Gamma L}$	3	$|2\cos 3\pi u|$	0.594	1.6 eV	0.95 eV
$2\vec{\Gamma T}$	1	$|2\cos \pi u|$	0.2	1.6 eV	0.32 eV
G_{19}–G_{24}	-1	$|2\cos \pi u|$	0.2	0.65 eV	0.13 eV
G_1–G_2	5	$|2\cos 5\pi u|$	0.963	0.65 eV	0.626 eV
G_{11}–G_{12}, G_{15}–G_{16}	3	$|2\cos 3\pi u|$	0.594	0.65 eV	0.386 eV

(b) The expression of the structure factor for $u = 0.5$ contains an odd multiple of $(\pi/2)$ for $(n_1 + n_2 + n_3)$ odd. This is the case for the vectors numbered (3),(1), or (5). The vanishing of the structure factor comes from the fact that for a simple cubic structure, these vectors are not reciprocal lattice vectors. In this case the only reciprocal lattice vectors are combinations of $(4\pi/a)(i, j, k)$.

(10°) (a) In accordance with the theory of quasi-free electrons:

$$E_{\pm 1}(L) = E_1(L) \pm |V_G| = 2.535 \pm 0.95 \text{ eV} \qquad (B.75)$$

Along ΓL the energy dispersion curve is that of free-electrons except in the vicinity of L. The variation is represented on Fig. B.13 (left) in a reduced zone scheme.

Likewise the degeneracy lifting, and the representations of $E_1(T)$ are (right of Fig. B.14 as follows):

$$E_{\pm 1}(T) = E_1(T) \pm |V_G| = 2.535 \pm 0.32 \text{ eV} \qquad (B.76)$$

(b) The levels being 6 fold degenerate one must diagonalize a 6x6 matrix.

(c) The relevant Fourier coefficient, for the vectors $G(19)$–$G(23)$ is, according to the table $E_{\pm 2}(L) = E_2(L) \pm 0.13$eV. (Each level is three-fold degenerate.) The corresponding dispersion curve is represented on Fig. B.15, left.

(d) As indicated in the text one determines separately the degeneracy lifting of E' and that of E''. They are respectively determined by the Fourier coefficients of $[G(1) - G(2)]$ and of

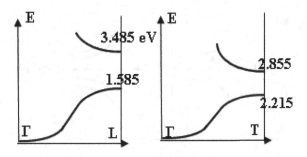

Fig. B.13.

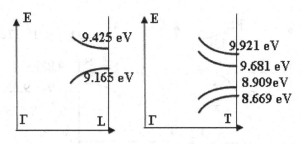

Fig. B.14.

$[G(11)G(12)G(15)G(16)]$ thus (Fig. B.14, right):

$$E_2'^{\pm}(T) = E_2(T) \pm 0.626 \text{ eV} = 9.295 \pm 0.626 \text{ eV} \qquad (B.77)$$

$$E_2''^{\pm}(T) = E_2(T) \pm 0.386 \text{ eV} = 9.2895 \pm 0.386 \text{ eV} \qquad (B.78)$$

(11°) (a) The undeformed lattice for $u = 0.5$ has a single atom per elementary unit cell. There is an odd number of electrons in the unit cell. This is the situation of alcaline metals in which the last band is partly empty (there are at least N_M empty states above the last occupied state). The behaviour is therefore that of a conductor.

(b) For the non-deformed lattice, the degeneracy liftings are on either sides of an identical average energy value, for both the points L and T. The spectrum is as represented in the next page.

The forbidden band has the width $\Delta E = 2|V_G|(L) = 0.26$eV. Below this forbidden band there are for the L point two non-degenerate bands and one three-fold degenerate band. Hence, there are five elementary bands. This is also the situation at the T point. These bands can be occupied by $10N_M$ electrons thus the anti-renumber of electrons. This is the situation of an insulator.

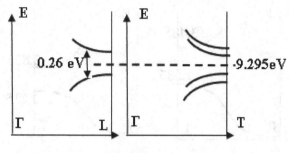

Fig. B.15.

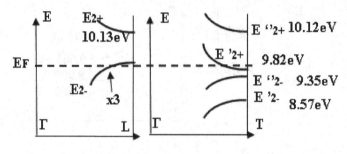

Fig. B.16.

(c) The displacements calculated in question (6°) lead to the representation of the spectrum as shown on Fig. B.16.

The three-fold degenerate band in L overlaps the sixth non-degenerate band in T. There is no global forbidden band. Bismuth is a conductor, *due to the deformation of the lattice.*

(12°)(a) The Fermi level cannot be above the sixth band at L because more than $10N_M$ electrons would be required to fill all the states. It cannot either be below the beginning of the sixth band at T. Indeed the fifth band in L being partly occupied, there would be less than $10N_M$ occupied states. Thus the Fermi level is situated as indicated on the figure. At L four elementary bands are totally occupied and one band partly occupied. AT T the sixth band is partly occupied.

(b) There is a band of *holes* in L because there are empty states at the top of the fifth band. There is an electron band in T because only the first states of the sixth band are occupied.

(c) The overlap of bands is small ($\approx 0.1\,\text{eV}$) with respect to the bands widths. The electronic properties are determined by the electrons which occupy partly the sixth band at T. They occupy a small fraction of this band. A similar situation occurs for the holes in L. In an ordinary metal the number of conduction electrons is comparable to N_M. The Hall voltage is inversely proportional to the number of carriers (cf. Chapter 1). Hence this voltage will be much higher than in a "good" metal as copper.

(d) Surprisingly for a simplified calculation as performed in the problem, one finds in the elaborate calculation results for the interval between E_1 and E_2 only differ by 10% from the present ones. The difference is 50% for the degeneracy lifting. The Fermi level intersects, as found here, the fifth band in L and the sixth band in T.

Constants Values

Angström 10^{-10}m
Micron 10^{-6}m
Avogadro number $N = 6.02 \times 10^{23}$
Barn (Scattering cross-section) 10^{-24}cm^2 (square centimeters)
Bohr Magneton: $\mu_B = 9.24 \times 10^{-24}$J.T^{-1} (Joule.Tesla^{-1})
Velocity of light $c = 2.99 \times 10^8$m.s^{-1}
Electron charge: $-e = -1.602 \times 10^{-19}$C (Coulomb)
Electron mass: $m = 9.11 \times 10^{-31}$kg (Kilogram)
Electron-volt (energy): eV $= 1.6 \times 10^{-19}$J (Joule)
Boltzmann constant: $k_B = 1.38 \times 10^{-23}$J.K^{-1} (Joule.degree^{-1})
Hydrogen atomic radius: 0.512 Å
Planck Constant: $h = 6.63 \times 10^{-34}$J.s (Joule.second)
$\hbar = h/2\pi 1.06 \times 10^{-34}$J.s
Proton mass: 1836 electron mass. $M = 16.72 \times 10^{-28}$kg

C.1 Notations

$\overrightarrow{\varepsilon}$	electric field
$f(\lambda, \theta)$	atomic scattering factor
$g(\overrightarrow{k})$	reciprocal space density of states
	(number of states in the unit reciprocal volume)
$g(E)$	energy density of states
	(number of states per unit energy interval)
$H(\overrightarrow{r}, \overrightarrow{p})$	Hamiltonian

$\widehat{H}$	Hamiltonian operator
$\vec{j}$	current density
$\vec{k}$	vector in reciprocal space
χ	magnetic susceptibility
$\vec{p}$	momentum
$\widehat{\vec{p}}$	momentum operator $= \frac{\hbar}{i}\vec{\nabla}_r$
$\vec{r}$	coordinate
$\widehat{\vec{r}}$	coordinate operator
ρ	electrical resistivity $\rho = 1/\sigma \times$ (Ohm.m)
$\rho(E)$	number of electrons which have their energy E in a unit energy interval
$\rho(\vec{r})$	electron density
σ	electrical conductivity $\vec{j} = \sigma\,\vec{\varepsilon}$
$V(\vec{r})$	potential energy
$\widehat{V}(\vec{r})$	potential energy operator
$\vec{v}$	velocity
Θ_F	fermi temperature

Index

Printed in the United States
by Baker & Taylor Publisher Services